Ecological Studies, Vol. 190

Analysis and Synthesis

Ecological Studies

Volumes published since 2002 are listed at the end of this book.

J.T.A. Verhoeven B. Beltman R. Bobbink D.F. Whigham (Eds.)

Wetlands and Natural Resource Management

With 91 Figures, 3 in Color, and 35 Tables

Springer

Prof. Dr. Jos T.A. Verhoeven
Dr. Boudewijn Beltman
Dr. Roland Bobbink
Landscape Ecology
Institute of Environmental Biology
Utrecht University
PO Box 80084
3508 TB Utrecht
The Netherlands

Dr. Dennis F. Whigham
Smithsonian Environmental Research Center
PO Box 28
Edgewater, MD 21037
USA

Cover illustration: Large picture: *Cladium jamaicense* lawns with tree islands, Everglades National Park, USA (Photo: J.T.A. Verhoeven)
Small pictures: La Pérouse Bay, Manitoba (Photos Hudson Pay Projekt Team): *Top* The effects of grubbing by lesser snow geese in early spring on the intertidal saltmarsh; *Middle* Death of willow bushes and exposure of the surface organic layer after goose grubbing in the supratidal marsh; *Bottom* Grazing exclosure indicating that in the absence of grubbing the vegetation remains intact on the intertidal marsh

ISSN 0070-8356
ISBN-10 3-540-33186-7 Springer Berlin Heidelberg New York
ISBN-13 978-3-540-33186-5 Springer Berlin Heidelberg New York

Springer is a part of Springer Science+Business Media
springer.com

Editor: Dr. Dieter Czeschlik, Heidelberg, Germany
Desk editor: Dr. Andrea Schlitzberger, Heidelberg, Germany
Cover design: WMXDesign GmbH, Heidelberg, Germany
Typesetting and production: Friedmut Kröner, Heidelberg, Germany

31/3152 YK – 5 4 3 2 1 0 – Printed on acid free paper

Preface

The two volumes on “Wetlands as a Natural Resource” in the book series Ecological Studies (Volumes 190, 191) are based on the highlights of the 7th INTECOL International Wetland Conference in Utrecht, 25–30 July 2004. This conference brought together about 900 participants from 61 countries, who discussed a very broad range of science-, policy- and management-oriented issues related to wetland ecology and hydrology, wetland conservation and creation, the impact of global change and wetlands as a resource in terms of food, flood protection and water quality enhancement. The participants were from different sectors of society, i.e., science and technology (scientists 45 %; PhD students 20 %), natural resource management (20 %) and policy (15 %). There were 38 symposia with invited speakers centered around the nine conference themes. We have given the organizers of these symposia the opportunity to produce one chapter for these books with the integrated content of their symposium. This has resulted in 25 chapters, of which 13 are included in Volume 190 under the heading “Wetlands and Natural Resource Management” and 12 in Volume 191 under the heading “Wetlands: Functioning, Biodiversity Conservation and Restoration”.

With these books, we had the aim to summarize the most important recent scientific results in wetland science, their applications in wetland and water resource management and their implications for the development of global, national and regional policies in the perspective of the ever-progressing deterioration of natural wetlands and the major impacts that future climate change will have. We hope that the integrated content of the chapters on such a wide scope of different fields in wetland science will serve as a valuable source of information, both for professionals in environmental science and natural resource management and for students and young professionals seeking to familiarize themselves with these fields. We also hope that the interaction between scientists from different disciplines, resource managers and policy makers will be stimulated by the content of these publications.

We as editors have worked according to a strict time schedule and we want to thank the authors for their timeliness in producing inspiring manuscripts and the scientists who have contributed to the peer reviews of the chapters for their active and prompt participation, which has enabled us to complete our task more or less according to this schedule. We acknowledge the series editor of the Ecological Studies book series, Prof. Dr. Ulrich Sommer, for his invitation to produce these volumes as one of the outcomes of the INTECOL Conference. We also thank Dr. Andrea Schlitzberger of Springer for her advice and help. We would like to take the opportunity to thank all key people who made the conference into such a success. In particular we want to thank Prof. Dr. Eugene Turner and the other members of the INTECOL Wetlands Working group, as well as the

International and National Scientific Committees for their support. We are most indebted to the team that organized the conference, in particular the inner circle, Fred Knol, René Kwant, Nienke Pot and Miranda Motshagen. The members of the Landscape Ecology Group at Utrecht University are thanked for their enormous efforts during the conference.

These two volumes are the most tangible, durable result of the conference. It is our wish that they will find their way to wetland professionals and students worldwide and will contribute to the wise use and conservation of the still large wetland resources that remain on our planet.

Utrecht, June 2006

The Editors

Jos T.A. Verhoeven,
Roland Bobbink,
Boudewijn Beltman,
Dennis F. Whigham

These two volumes are major contributions from a well-run meeting inspired by the collegiality and good will of the hosts. This meeting sparked professionalism through the expression of the finer parts of Dutch culture and, indeed, of all cultures. The 7th International Wetland Conference, like the preceding meetings, are successful because people care about living systems – i.e., people, landscapes, science culture, political structures, birds, etc. – as they go about trying to make things a little better and a little sooner than when wetlands were first appreciated in their collective minds. The successes from the meetings, exemplified by these two volumes, is partly because they enhance the possibilities for clarity and develop a strong scientific enterprise amidst the interactions of people in neutral spaces and a sometimes strong gradient of personalities and cultures. We never quite know ahead of time what the results of the meetings will be, although it has always been wonderful to see them evolve to closure.

It is humbling to know how small things influence others, which is a lesson in being careful, thoughtful and open. These efforts and successes are an explicit recognition of the interdependency of our discipline interests, but also the fabric of human interactions through politics, science, economics, etc. This interdependency suggests that being involved in wetland science and management is a great way to improve the quality of the natural world, but also society. The world needs, whether it knows it or not, the expertise and clear thinking of experts of general and detailed understanding to contribute to the social good. These two volumes do exactly that. Kudos to the Editors!

R. Eugene Turner, Chair
On behalf of the INTECOL Wetland Working Group

Contents

Section II Wetland Science for Environmental Management

Section IV Wetlands and Climate Change Worldwide

Contributors

CLAUDE AMOROS

Université Claude Bernard, Lyon I, 69622 Villeurbanne Cedex, France,
e-mail: amoros@univ-lyon1.fr

ANGELA ARTHINGTON

Centre for Riverine Landscapes, Faculty of Environmental Sciences,
Griffith University, Brisbane, Australia

JAN P. BAKKER

Community and Conservation Ecology Group, Centre for Ecological and
Evolutionary Studies, University of Groningen, PO Box 14, 9750 AA Haren,
The Netherlands, e-mail: j.p.bakker@rug.nl

BOUDEWIJN BELTMAN

Landscape Ecology, Institute of Environmental Biology, Utrecht University,
PO Box 80084, 3508 TB Utrecht, The Netherlands,
e-mail: b.beltman@bio.uu.nl

CHRISTOPHE BENE

WorldFish Center – Cairo Office, 3 Abou El Feda Street, PO Box 1261, Maadi,
Cairo, Egypt, e-mail: cbene@worldfish-eg.org

WLADIMIR BLEUTEN

Department of Physical Geography, Utrecht University, PO Box 80115,
3508 TC Utrecht, The Netherlands, e-mail: w.bleuten@geo.uu.nl

ROLAND BOBBINK

Landscape Ecology, Institute of Environmental Biology, Utrecht University,
PO Box 80084, 3508 TB Utrecht, The Netherlands,
e-mail: r.bobbink@bio.uu.nl

PAUL L.E. BODELIER

Centre for Limnology, Department of Microbial Ecology, Netherlands Institute of Ecology (NIOO-KNAW), PO Box 1299, 3600 BG Maarssen, The Netherlands, e-mail: p.bodelier@nioo.knaw.nl

WIEBE BORREN

Centre for Limnology, Department of Microbial Wetland Ecology, Netherlands Institute of Ecology (NIOO-KNAW) Rijkstraatweg 6, 3631 AC Nieuwersluis, The Netherlands, e-mail: w.borren@geo.uu.nl

S. BRIDGHAM

University of Oregon, Eugene, USA, e-mail: Bridgham@uoregon.edu

HANS BRIX

Department of Biological Sciences, University of Århus, Ole Worms Allé, Building 135, 8000 Århus C, Denmark, e-mail: hans.brix@biology.au.dk

CATE A. BROWN

Southern Waters Ecological Research and Consulting, Zoology Department, University of Cape Town, Private Bag Rondebosch, South Africa, e-mail: cbrown@southernwaters.co.za

R.E. BRUMMETT

The WorldFish Center, Cameroon

DONALD R. CAHOON

US Geological Survey, Patuxent Wildlife Research Center, Laurel, Maryland, USA, e-mail: don_cahoon@usgs.gov

HUGO COOPS

RIZA, PO Box 17, 8200 AA Lelystad, The Netherlands, e-mail: h.coops@riza.rws.minvenw.nl

R. CORSTANJE

University of Florida, Gainesville, Florida, USA, e-mail: ron.corstanje@bbrsc.ac.uk

ANDRES CÓZAR

Dipartimento Scienze e Tecnologie Chimiche, University of Siena, 53100 Siena, Italy, e-mail: cozar@uca.es

Sanjini de Silva

International Water Management Institute, PO Box 2075, Colombo, Sri Lanka, e-mail: sanjini.desilva@cgiar.org

P. Denny

UNESCO-IHE, Institute for Water Education, Delft, The Netherlands

Harold L. Drake

Department of Ecological Microbiology, University of Bayreuth, 95440 Bayreuth, Germany, e-mail: HLD@uni-bayreuth.de

Rudi H. Drent

Animal Ecology Group, Centre for Ecological and Evolutionary Studies, University of Groningen, PO Box 14, 9750 AA Haren, The Netherlands, e-mail: r.h.drent@rug.nl

Patrick Dugan

WorldFish Center – Cairo Office, 3 Abou El Feda Street, PO Box 1261, Maadi, Cairo, Egypt

Peter Frenzel

Max Planck Institut für terrestrische Mikrobiologie, Karl-von-Frisch-Straße, 35043 Marburg, Germany, e-mail: peter.frenzel@mpi-marburg.mpg.de

Paul H. Glaser

Department of Geology & Geophysics, University of Minnesota, Minneapolis, MN 55455, USA, e-mail: glase001@umn.edu

Margaret Greenway

School of Environmental Engineering and Cooperative Research Centre for Catchment Hydrology, Griffith University, Nathan, Queensland 4111, Australia, e-mail: m.greenway@griffith.edu.au

M.R. Hasan

FAO of the United Nations, Rome, Italy

Thomas Hein

University of Vienna, Institute of Ecology and Conservation Biology, Althanstrasse 14, 1090 Vienna, Austria. Present address: Institute of Hydrobiology and Aquatic Ecosystem Management, Department of Water – Atmosphere – Environment, University of natural Resources and Applied Life Sciences, Vienna, Max Emanuel Strasse 17, A 1180 Vienna, Austria, e-mail: thomas.hein@boku.ac.at

Philippe F. Hensel

US Geological Survey, Patuxent Wildlife Research Center, Laurel, Maryland, USA, e-mail: philippe_hensel@usgs.gov

Thomas Hurek

Laboratory for General Microbiology, Faculty of Biology, University of Bremen, PO Box 33.04.40, Germany, e-mail: thurek@uni-bremen.de

Robert L. Jefferies

Department of Botany, University of Toronto, 25 Willcocks Street, Toronto, Ontario M5S 3B2, Canada, e-mail: jefferie@botany.utoronto.ca

Hans Joosten

Ernst-Moritz-Arndt Universität, Botanisches Institut, Greifswald, Germany, e-mail: joosten@uni-greifswald.de

Rose C. Kaggwa

PO Box 16080, Kampala, Uganda, e-mail: rose.kaggwa@nwsc.co.ug

Frank Kansiime

Institute of Environment & Natural Resources, Makerere University, Kampala, Uganda, e-mail: fkansiime@muienr.mak.ac.ug

Peter Kelderman

UNESCO-IHE Institute for Water Education, Delft, The Netherlands, e-mail: p.kelderman@unesco-ihe.org

Jackie M. King

Southern Waters Ecological Research and Consulting, Zoology Department, University of Cape Town, Private Bag Rondebosch, South Africa

J. Kipkemboi

Department of Zoology, Egerton University, PO Box 536, Njoro, Kenya

Kirsten Küsel

Limnology Research Group, Institute of Ecology, University of Jena, 07745 Jena, Germany, e-mail: kirsten.kuesel@uni-jena.de

Elena D. Lapshina

Institute of Biology and Biophysics, Tomsk State University,
Prospekt Lenina 36, 634050 Tomsk, Russia, e-mail: edlapshina@hotmail.com

Steven Loiselle

Dipartimento Scienze e Tecnologie Chimiche, University of Siena,
53100 Siena, Italy, e-mail: loiselle@unisi.it

Charles Lovell

Department of Biological Sciences, University of South Carolina, Columbia,
SC 29208, USA, e-mail: lovell@biol.sc.edu

Markku Mäkilä

Geological Survey of Finland, PO Box 96, FIN-02151 Espoo, Finland,
e-mail: markku.makila@gsf.fi

Ülo Mander

Institute of Geography, University of Tartu, 46 Vanemuise St., EE-51014,
Tartu, Estonia, e-mail: mander@ut.ee

Zhanpo Mao

Research Center for Eco-Environmental Science, CAS, Beijing 100085,
P.R. China

N.S. Mattson

Mekong River Commission, Fisheries Programme, PO Box 7035, Vientiane,
Lao PDR

Karen L. McKee

US Geological Survey, National Wetlands Research Center, Lafayette,
Louisiana USA, e-mail: karen_mckee@usgs.gov

Patrick Megonigal

Smithsonian Environmental Research Center, PO Box 28,
Edgewater, MD 21037, USA, e-mail: megonigal@serc.si.edu

Aaron Mills

Kennedy Space Center, USA, e-mail: aaron.l.mills@nasa.gov

TATIANA MINAYEVA

Central Forest Biosphere Nature Reserve, Tver region, Nelidovo District, p/o Zapovednik 172513, Russia, e-mail: tminaeva@wwf.ru
S. NEWMAN

South Florida Water Management District, West Palm Beach, Florida, USA, e-mail: snewman@sfwmd.gov

A. OGRAM

2169 McCarty Hall, University of Florida, Gainesville, FL 32611-0290, USA, e-mail: avo@mail.ifas.ufl.edu

K. PORTIER

University of Florida, Gainesville, Florida, USA, e-mail: portier@ufl.edu

GERRY QUINN

School of Biological Sciences, Monash University, Victoria, 3800, Australia. Present address: Deakin University, PO Box 423, Warrnambool, Victoria, 3280, Australia, e-mail: gerry.quinn@deakin.edu.au

DENISE J. REED

Department of Geology and Geophysics, University of New Orleans, New Orleans, Louisiana, USA, e-mail: djreed@uno.edu

BARBARA REINHOLD-HUREK

Laboratory for General Microbiology, Faculty of Biology, University of Bremen, PO Box 33.04.40, Germany, e-mail: breinhold@uni-bremen.de

FRANK RIJSBERMAN

International Water Management Institute, PO Box 2075, Colombo, Sri Lanka, e-mail: f.rijsberman@cgiar.org

NEIL SAINTILAN

Rivers and Wetland Unit, Science and Policy Division, Department of Environment and Conservation, 59-61 Goulburn St, Sydney, PO Box A290 Sydney South NSW 1232, Australia, e-mail: neil.saintilan@environment.nsw.gov.au

MATT SAUNDERS

Department of Botany, Trinity College, Dublin, Ireland, e-mail: saunderm@tcd.ie

BAOQING SHAN

Research Center for Eco-Environmental Science, CAS, Beijing 100085, P.R. China

DON SIEGEL

Department of Earth Science, Syracuse University, Syracuse, NY 13244, USA, e-mail: disiegel@mailbox.syr.edu

SILVIO SIMONIT

Environment Department, University of York, York, UK, e-mail: silvio.simonit@undp.org

A. SIRIN

Laboratory of Peatland Forestry and Hydrology, Institute of Forest Science Russian Academy of Sciences, Uspenskoye 143030, Moscow Region, Russia, e-mail: sirin@proc.ru

BRIAN SORRELL

National Institute of Water and Atmospheric Research, PO Box 8602, Christchurch, New Zealand, e-mail: b.sorrell@niwa.cri.nz

TOM SPENCER

Depatment of Geography, University of Cambridge, Cambridge, UK, e-mail: ts111@hermes.cam.ac.uk

VASU SUGUNAN

WorldFish Center – Cairo Office, 3 Abou El Feda Street, PO Box 1261, Maadi, Cairo, Egypt

KLEMENT TOCKNER

EAWAG/ETH, 8600 Dübendorf, Switzerland, e-mail: tockner@eawag.ch

KARIN TONDERSKI

Department of Physics, Measurement Technology, Chemistry and Biology, Linköping University, SE-58183 Linköping, Sweden, e-mail: karsu@ifm.liu.se

TAKEO TSUCHIHARA

Satoshi Ishida, Masayuki Imaizumi: 2-1-6 Kannondai, Tsukuba, Ibaraki 305-8609, Japan, e-mail: takeo428@nkk.affrc.go.jp

Anne van Dam

UNESCO-IHE Institute for Water Education, Delft, The Netherlands,
e-mail: a.vandam@unesco-ihe.org

K.K. Vass

National Research Centre on Coldwater Fisheries (ICAR), Bhimtal, Nanital,
Uttranchal, India

Jos T.A. Verhoeven

Landscape Ecology, Institute of Environmental Biology, Utrecht University,
PO Box 800.84, 3508 TB Utrecht, The Netherlands,
e-mail: j.t.a.verhoeven@bio.uu.nl

Jan Vymazal

ENKI, o.p.s., Říčanova 40, 169 00 Praha 6, Czech Republic and
Nicholas School of the Environment and Earth Sciences, Duke University
Wetland Center, Box 90333, Durham, NC 27708, USA,
e-mail: vymazal@yahoo.com

Martin J. Wassen

Utrecht University, Copernicus Institute, Box 80.115, 3508 TC Utrecht,
The Netherlands, e-mail: m.wassen@geog.uu.nl

Robin L. Welcomme

Renewable Resources Assessment Group, Department of Environmental
Science and Technology, Imperial College, London, UK,
e-mail: welcomme@dsl.pipex.com

R. Wetzel

(deceased)

Dennis F. Whigham

Smithsonian Environmental Research Center, PO Box 28, Edgewater,
MD 21037, USA, e-mail: whighamd@si.edu

Chengqing Yin

Research Center for Eco-Environmental Science, CAS, Beijing 100085,
P.R. China, e-mail: cqyin@mail.rcees.ac.cn

1 Wetland Functioning in a Changing World: Implications for Natural Resources Management

Jos T.A. Verhoeven, Boudewijn Beltman, Dennis F. Whigham, Roland Bobbink

1.1 Introduction

Wetland ecosystems are a natural resource of global significance. Historically, their high level of plant and animal (especially bird) diversity is perhaps the major reason why wetland protection has become a high priority worldwide, supported by international agreements, such as the Ramsar Convention and the International Convention of Biological Diversity (Fig. 1.1). More recently, a number of goods and services provided specifically by wetland ecosystems have been identified that may even outweigh biodiversity in terms of their importance for human welfare and sustainable natural resource management worldwide. Wetlands, as transitional zones between land and water, provide a natural protection against extreme floods and storm surges. They may also store freshwater to be used for drinking-water preparation or for irrigation. Wetlands bordering streams, rivers and lakes have a water quality enhancement function that is increasingly recognized. As natural habitats for fish, riverine wetlands, shallow lakes and coastal wetlands have the potential to produce large fish stocks, which are exploited commercially in some regions but could be enhanced by restoring wetlands in degraded areas. Because wetlands often provide spawning habitats, their importance as a source of juvenile fish for large aquatic lakes and river channels should not be underestimated. In addition to these local and regional benefits, wetlands as a global resource provide a net sink of carbon dioxide. The world's peatlands are the only type of terrestrial ecosystem with a long-term net carbon storage function. However, the large amounts of carbon that have accumulated historically in peatlands may be released as a result of drainage or excavation.

Wetlands do produce a striking variety of goods and services and it is no wonder that, more often than any other terrestrial ecosystem, they are used by

Ecological Studies, Vol. 190
J.T.A. Verhoeven, B. Beltman, R. Bobbink, and D.F. Whigham (Eds.)
Wetlands and Natural Resource Management

environmental economists to illustrate ecosystem functions and their value to mankind. However, in spite of the high biodiversity and the high importance of the goods and services of wetland ecosystems, their global status is poor. Many wetlands, particularly river floodplains, deltas and estuaries, have been strongly degraded because of human impacts. Early civilizations were particularly successful in these areas, where agriculture thrived because of the natural fertility of the soils and transport was favoured by the river channels. In the industrial era, these impacts became dramatically negative as a result of floodplain reclamation, poldering, construction of flood control structures, drainage for agriculture, excavation of peat for fuel and modification and straightening of river channels in favour of navigation. Worldwide, more than 50 % of the wetland resource has been lost because of these reasons. In some densely populated regions in Europe, North America and East Asia, more than 80 % of the wetlands have been lost or severely degraded.

This volume, containing an integrated account of a number of major symposia presented at the 7th INTECOL International Wetlands Conference in Utrecht, investigates the major natural resource management issues involved in the protection of the remaining wetland resource, the enhancement of the goods and services arising from this resource and the restoration of degraded wetlands and wetland functions. In this introductory chapter, we will give an overview of recent advances in the comprehension of how both wetland biodiversity and the wetland ecosystem goods and services can be enhanced by management decisions, as treated in more detail in the other chapters of this volume. We will also identify remaining gaps in scientific knowledge and understanding that need to be addressed to optimize the decision-making process on wetland land use and management.

1.2 Clarity on Wetlands and Water Use

It is widely recognized that the limited availability of clean freshwater will increasingly become a matter of controversy between local communities in many semi-arid regions of the world. Access to healthy freshwater resources has even been identified as a fundamental human right. The relation between wetlands and the availability of freshwater recently led to confusion among natural resource managers. As Rijsberman and De Silva point out in Chapter 3, one of the services of wetland ecosystems was described as the 'providing' or 'provision' of water. This service would suggest that wetlands are sources of water and do not compete with other water-demanding sectors such as agriculture or water use for sanitation or industry. In this view, wetlands would even be potential sources of water. However, in reality, wetlands are as much dependent on water as these other sectors. Being systems with a high water table, they can only maintain their characteristic biota and functioning if

Fig. 1.1 Species-rich wet meadow with *Dactylorhiza majalis* (broad-leaved marsh orchid) bordering brackish pools on the island of Texel, The Netherlands

Fig. 1.2 *Cladium jamaicense* (sawgrass) lawns with tree islands in Shark Slough, Everglades National Park

Fig. 1.3 A stand of *Nelumbo nucifera* (water lotus) in Kakadu National Park, Australia. The water lotus is a typical wetland plant with aerenchyma

water outputs are balanced by water inputs, in a way typical for the water regime of the wetland in question. Most wetland types require inputs of surface water or groundwater in addition to the inflow through precipitation. In practice, wetlands often compete for water with agriculture or drinking-water preparation, in particular in semi-arid regions.

A first example is the use of water in many (sub)tropical countries for irrigation, which has led to the drying-out of vast wetland resources. Water resource managers often are mainly involved with the so-called 'blue' water resources, mainly to be used for urban and industrial use. The water used for irrigation, mostly present as soil moisture, also leads to a major regional water loss which is equal to the amount of water evapotranspired by the crops. This so-called 'green' water use is now increasingly being addressed in an integrated way with other water uses by water resource managers. The notion that wetlands 'provide' water has been nuanced: water conservation may be optimized by using wetlands as reservoirs where water can be temporarily stored.

A second example of a controversy on the uses of limited water resources is the situation in the Everglades, Florida. Here, a large freshwater surface resource flowing over the land surface south of Lake Okeechobee is used increasingly for urban, agricultural and industrial purposes by the metropolitan region surrounding Miami. As a result, the large wilderness area of the Everglades wetlands (see Fig. 1.2), partly protected in nature reserves (e.g. Everglades National Park), is suffering from water shortage and is threatened on the longer term. A multi-million dollar Comprehensive Everglades Restoration Plan (CERP; http://www.evergladesplan.org/index.cfm) has been designed to mitigate water shortages in the future. It remains to be seen whether this will provide a sustainable solution to the protection of the Everglades and its many biota and other values.

1.3 Wetlands and Environmental Flows

The definition and implementation of 'environmental flows' have become a major management tools in river catchments worldwide, particularly in semi-arid regions. The idea is that so much water in river systems is diverted and used for agriculture, cities or industry that rivers can no longer function naturally. River flow and flood events are increasingly limited to a narrow zone bordering the river channel, while the lateral interactions with the often extensive floodplains become diminished. This has drastic negative consequences for the biota characteristic for floodplains and for the goods and services provided by the floodplain habitat. Environmental flows (EFs) are defined as a minimum river discharge needed to meet certain targets in terms of biodiversity and/or ecosystem goods and services. Restoration of lateral connectivity by bringing floodwater only to selected parts of the original floodplain may help in restoring the intensity and temporal dynamics of typical flooding events, rather than allowing the water to create too small a flooding event in the total floodplain. This is illustrated for rivers in Australia by Coops et al. (Chapter 2) and for tropical rivers by Welcomme et al. (Chapter 7). The EF concept is in the stage of becoming widely accepted among water resource managers as a tool to maximize the quality of biodiversity restoration and associated fisheries in large river floodplains with diminished river discharge. Hopeful developments have been initiated in several large European river catchments, where river rehabilitation projects have focused on: (1) the restoration of lateral connections by removing so-called 'summer dikes', which has resulted in a higher frequency of flooding of floodplain habitats, (2) the restoration of side-channels and river dunes and (3) the enhancement of river water quality. Some of the projects described in Chapter 2 illustrate the successful restoration of floodplain habitats in the basins of the rivers Rhine, Rhone and Danube.

1.4 Wetlands and Water Quality

The role of wetlands in river and lake catchments in enhancing water quality is well established. A recent review gives a global perspective of this ecosystem service in areas of the world with high intensity of agricultural use (Verhoeven et al. 2006). Riparian wetlands bordering lower-order streams and floodplains of mid-size and larger rivers have a great potential to remove nutrients and pollutants from through-flowing water. Nitrate in surface and subsurface runoff from agricultural fields and pastures, when exposed to superficial soil layers in the riparian zone, is transferred to gaseous nitrogen species and emitted to the atmosphere, while phosphate and ammonium are retained in vegetation or bacterial biomass, adsorbed to soil particles or laid down in sediments. Long-term loading of these zones, however, enriches these riparian wetlands, which often leads to the loss of characteristic species. Critical loading rates for N and P have been established for freshwater wetlands, beyond which losses of plant and animal species are to be expected. Riparian zones have also been shown to be only effective at the catchment scale if they are sufficiently large and continuous in the landscape. Only when their total area comes near 5 % of the total catchment area can they really make a difference to water quality in the catchment. Wetlands restoration schemes in agricultural areas should take into account these limitations.

The chapter by Yin et al. (Chapter 4) deals with a rural system for water resources management which has been developed in the southern part of China. This so-called 'multipond' system, a system of many shallow ponds in the landscape connected by ditches, is the result of 2000 years of engineering experience and combines benefits such as water storage, flood protection and water quality enhancement. Another example of the pivotal role of natural wetlands in this respect is given by Loiselle et al. (Chapter 6) for the extensive papyrus wetlands around Lake Victoria in Africa. These wetlands are enormously important to halt the eutrophication of the large lake. Many local communities around the lake depend on the fisheries as their main source of protein. In addition to the nutrient removal service, these wetlands provide a number of other important goods and services, such as papyrus stems, protection from damage by storm surges and a habitat for juvenile fish.

Another application of the water quality enhancement service of wetland ecosystems is the construction of wetlands just for this purpose. Vymazal et al. (Chapter 5) give a thorough review of the latest knowledge on the performance, efficacy and application of different kinds of constructed wetlands. The past 40 years of experience with these systems has taught us that the biogeochemical processes in wetlands can be used most optimally by designing and managing them in a 'tailor-made' way, according to the nature and concentration of the pollutants, the amount of discharge to be treated and the climatic conditions. Designs that combine different types of constructed wetlands in sequence ('hybrid systems') have been shown to be most effective. An example of such a system is a combination of a vertical-flow wetland, in which organic matter is broken down and nitrification can take place, followed by a horizontal-flow wetland in which denitrification reduces the N content of the water. In almost all cases, constructed wetlands form an inexpensive and sustainable alternative to more technological solutions which require higher energy inputs and more expensive investments.

A disadvantage of constructed wetlands is that they emit relatively high quantities of the greenhouse gases nitrous oxide and methane. This is particularly true for highly nutrient-loaded wetlands, which mostly have only a limited surface. Better understanding of the factors controlling emission rates of these gases may give additional guidelines for the management of these systems.

1.5 Biogeochemical Insights

The biogeochemical functioning of wetlands is a very challenging and complex research subject. The high water table near the soil surface and the temporal patterns of water level fluctuations create many redox gradients in space and time, which leads to complex interactions among plants, microbes and geochemical processes. Redox cycles are often driven by the activity of aerobic and anaerobic bacteria and by the availability of alternative electron acceptors such as nitrate, iron, manganese, sulphate and carbon dioxide. The role of the plants is that they produce organic substances in the rhizosphere through leakage of dissolved substances and death of plant parts, summarized as 'rhizodeposition'. Further, wetland plants have aerenchyma, a system of air-filled cavities bringing oxygen to the roots and rhizomes (Fig. 1.3). Oxygen leaks into the rhizosphere to a certain degree.

Recent studies have focused on the role of bacteria in wetland rhizospheres. Bodelier et al. (Chapter 10) give a fascinating overview of recent results in this field. New molecular methodologies such as micro-array technology have enabled estimations of the diversity of various functional groups of bacteria. Large differences in diversity were found. Diazotrophic microbes, which are capable of fixing atmospheric nitrogen, show a remarkable high diversity in wetland rhizosphere systems. This high degree of functional redundancy ensures the sustaining of N_2 fixation under a wide range of different environmental conditions. This indicates the great importance of this ecological function for the ecosystem. Microbes active in the cycling of iron in wetland soils have been found to show a rather poor diversity. Results of new studies of these bacteria has revealed that the redox cycles involving iron in wetland soils are primarily mediated by microbes and not by physicochemical processes, as previously generally assumed. Much progress has also been made in the study of methanogenic and methane-consuming bacteria in wetlands. The balance between methanogenesis and methane oxidation determines the emission of methane, an important greenhouse gas. Knowledge on the ecophysiology and identity of the groups of bacteria involved in the two processes is urgently needed to understand the conditions favouring methane emission and to develop management approaches to minimize such emissions from wetlands.

A major bottleneck for linking this new information of the functional ecology of microbial taxa to the fluxes of matter and energy at the ecosystem level is the enormous difference in the scales in space and time at which the microbes operate, in comparison to the vegetation or the environmental dynamics such as flooding or sedimentation. New integrative approaches combining the expertise of microbiologists, geochemists, landscape ecologists and statisticians is needed for such linkages. Ogram et al. (Chapter 11) explain some of these approaches and apply a combination of a clustering method, stepwise discrimination and canonical discrimination to a large data set of biogeochemical studies made in the Northern Everglades. They were able to link microbial activity data (e.g. alkaline phosphatase activity) to gradients in total phosphorus richness; and they concluded in a more general way that microbial activities are demonstrably responsive to environmental variables. Together with the new methodologies in microbial ecology, these new approaches at the landscape-scale to investigate the implications of microbial activity at the system and landscape level, form a very promising new field in wetland ecology that needs further exploration by interdisciplinary teams.

The vast peatland areas in the northern hemisphere are to a large degree degraded because of human-induced drainage and/or peat cutting in the more densely populated, more southern areas of the peatland regions. Very large areas remain in a more or less pristine state in the remote, more northern areas in Siberia and Canada. Detailed study of the functioning of these large, pristine peatlands has lagged behind that of the degraded systems because they are so remote, inaccessible and large-scale. Bleuten et al. (Chapter 9) describe the ecohydrological functioning of large peatland complexes in Siberia and Canada and show that bog–fen gradients in these very extensive areas do not always follow standard patterns, because the hydrology in the top layer of the peatlands is so dominated by horizontal, rather superficial throughflow, transporting rain-water or groundwater over quite large distances. These hydrological processes have been modelled, revealing a better understanding of these vast wetlands which are so important globally because of their carbon storage function.

1.6 Wetlands and River Fisheries

One of the major goods produced by wetlands which are unambiguously valued by humans is fish. Freshwater wetlands associated with river or lake systems are a major spawning and feeding habitat for a whole range of fish species which are, often commercially, exploited by local fisheries. The same holds for brackish and saltwater wetlands which have a function as breeding grounds for juvenile fish later dispersing toward coastal seas or the oceans. River fisheries have decreased in importance in regions of the world where the large river catchments have been used for intensive agriculture and industry and the river channels have been diked and dammed for navigation and the production of electricity. While rivers in many temperate regions have lost their integrity, more naturally functioning rivers do still occur in north-temperate, boreal and (sub)tropical regions. Particularly in developing countries, the extra protein source provided by river fisheries is very important to support rural economies and their welfare. Welcomme et al. (Chapter 7) indicate the threats to river fisheries in tropical regions because of the gradual damage to the natural functioning of rivers because of human disturbances. The diversion of river water for irrigation, changes of the river channel for navigation and the building of dams become more and more prominent. They also point to the concept of EFs to help solve questions associated with unexpected negative effects of changes to river functioning for river fisheries. Based on the current understanding of the relation between river hydrology and the occurrence and productivity of major fish species, EFs can be determined which are required to maintain a certain level of fish production. This information on EFs can be used in the decision-making process to weigh the benefits of irrigated agriculture versus river fisheries. The authors describe methodologies to assess the types of EF necessary in a particular case and give examples where this approach has worked well in countries in southern Africa (Mozambique, Tanzania, Zambia). The chapter clearly shows how important water issues are in wetland resources management.

Apart from improving the hydrology, there are other ways to enhance the success of river fisheries in developing countries. In Chapter 8, Welcomme et al. give an overview of approaches to improve fish catches based on traditional experience of rural communities. Such methods include measures to optimize the landscape for fish catch, e.g. the construction of 'drain-in ponds', parallel depressions in floodplains to keep fish in the floodplain as the floods recede. Fish stocking in combination with the use of fertilizer is another well known method to enhance fish production. The chapter gives detailed examples of stocking procedures and investigates their opportunities and limitations. Often, stocking is a very successful method to increase fish catch, but caution should be used with respect to the use of non-native species and the level of fertilization. Species introductions have led to large-scale invasions, with negative consequences for the native fish populations. Eutrophication is the risk associated with too high a level of fertilization. A number of in-site measures are also discussed, e.g. 'brush parks', like the use of aquatic plants or reed screens to provide spawning habitat and shelter, or the use of water level control structures to manipulate water levels to facilitate both fish growth and fish catch. The wise use of such measures strongly contributes to riverine fish catch. Communication of this knowledge to wetland users around the world is very important.

1.7 Wetlands and Climate Change

All terrestrial ecosystems are anticipated to react strongly to the predicted climate change that will take place in the next century as a result of the enhanced greenhouse effect. Projected changes in climate mainly encompass an increase in average global temperatures and very distinct changes in rainfall patterns, with more extreme episodes of heavy rainfall and/or severe drought. Further, the mean sea level is expected to rise at least 30 cm, with some estimates predicting even 80 cm. The impacts of these climatic changes are expected to be exceptionally large for the world's wetland ecosystems. The balance between water inputs and outputs, which determines the water regime and ecological characteristics of all wetland ecosystems, will change, with enormous consequences, such as changes in the flooding frequency, amplitude and duration of large river–floodplain systems, paludification of low-lying areas, saltwater intrusion into coastal habitats, more complete and longer desiccation of wetlands and shrinking of peatlands.

Clues for understanding the impact of climatic changes on the rate of peat accrual in bogs and fens can be derived from stratigraphic research in peatlands. A study of Russian mires within the southern taiga belt in the northern hemisphere by Minayeva et al. (Chapter 13) clearly demonstrates substantial changes in peat growth and carbon accumulation in response to variations in palaeoclimatic conditions. Bog sites that had little or no contact with groundwater were more sensitive to palaeoclimate fluctuations than groundwater-fed mires.

Peat accumulation rates at the Malaya–Icha rain-fed bog site in West Siberia varied especially during the early Holocene as a result of changes in the sequestration rate, which can be attributed to climate changes. Within the Holocene, there was a steady growth of peatland area in this region, resulting in an expansion of peat deposits and increased carbon storage. Many Eurasian mires tended to grow faster in cold periods and to slow down in warmer periods during the past three millennia. Even if the vegetation of specific mire did not react to minor climatic changes (according to data from macrofossil analysis), such changes were reflected in the rates of carbon accumulation. A future climate change resulting in a temperature rise (IPCC 2001) can enhance primary production and decomposition in mires, with a net negative effect on carbon sequestration, particularly when the climate becomes drier. However, if precipitation increases more than evapotranspiration combined with higher temperatures, a rapid peat growth and higher carbon sequestration may occur locally.

Coastal wetlands such as saltmarshes and mangroves are, more than any other wetland type, sensitive to changes in mean sea level. Through the interaction of biotic and abiotic processes, their elevation often gradually increases and can keep up with sea level rise. Plant growth and associated organic matter accumulation, sedimentation, algal mat formation and animal burrowing all contribute to slowly increasing ground levels and a lateral expansion in coastal wetlands. It is clear that increased rates of global sea level rise as a result of climate change can be expected to have drastic impacts on the extent and functioning of these wetlands. However, quantitative relationships between sedimentation and erosion and the role of plants and animals in coastal wetlands, as influenced by sea level rise, are still almost unknown. Cahoon et al. (Chapter 12) report on the SET-MH methodology as an example to investigate these relationships. SET stands for *surface elevation tables*, permanently installed measuring devices for assessing marsh elevation, whereas MH means *marker horizons*, layers of dyed clay which can be buried at a known depth to measure sediment accretion or erosion.

Currently, there is an informal SET-MH network of scientists in different parts of the world who have installed these devices in coastal wetlands and exchange information on their results. In some areas of the United States and Europe, there are already more than 15 years of data available. It has become clear that many coastal wetlands do increase their elevation at the same pace as the current sea level rise. Large differences in rates of surface accretion have been found in relation to sediment inputs, erosion phenomena, macrophyte productivity and algal growth. These data are being used in models describing soil accretion in relation to environmental factors. The models are of great value for predictions of the effects of future enhanced sea level rise scenarios associated with climate change. First results of such analyses indicate large losses of coastal wetland area as a result of increased sea level rise. More research is needed to evaluate the more indirect effects of climate change on marsh accretion, e.g. through modified river discharge and groundwater flow, changes in plant growth, shifts in the latitudinal distribution of plant communities and the way these interact with the effects of sea level rise.

Another way to evaluate the consequences of climate change on wetland ecosystems is to investigate the response of key species or functional groups. The chapter by Jefferies et al. (Chapter 13) gives an overview of the responses of goose species to changes in agricultural land use and climate in the northern hemisphere. It is clear that these changes have strongly influenced the population numbers and migration behaviour of goose species frequenting wetlands and, increasingly, also crop fields and pastures. The Arctic-breeding goose species (i.e. barnacle goose and Brent goose) breed in northern tundra regions and traditionally used to winter-feed in wetlands, such as saltmarshes and seagrass beds. In the past 50 years, they have increasingly started to feed on agricultural croplands and pastures while wintering. As agriculture has become much more intensive, with higher crop production and higher food quality, the population numbers of most goose species have increased strongly. This has become quite apparent in the numbers of wintering geese in warm/temperate regions on both sides of the Atlantic. Apart from this land-use-driven change, global climate change will also influence the population dynamics of goose species. During the spring migration from their wintering grounds to the breeding areas, the Arctic-breeding species, at least in Europe, need to stage at grassland or wetland sites for extra foraging during the long migration flight. They move northward just in time to reach staging sites where the air temperature is between 3 °C and 6 °C, the temperature at which the onset of the spring growth of grasses occurs. In the event of increased global warming, it might be expected that the northerly spring migration will occur earlier. However, the presence of the boreal forest with its near absence of grass and croplands limits the ready accessibility of forage plants that respond to warmer temperatures early in the season. There are already indications in the migration behaviour of Brent goose and lesser snow goose that the subtle interplay between the direct and indirect effects of climate on plant growth and the foraging responses of geese can result in winter-range extensions and changes in the migration routes of birds. It is hard to predict whether these changes will lead to the selection of new staging sites and/or new breeding grounds at different locations, which may have further drastic impacts on population sizes. As geese, through their grazing activity, often have a major impact on the vegetation of the wetlands where they feed, these changes may drastically impact the wetland ecosystem functioning as well.

1.8 Further Developments in Wetland Science and its Applications

The importance of ecosystem services provided by wetlands justifies a high priority for scientific studies of wetland ecosystem functioning. Water quality improvement, enhancement of freshwater fish production, sustainable use of water for irrigation and river management for flood protection are natural resource management issues increasingly recognized by regional and national authorities. The scientific evidence for these ecosystem services is growing. The most important further scientific priorities remain: (1) the understanding of how these services can be optimized in a sustainable way, simultaneously ensuring the conservation and restoration of natural habitat for plants and animals and (2) the anticipation of changes that will be provoked by future climate change, by using scientific information on ecosystem functioning in predictive models. It will be a challenge to try to work with the changes in climate and associated regional water regimes to further approach the above-mentioned targets of sustainable wetland and water resource management.

References

IPCC (2001) Climate change 2001: the scientific basis. The carbon cycle and atmospheric carbon dioxide. Cambridge University Press, Cambridge, pp 183–237

Verhoeven JTA, Arheimer B, Yin C, Hefting MM (2006) Regional and global concerns over wetlands and water quality. Trends Ecol Evol 21:96–103

Section I

The Role of Wetlands for Integrated Water Resources Management: Putting Theory into Practice

2 Restoring Lateral Connections Between Rivers and Floodplains: Lessons from Rehabilitation Projects

Hugo Coops, Klement Tockner, Claude Amoros, Thomas Hein, Gerry Quinn

2.1 Introduction

Most large rivers have been strongly modified by human activities, including flow regulation and floodplain embankment. Out of the largest river systems in the northern hemisphere, at least 77 % can be considered to be moderately to strongly affected by dams (Dynesius and Nilsson 1994), having severe impacts on natural stream flow and floodplain hydrology. Additionally, inputs of pollutants are very high, mainly because river corridors are traditionally the places where urban and industrial development and agricultural production occur. The negative effects include loss of biodiversity, reduced biogeochemical processing and declining water provision for human use, recreation and ecosystems. For various large rivers in Europe and North America, rehabilitation plans have been formulated and initiated to achieve partial recovery from past degradation (Buijse et al. 2005). However, such initiatives often have started in isolation and without the full use of experiences gained in completed rehabilitation projects.

Despite growing information on their ecological functioning, many aspects of river–floodplain ecosystems are rather poorly understood. Limited understanding prohibits accurate forecasts of consequences of ecosystem manipulations and the attainability of restoration targets – even when a complete description of their patterns and processes has been made. Ecosystem responses to environmental modifications are usually not linear but stochastic and delayed, governed by catchment climate and hydrology, spatially and temporally interacting, and operating differently on ecological functions at widely varying scales. The complex spatio-temporal dynamics of floodplain ecosystems are reflected in their patterns of biodiversity (Ward et al. 1999). Intermittent flooding plays a key role in maintaining different successional stages within a variety of lentic, lotic and semi-aquatic habitat types. In these

Ecological Studies, Vol. 190
J.T.A. Verhoeven, B. Beltman, R. Bobbink, and D.F. Whigham (Eds.)
Wetlands and Natural Resource Management

dynamic environments, many biota depend on boundaries (transition zones between habitat patches) and different degrees of connectivity (interactions between habitat patches).

Until recently, the main river channel has been the key focus of river research. Lateral (semi-)aquatic habitats – ponds, backwaters, and tributary confluences – have been widely ignored or studied in isolation. A reason for the underestimation and undervaluation of lateral water bodies in river research is their almost complete absence in small headwater streams (where most river research has been carried out) and in heavily modified downstream sections. Lateral aquatic habitats are among the first landscape elements that disappear as a consequence of river regulation and flow control (comparable to vegetated islands and gravel bars). However, the understanding of the functional and structural role of these habitats along river corridors forms a prerequisite for successful and sustainable river management (Karaus 2005).

In this chapter, we present several case studies that highlight the restoration potential of interactions between the river and its floodplain and address their value for conservation and management of large rivers. First, we address the aquatic–terrestrial interface dynamics and the resulting biodiversity of lateral habitats within river corridors, as illustrated by riparian arthropods. Then, two local restoration cases are presented in more detail, in which river–floodplain connectivity was restored, aiming at enhancing species richness and nutrient processing, respectively. We also explore the concept of 'environmental flows' in semi-arid regions, an increasingly important tool for restoring ecological function and biodiversity of floodplain wetlands on large regulated rivers. We finish by pointing out what lessons should be drawn from these case studies, to the benefit of planning and design of new restoration projects and, eventually, for upscaling measures to the scale of river sections or entire river systems.

2.2 Threatened Life at the Aquatic–Terrestrial Interface

Only recently have ecologists begun to better understand the flow of matter and organisms across habitat boundaries, which is particularly tight in river–floodplain ecosystems (Polis et al. 2004). Floodplains are pulsed systems; therefore, organisms should be able to cope with large changes in the availability of resources. In particular, communities living at the aquatic–terrestrial interface are subject to pulsed resources, such as stranded organic matter and pulses of emerging aquatic insects. Recent investigations of the aggregative response of arthropods to short-term resource pulses demonstrated the different availability of riparian arthropods to react to changes in resource availability. Their response clearly depends on their general feeding strategy (Paetzold 2005). Alterations of the flow regime, channel modification, and removal of buffer strips can alter the riparian community indirectly by impeding the energy flow across the boundary (Paetzold and Tockner 2005).

Terrestrial shoreline communities consist of an abundant and often endangered arthropod community that is very susceptible to human impacts. Riparian arthropods can be used as sensitive indicators to assess the ecological integrity of river corridors, as already successfully applied in the Rhône–Thur river restoration project in Switzerland (A. Peter, personal communication). The great advantage of shoreline communities as indicators is that they can be applied in small streams and large rivers and they provide information not only on local habitat conditions but also on the functional linkage between water and land, a key advantage to the more traditional indicators (e.g. benthic invertebrate community structure).

In a recent study, Karaus (2005) quantified the species diversity of benthic invertebrates along three Alpine river corridors (Swiss Rhone, Thur, Tagliamento) by including the lateral dimension along each corridor. Between 14 and 17 1-km reaches along each corridor and four aquatic habitat types (when present) within each reach were sampled. Results clearly demonstrated that lateral habitats disproportionately contributed to longitudinal diversity (Fig. 2.1). Overall, 162 taxa of Ephemeroptera, Plecoptera, and Trichoptera were identified in 119 composite samples, which was between 73 % and 77 % of the total expected richness along each corridor. Lateral habitats con-

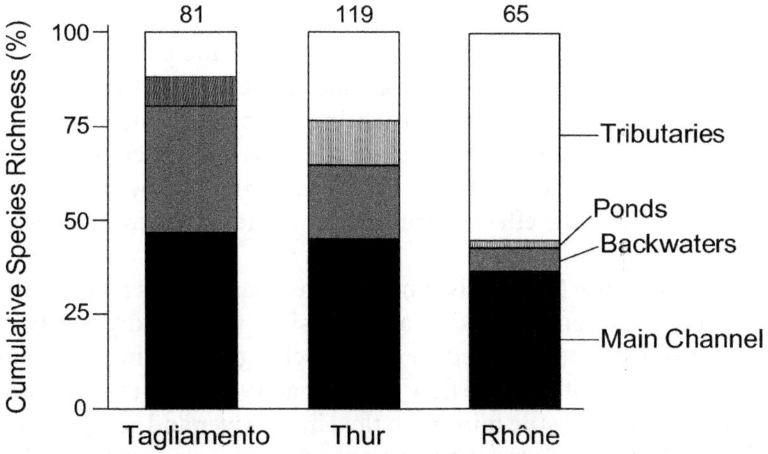

Fig. 2.1 Cumulative (%) and total species richness of Ephemeroptera, Plecoptera, and Trichoptera along three Alpine river corridors. For backwaters, ponds, and tributary confluences, only those taxa that did not occur in previous habitat types were added. Jack-knife analyses were used to calculate total expected taxon richness for each individual river corridor (source: U. Karaus and K. Tockner, unpublished data).

tributed >50% to total corridor species richness, although they covered <10% of the aquatic surface area (Karaus 2005). Further, diversity was hierarchically partitioned into its components (alpha, beta, gamma diversity) to quantify the relative contribution of individual samples, habitats, and corridors to the overall diversity of the three Alpine river corridors. Among-sample and among-corridor diversity components contributed most to total taxa richness, while <15% was due to within-sample and among-habitat components. This study clearly emphasised the importance of lateral aquatic habitats for maintaining high aquatic biodiversity along river corridors. Consequently, these habitats need to be fully integrated in future conservation and restoration projects; particularly since these are the first habitats that disappear as a consequence of river regulation and flow control.

2.3 Reconnecting Side-Channels Along the Rhône (France)

Existing side-channels along large lowland rivers have been re-opened in a number of cases to restore lateral connectivity. Side-channels along many large rivers in industrialised countries have been heavily impacted by construction and embankment to improve navigation. Embankments concentrate the flow in the riverbed into a single straight channel and result in riverbed incision. The important functions of navigation and flood control make it no longer feasible to restore the original state of ecosystems before degradation occurred. As an alternative target for ecosystem rehabilitation, the increase in biodiversity through the increase of habitat diversity between and within ecosystems may be used as a guiding principle to define targets in experimental rehabilitation (Amoros 2001). This principle will be illustrated by a project to reconnect side-channels along the Rhône River. In this case, habitat diversity was enhanced by orienting the project towards the restoration of: (a) hydraulic connectivity (including both surface connectivity to the main river and groundwater connectivity) and (b) flooding (including water level rise and scouring effects resulting in some cases from flow velocity increase).

In the sector of the Rhône River involved, embankments made in the nineteenth century had led to considerable incision, while hydro-electric works during the twentieth century reduced the discharge to a minimum flow varying from 1% to 30% of the base flow (Roux et al. 1989). The rehabilitation project included increasing the minimum flow in the river and re-opening several cut-off side-channels. Three different approaches were evaluated: (1) GRW (GRoundWater) – a channel selected for its groundwater supply and designed with a slope to enhance groundwater drainage, (2) BIC (BIConnected) – a channel close to the river and reconnected at both ends, (3) BKF (BacKFlow) – a channel reconnected at its downstream end in order to allow river backflow. The rehabilitated channels were compared with a CTR channel (ConTRol) that had been permanently connected at its downstream end. Bed excavation and sediment removal were carried out between February and August 1999, while the increase in the minimum flow of the river from 10–20 $m^3 s^{-1}$ to 100 $m^3 s^{-1}$ was achieved in July 2000.

Physico-chemical monitoring results demonstrated that, according to the expectations formulated before the project started, water supply was by river water in the BIC and BKF channels, whereas the GRW channel was mostly supplied by groundwater except during floods (Amoros et al. 2005). Sedimentation rates (monitoring the thickness of deposited sediment at 130 sampling plots) showed the highest sedimentation rate in BKF (7.59±4.56 cm $year^{-1}$) and the lowest in BIC (2.28±2.76 cm $year^{-1}$), with GRW in an intermediate position (4.17±2.91 cm $year^{-1}$). The differences in sedimentation rate of the three rehabilitated side-channels may be related to their connectivity and flooding effects. Evidently, the backwater flow of BKF traps sediments which are supplied by the Rhône River during high water levels and floods, with most of the deposition occurring at its downstream end (Amoros et al. 2005); disconnection of this channel is expected within 5–10 years. The low sedimentation rate observed in BIC (owing to permanent running water and high water velocities during floods) leads to the expectation of consolidated running-water conditions over a long period of time (more than 100 years). GRW is expected to be silted up in within 25–40 years; the intermediate position of GRW resulted from its length, reducing the effects of backflow, and upstream overbank flow which allowed scouring of some of the sediment deposited previously.

In the control channel (CTR), the number of aquatic plant species remained at eight to 12 after the works, whereas it increased in the three rehabilitated side-channels (Fig. 2.2A). The highest colonisation rate occurred in GRW between 1999 and 2000, but in later years the species number decreased. A slower colonisation in BKF continued until 2002. The decrease of species numbers in BIC and GRW during 2001 and 2002 followed the disappearance of species that colonised very sparsely just after the rehabilitation works; their extinction may have resulted from either unsuitability of habitat conditions within these side-channels or competitive exclusion, as indicated by high overall cover percentages. Similar patterns of early colonisation followed by reduction in species number resulting from competition have been reported in previous investigations (Barrat-Segretain 1996; Barrat-Segretain et al. 1999).

The numbers of species unique to one side-channel (Fig. 2.2B) were highest in GRW in the first three years after rehabilitation, remaining rather high afterwards and constituting 29% of the total species number in this side-channel in 2003. In BIC, unique species constituted 35% of the total species number in 2003. The numbers were lowest in BKF and in CTR.

The different connectivities resulted in a clear difference in vegetation composition between GRW and BIC (Amoros et al. 2005). Species characteris-

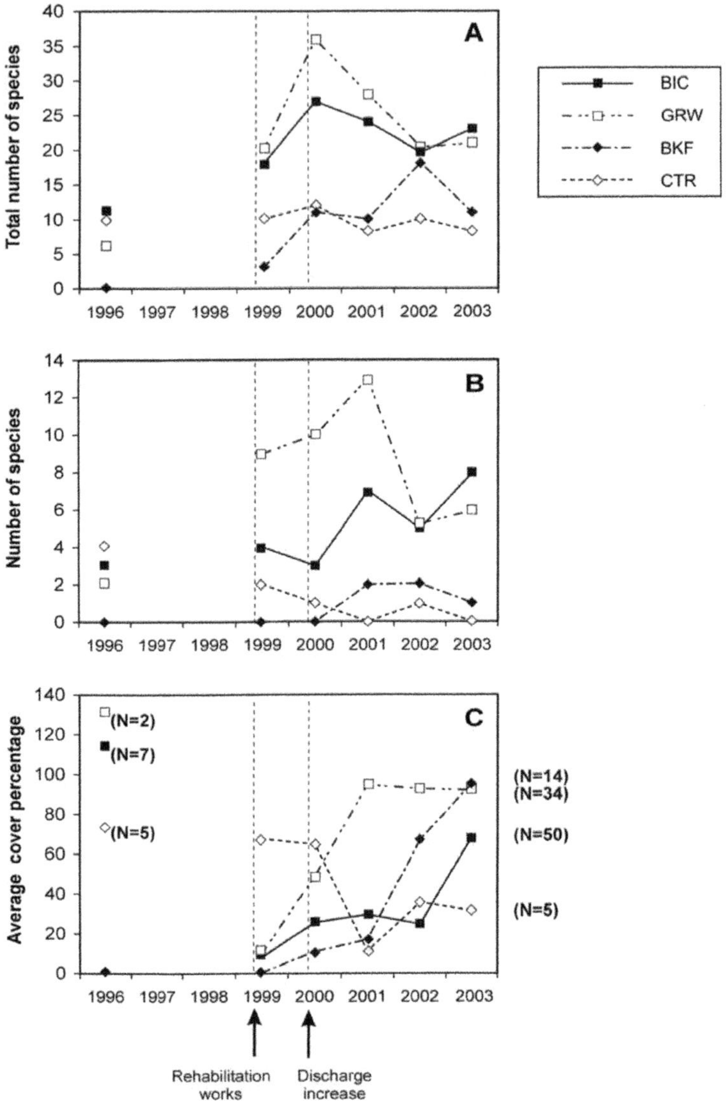

Fig. 2.2 Aquatic vegetation dynamics before (1996) and after (1999) rehabilitation works and the discharge increase in the by-passed river (2000). **A** Total species numbers in each side-channel. **B** Number of species occurring only in a single channel. **C** Total vegetation cover (%) averaged per sampling plot. *BIC* Biconnected channel, *GRW* flood-scoured and downstream-connected channel, *BKF* downstream-connected channel, *CTR* non-rehabilitated channel. Aquatic vegetation was surveyed in 1996, then annually in August–September after the works (from 1999 to 2003). Vegetation cover was recorded in 2-m sampling plots crossing the channel and distributed every 25 m along the rehabilitated side-channels (50 plots along BIC, 33 along GRW, 17 along BKF, five along CTR). Species cover was visually estimated as the percent cover of species on the plot, using the following cover classes: 1 =5 %, 2 =25 %, 3 =50 %, 4 =75 %, 5=100 % (Henry et al. 1996). Since species can occur in several strata, the cumulated vegetation cover on a plot may exceed 100 %.

tic for GRW were: *Berula erecta*, *Callitriche* sp., *Potamogeton crispus*, *Chara* sp., *Ranunculus circinatus*, *Veronica anagallis-aquatica*, *Juncus articulatus*, and *Alisma plantago-aquatica*, and for BIC: *Carex* sp., *Juncus subnodulosus*, *Veronica beccabunga*, *Najas marina*, and *Ranunculus fluitans*.

The role of connectivity was also demonstrated in the colonisation rate, with the two channels that were upstream-connected during floods (GRW and BIC) exhibiting a higher colonisation rate than BKF, which was only downstream-connected. Average vegetation cover (Fig. 2.2C) was very high in GRW and BIC before rehabilitation because the remaining standing-water bodies were very small and shallow, resulting in very crowded aquatic vegetation. The strong increase in average cover in GRW indicated a high colonisation rate from 1999 to 2001. BKF showed the smallest increase from 1999 to 2001, but increased sharply between 2001 and 2003, indicating a delayed response. The increase in BIC was moderate between 1999 and 2000 and cover remained relatively low (20–30 %) until a sharp increase occurred in 2003, probably associated with drought and subsequent low flow. Average cover in CTR was constant till 2000, but decreased sharply in 2001, suggesting a high scouring effect of spring floods.

The role of propagule drift during floods in the colonisation of the river side-channels, in relation to their connectivity, has been previously reported (Henry et al. 1994; Cellot et al. 1998; Amoros and Bornette 1999; Ward et al. 1999). However, a permanently high flow velocity, like in BIC (where it may reach 1.4 m s^{-1} in several stretches), may limit or delay the development of vegetation biomass (Chambers et al. 1991; Boeger 1992; French and Chambers 1996; Madsen et al. 2001).

2.4 Rehabilitation of Side-Channels of the River Danube (Austria)

The main focus of this section is to demonstrate the effects of rehabilitation on biogeochemical functions such as nutrient retention (phosphorus deposition, nitrate removal) and organic matter processing. In this section, we introduce the Danube Restoration Project (Austria) as an example of the reconnection of side-channels. The rehabilitation project downstream of Vienna (Haslau–Regelsbrunn floodplain) was pivotal for initiating further restoration measures within the 'Alluvial Zone National Park' and was an important step forward for the development of similar projects along other large braided rivers (Tockner et al. 1998). Its value stems mainly from the large size and rich ecological structuring of the affected area, the long-term interdisciplinary planning phase, and the comprehensive monitoring programme, which all provided advanced insights to the understanding of large-river floodplain ecosystems (Schiemer et al. 1999; Schiemer and Reckendorfer 2004). The aim has been to re-initiate key environmental processes – which once were typical for this section of the river – and thereby shifting the floodplain from a lentic backwater towards a more connected and lotic side-channel system.

In their pristine state, side-channels along the Upper Danube exhibited lotic conditions for most of the year. They could be arranged along a flow gradient from low to high velocities. The Danube exhibited a braided style, with shallow waters and highly unstable bars and islands. High sediment dynamics created large alluvial fans in this unconstrained section and the width of the floodplain was up to 10 km and more (Schiemer et al. 2001). After the major regulation scheme in 1875, long-term development led to reduced hydrologic connectivity and a dramatic loss of riverine habitats. Less than 20% of the original inundation area remained after the construction of flood protection dams. Embankments reduced surface connectivity and geomorphic dynamics of the active channel. The remaining high natural values and the potential for restoring key ecological processes led to the declaration of a national park in 1996 (Alluvial Zone National Park; Schiemer et al. 1999).

The Regelsbrunn floodplain area had a high potential for hydrologic exchange prior to rehabilitation (Tockner et al. 1998; Schiemer et al. 1999). The restoration programme, carried out between 1993 and 1997, included the re-connection of the side-channel to the main Danube channel at six inlets over an entire stretch of 10 km, the lowering of check-dams, and the creation of culverts within the floodplain system (Schiemer et al. 1999). Surface connectivity increased between mean water and bank-full level from $<10\ m^3\ s^{-1}$ prior to the measures to a maximum of more than $200\ m^3\ s^{-1}$. At low and high water, the surface connectivity was not changed by the measures at all. After rehabilitation, the relative proportion of river flow passing through the side-channel system increased non-linearly from less than 0.5% of the total river flow at low water ($<6\ m^3\ s^{-1}$, out of $1200\ m^3\ s^{-1}$ river flow) up to 12% (about $650\ m^3\ s^{-1}$) at high water.

Monitoring results, including biogeochemical processes, geomorphic dynamics, aquatic vegetation, and changes in biodiversity, have been compiled by Schiemer and Reckendorfer (2004). Key messages in relation to the hydro-geomorphic effects are:

1. The project was very successful in stimulating hydrological and geomorphological dynamics and in increasing the availability of semi-aquatic habitats.
2. The reconnection of floodplain waters to the main river significantly increased the variability of discharge and water velocity in the main side-channel. A visual inspection showed local bank erosion in areas with increased stream power.
3. The average shoreline length (interface between aquatic and terrestrial habitats) and the availability of shallow-water habitats increased significantly. Generally, the entire area became hydrologically more heteroge-

neous, exhibiting a gradient from dynamic to stagnant water bodies (Reckendorfer and Steel 2004).

4. An important task of the project was the development of prognostic models, especially an eco-hydrological model, to improve evaluation of the ecological consequences. The key parameters turned out to be water age and days of surface connectivity (e.g. Baranyi et al. 2002; Reckendorfer, unpublished data).

The relative contribution of the side-channel to the nutrient and organic matter dynamics of the whole river corridor was demonstrated by mass balance calculations. Based on the results of the hydrological model and data sets for 1997–2000, the input and output of all major species of phosphorus and nitrogen were calculated. The phosphorus input was mainly of particulate forms and was significantly related to the inorganic suspended solid load (r^2=0.55, n=55, P<0.01) and therefore comparable to the suspended solids dynamics (Tockner et al. 1999). The phosphorus balance was mainly controlled by deposition during floods, when terrestrial components were the most decisive elements (Fig. 2.3). Thus, increased connectivity did not change the total phosphorus balance; and only for soluble reactive phosphorus was

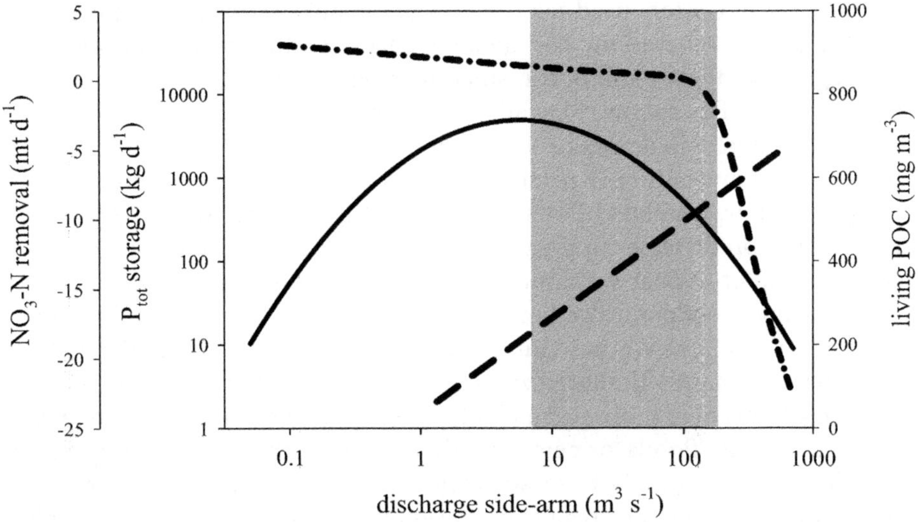

Fig. 2.3 Relationships between discharge and total phosphorus (P_{tot}; *broken line*), nitrate (*broken/dotted line*) and living particulate organic matter (*POC*; *solid line*) in the side-channel system (data for 1997 to 2000; Hein et al. 2003, 2005). Nitrate removal was calculated on the basis of input and removal efficiencies. Significance of regressions for phosphorus, nitrate, and living POC are, respectively: r^2 = 0.74, n=87, P<0.01; r^2 = 0.69, n=55, P<0.01; r^2 = 0.21, n=34, P<0.01. The *shaded area* marks the increased discharge through the side-channel system due to the rehabilitation measures.

an increased biological (phytoplankton) uptake detected (Hein et al. 2005). The effect of the local erosion processes was not quantitatively assessed, although we expected an increased remobilisation of deposited phosphorus. The removal of nitrate decreased with increasing discharge. During flood events, nitrate was even released from the floodplain (Fig. 2.3). In average, 0.5 % of the total flux through the river corridor was retained in the large (520 ha) floodplain area. Organic matter processing was demonstrated by the amount of living particulate organic matter (POC) produced and exported from the side-channel system, which was especially important during medium flow periods (Fig. 2.3; Hein et al. 2003). After rehabilitation, export of living POC increased by ~100 %.

2.5 'Environmental Flows' for Rehabilitating Floodplain Wetlands (Australia)

Flow regulation has been applied to many rivers in semi-arid areas. Australia has some of the most hydrologically variable rivers in the world (Puckridge et al. 1998) and the majority of lowland rivers in south-eastern Australia have been subjected to some form of flow regulation (Walker 1985). In most cases, a storage dam traps water upstream during the wet months and this water is released for irrigation, used for urban water supply, or processed for hydroelectric power generation. The general effects of this regulation are that, although large floods still occur, small to moderate floods are now less frequent and the floodplain is usually disconnected from the river channel. In many rivers, water in dams is released for irrigation during normally dry periods (summer and autumn), resulting in bank-full flows when the river should be at base-flow. Regulated flows are designed to provide water users with a reliable and predictable supply of water, so regulation also reduces both seasonal and inter-annual variability in flows. As a result of the regulation of flows, many floodplain wetlands in south-eastern Australia have been reduced in size and are now inundated less frequently than under natural flows, with substantially altered wetting–drying cycles. In particular, wetlands are likely to become more permanent, either permanently dry due to lack of small to medium floods, or permanently wet if they are 'topped-up' by irrigation flows in the normal dry seasons (Quinn et al. 2000). It is well known that rewetting of dry wetlands results in pulses of carbon, nitrogen, and phosphorus, both from riverine inputs and from litter and soils, that can drive food webs and increase productivity (Baldwin and Mitchell 2000; Scholz et al. 2000).

Changes in the wetting and drying cycles of floodplain wetlands due to regulated flow regimes are likely to have major impacts on productivity and can alter biotic assemblages. In south-eastern Australia, Quinn et al. (2000)

showed that invertebrate assemblages were markedly different between permanent and temporary wetlands on the regulated Murray River after flooding, but were indistinguishable between permanent and temporary wetlands on the unregulated Ovens River. They suggested that regulation changes not only the duration of wet and dry periods in floodplain wetlands but also the physical nature of inundation, with more extensive overbank floods on unregulated rivers resetting wetland biota.

To alleviate the problems caused to floodplain wetlands by flow regulation, environmental water allocations are being targeted to key floodplain sites and their associated wetlands, especially along the regulated Murray River. One such site is the Barmah Forest, where river regulation has reduced the frequency of successful bird breeding episodes (Leslie 2001). To address this issue, up to 100 Gl $year^{-1}$, accumulated up to four years, is stored in Hume Dam as an environmental water allocation to enhance flooding of the forest. This water has been released to top-up natural high flows in 1998 and 2000. The latter allocation (October 2000 to January 2001) resulted in increased waterfowl breeding success: egrets bred for first time since 1993 (Great), 1992 (Intermediate), and 1975 (Little). Components of the macrophyte assemblages in these wetlands also responded to the extended flooding event (Reid and Quinn 2004), although the relative importance of longer-term historical flood patterns compared to recent flood events varied among macrophyte taxa.

Floodplain wetland conditions in regulated rivers can be improved by the so-called 'environmental flows', which have become a major component of river restoration strategies used by water resource managers in south-eastern Australia. There are many (>200) approaches used for setting environmental flows (Tharme 2003), with a recent emphasis on holistic methods that explicitly link the flow regime to the biophysical components of the riverine environment. Within Australia, however, the amount of water allocated to environmental flows has rarely been enough to provide for overbank flooding and the focus of environmental flow assessments has been mainly on in-channel biota and ecological processes. More recently, the problems caused to floodplain wetlands by their hydrological disconnection from the main river channel have resulted in specific environmental flow allocations to 'significant ecological assets' (icon sites) along the Murray River, five of which are major floodplain wetland sites.

2.6 Lessons from Rehabilitation Projects

Past and ongoing anthropogenic modifications of large rivers, i.e. increased levels of pollutants, floodplain embankment, and hydrological regulation, have resulted in large-scale degradation of riparian corridors. Reduction of pollutant loads has often been the first step towards reversal of this trend. It is recognised, however, that recovery of the patterns and processes that naturally operate in river–floodplain systems would require an at least partial re-establishment of their natural hydromorphological regime (Hein et al. 2005; Tockner et al. 2005). Habitats that depend on river–floodplain interfaces may thus recover and be continuously re-created. Understanding the responses of river–floodplain systems and their biota to hydromorphological dynamics, however, is a prerequisite for successful restoration. This understanding should not focus on the functioning of either river channel or floodplain in isolation, but should include the interactions between rivers and their floodplains.

Key processes in river–floodplain systems, such as production, consumption, and decomposition of organic matter, are driven by the flood pulse (Bayley 1991; Tockner et al. 2000). Variability in flow, water-level fluctuations, sedimentation and erosion processes, and dispersal of organisms are the key determinants of ecological succession in these systems. Hence, interference with lateral connectivity affects the availability of physical habitats and the viability of populations.

The examples introduced in this chapter emphasise the importance of restoring lateral connectivity in river–floodplain systems for several reasons:

- Lateral connections are critical for aquatic organisms that use channel, riparian, and floodplain habitats during different stages of their life cycle (Bunn and Arthington 2002).
- Enhanced lateral flow may create hydrologically and geomorphologically dynamic conditions that are important in maintaining a high habitat heterogeneity and biodiversity, as illustrated for riparian arthropods occurring along braided rivers (see Section 2.2) and for vegetation in reconnected side-channels (see Section 2.3).
- Lateral interactions between river channel and floodplain wetlands allow enhanced processing of organic matter and nutrients (illustrated in Section 2.4).

As more and more projects are completed, there is a growing experience that (partially) re-established lateral connectivity in large river–floodplain systems potentially leads to successful restoration. Likewise, a more directed approach of delivering water to selected floodplain wetlands, as advocated by the concept of environmental flows (Section 2.5), may be the only possible way to conserve or restore the biodiversity of riverine wetlands in semi-arid regions. Here, river regulation schemes have changed the patterns of permanency of water bodies and wetlands in floodplains, with more wetlands remaining permanently wet or dry; and environmental flows can be targeted to restore connectivity between key floodplain wetlands and the main river channel, resulting in partial restoration of water level dynamics.

Increased lateral connectivity between channel and floodplain habitats may be achieved by either re-opening or creating side-channels, or by restoring natural overbank flood regimes. Side-channel reconnection along the Rhône River, for example, has demonstrated that vegetation diversity benefits from re-opening 'unplugged' side-channels. Flow through these channels re-introduces patterns of scouring and sediment deposition and creates new habitats for aquatic species. In addition, dispersal of organisms and propagules among sites may be enhanced. Four years after re-connecting, the numbers of species in the channels with permanent flow (GRW and BIC) were 21 and 23, respectively, which is higher than the average species numbers reported by Bornette et al. (2001) for side-channels of the Upper Rhône River (17±8) and its tributary, the Saône River (15±3). Hence, it can be concluded that rehabilitation by enhancing connectivity of side-channels was successful in terms of vegetation re-colonisation. It can generally be assumed that faunal communities may respond in a similar way.

Increased lateral connectivity with the floodplain also has an impact on the processing of transported matter. Mass balance estimations developed for the Danube Restoration Project have demonstrated that, on average, the value of the floodplain as a sink for nitrogen as well as a source for autochthonous particulate organic matter has increased after backwater reconnection. Hence, rehabilitation may affect the nutrient and organic matter dynamics of the entire river, provided the scale of rehabilitation is sufficiently large. Connected side-channels increase both transformation capacity and autochthonous organic matter production. Results from the Danube River showed the potential effect of local measures on river corridor transport and on downstream reaches. Consequently, naturalised lateral exchange processes between rivers and their floodplains can support other engineering-based measures to achieve water quality goals (McClain 2002). Therefore, evaluation of rehabilitation measures should not focus only on the effects on biodiversity, but may be improved by applying morphologic or hydraulic metrics. The latter have a high potential to forecast changes of key ecosystem functions like nutrient processing and organic matter production (Schiemer et al. 1999; Hein et al. 2004).

Current river and floodplain rehabilitation initiatives are usually restricted to the scale of a local reach or site. The more holistic approach involves restoring natural flood patterns and is often unfeasible. In most cases, it is highly unlikely that complete restoration, i.e. reversing all disruptions, is possible: it is not feasible to restore ecosystems back to conditions prior to degradation. Restoration options may be restricted because of the potentially conflicting need to control flooding (as exemplified by the Sacramento River, USA, by Alpert et al. 1999), or because of other schemes that require some form of flow regulation (e.g. for agriculture, such as in the case of Australian rivers presented in the previous section). Moreover, large-scale restoration requires planning at a catchment scale, which may have limited feasibility due to institutional complexity (Adams and Perrow 1999).

The initial success of catchment-scale water quality improvement and various local-scale river and floodplain restoration projects demonstrates the feasibility of rehabilitating riverine ecosystems. There is ample evidence that ecological rehabilitation of river–floodplain systems will be strongly beneficial for various societal and economic functions. First of all, floodplains have a major function for flood retention and water storage. The designation of flood detention reservoirs along the Rhine, for example, actually means the creation of new floodplain habitats (Siepe 1994). Increased nutrient storage and removal also present a green service of socio-economic importance for the river corridor (Gren et al. 1995). Several other values may be relevant as well, such as riverine areas being a resource for local food and crafts, and the provision of recreational space where rivers cross urbanised areas.

Despite the advantages of enhancing or restoring natural floodplain functions, river managers are still facing several problems regarding the design, planning, and implementation even of small-sized rehabilitation projects. It is to be expected that the up-scaling of river restoration will face even more challenging problems. Following Adams and Perrow (1999), four major issues that currently may constrain the implementation of river restoration actions may be identified: (1) limited scientific understanding of the complexity of floodplain ecosystems, (2) little understanding of the effects of management interventions, (3) difficulties of co-ordination between many different organisations and interest groups, and (4) different perceptions of rules, both formal (legislation) and informal (cultural ideas, traditions).

A major gap potentially constraining effective rehabilitation is often the highly abstract level at which restoration targets are formulated, in contrast to individual restoration actions that generally cover a much smaller scale. Table 2.1 indicates that different scales of restoration should address differently formulated targets and, consequently, require different indicators for success and schemes for monitoring and assessment. The challenge of linking different scales may be hindered by the top-down hierarchy of forcing factors (e.g. basin hydrology > local hydrology > sedimentation > vegetation) as opposed to a bottom-up approach of rehabilitation (many small-scale projects together should contribute to basin rehabilitation).

In practice, river rehabilitation may be a combination of enhancement, mitigation, and local reinstatement of river–floodplain connectivity. The question then is: to what extent do local projects contribute to whole-river rehabilitation? Increasing habitat and species diversity between and within ecosystems must be the guiding principle in assessing the effects of rehabilitation works. Ultimately, the success of rehabilitation will be expressed by the recovery of viable populations of riverine organisms at all trophic levels.

Restoring natural flooding conditions locally may work out differently on different scales and for different ecotopes; in some cases, conservation values have developed as the result of regulation or transformation of river floodplains. Also, increasing the intensity of river–floodplain exchanges could be detrimental to floodplain lake communities when the water quality of the flood water is poor (Van den Brink 1994) and the regenerative capacity of

Table 2.1 Characteristic gradients, applicable rehabilitation targets and management actions, and ecological indicators for river ecosystems, depending on spatial scale.

Scale	Gradients	Rehabilitation targets	Management actions	Biological indicators
River system (100–10 000 km²)	Longitudinal gradient, stream/river order, river to estuary transition	Improved water quality, reinstated upstream–downstream connectivity (corridor and flyway protection)	Emission control (legislation), dam removal, designation of conservation areas	Regional biodiversity, migratory fish, birds
River section (10–100 km²)	Transverse gradient, river to hillslope transition, flooding intensity, minimum discharge	Undisturbed river–floodplain ecotone, landscape complexity	Mitigation of hydrological alterations (e.g. dam or weir operations, removal of minor embankments)	Resident fish, mammals, birds, flora
Local reach or floodplain (1–10 km²)	Geomorphic variability, surface and subsurface flow, flow velocity (flow-reduced areas)	Enhanced connectivity (geomorphology, hydrology) channels, naturalised vegetation dynamics	Introduction of large herbivore grazers, sediment displacement, creating side-channels	Vegetation patterns, reptiles, amphibians, birds
Site or ecotope (0.01–1 km²)	Patterns in soil and vegetation structure	Maintenance/creation of local structures and ecotones	Localised management (mowing, tree-cutting), reconnection of single oxbows or pools	Plant species composition, invertebrates, algal communities
Eco-element (<0.01 km²)	Microgradients related to humidity, nutrient availability, flow velocity, water depth, etc.	Maximum habitat availability, species conservation	Local management (mowing, fertilisation, tree-cutting), naturalising structures, pool maintenance	Microvegetation patterns, macroinvertebrates

desired organisms may have been reduced by loss of seed banks, lack of sedimentation sites, etc.

Despite these limitations, the cases reported here for the Lower Rhône (France), the Danube Restoration Project (Austria), and Barmah Forest (Australia), as well as various others, demonstrate a highly positive and encouraging initial response of floodplain ecosystems to restoring hydrological connectivity between river and floodplain. In some cases, rehabilitation is part of flood protection measures (Nienhuis and Leuven 2001), when these measures benefit both the allocation of larger flood volumes and nature restoration. Large-river managers and conservationists should benefit from these projects for the lessons they provide, as they improve our understanding of river ecosystems (Schiemer et al. 1999; Buijse et al. 2002). This provides a strong basis for the planning and design of new projects and, eventually, for up-scaling measures to the scale of river sections or entire river systems. Adequate monitoring and assessment of river rehabilitation projects is essential to learn from both their successes and their failures, and to allow integrated analyses across different rivers.

References

Adams WM, Perrow MR (1999) Scientific and institutional constraints on the restoration of European rivers. In: Marriott SB, Alexander J (eds) Floodplains: interdisciplinary approaches. Geological Society, London, pp 89–97

Alpert P, Griggs FT, Peterson DR (1999) Riparian forest restoration along large rivers: initial results from the Sacramento River. Restor Ecol 7:360–368

Amoros C (2001) The concept of habitat diversity between and within ecosystems applied to river side-arm restoration. Environ Manage 28:805–817

Amoros C, Bornette G (1999) Antagonistic and cumulative effects of connectivity: a predictive model based on aquatic vegetation in riverine wetlands. Arch Hydrobiol Suppl 115:311–327

Amoros C, Elger A, Dufour S, Grosprêtre L, Piégay H, Henry C (2005) Flood scouring and groundwater supply in rehabilitated side-channels of the Rhône River, France: sedimentation and aquatic vegetation responses. Arch Hydrobiol Suppl. 155:147–167

Baldwin DS, Mitchell AM (2000) The effects of drying and reflooding on the sediment and soil nutrient dynamics of lowland river–floodplain systems: a synthesis. Regul River Res Manage 16:457–467

Baranyi C, Hein T, Holarek C, Keckeis S, Schiemer F (2002) Zooplankton biomass and community structure in a Danube river floodplain system: effects of hydrology. Freshwater Biol 47:473–482

Barrat-Segretain MH (1996) Strategies of reproduction, dispersion, and competition in river plants: a review. Vegetatio 123:13–37

Barrat-Segretain MH, Henry CP, Bornette G (1999) Regeneration and colonization of aquatic plant fragments in relation to the disturbance frequency of their habitats. Arch Hydrobiol 145:111–127

Bayley PB (1991) The flood pulse advantage and the restoration of river–floodplain systems. Regul River Res Manage 6:75–86

Boeger R (1992) The influence of substrate and water velocity on growth of *Ranunculus aquatilis* L. Aquat Bot 42:351–359

Bornette G, Amoros C (1996) Disturbance regimes and vegetation dynamics: role of floods in riverine wetlands. J Veg Sci 7:615–622

Bornette G, Piégay H, Citterio A, Amoros C, Godreau V (2001) Aquatic plant diversity in four river flood plains: a comparison at two hierarchical levels. Biodiver Conserv 10:1683–1701

Buijse AD, Coops H, Staras M, Jans LH, Van Geest GJ, Grift RE, Ibelings BW, Oosterberg W, Roozen FCJM (2002) Restoration strategies for river floodplains along large low-land rivers in Europe. Freshwater Biol 47:889–907

Buijse AD, Klijn F, Leuven RSEW, Middelkoop H, Schiemer F, Thorp JH, Wolfert HP (2005) Rehabilitation of large rivers: references, achievements and integration into river management. Arch Hydrobiol Suppl 155:715–738

Bunn SE, Arthington AH (2002) Basic principles and ecological consequences of altered flow regimes for aquatic biodiversity. Environ Manage 30:492–507

Cellot B, Mouillot F, Henry C (1998) Flood drift and propagule bank of aquatic macro-phytes in a riverine wetland. J Veg Sci 9:631–640

Chambers PA, Prepas EE, Hamilton HR, Bothwell ML (1991) Current velocity and its effects on aquatic macrophytes in flowing waters. Ecol Appl 1:249–257

Dynesius M, Nilsson C (1994) Fragmentation and flow regulation of river systems in the northern third of the world. Science 266:753–762

French TD, Chambers PA (1996) Habitat partitioning in riverine macrophyte communi-ties. Freshwater Biol 36:509–520

Gren IM, Groth KH, Sylven M (1995) Economic values of Danube floodplains. J Environ Manage 45:333–345

Hein T, Baranyi C, Herndl GJ, Wanek W, Schiemer F (2003) Allochthonous and autochthonous particulate organic matter in floodplains of the River Danube: the importance of hydrological connectivity. Freshwater Biol 48:220–232

Hein T, Baranyi C, Reckendorfer W, Schiemer F (2004) The impact of surface water exchange on the hydrochemistry and particulate matter dynamics in floodplains along the River Danube, Austria. Sci Total Environ 328:207–218

Hein T, Reckendorfer W, Thorp J, Schiemer F (2005) The role of slackwater areas for bio-geochemical processes in rehabilitated river corridors: examples from the Danube. Arch Hydrobiol Suppl 155:425–442

Henry CP, Amoros C, Bornette G (1994) Differential effects of floods on the aquatic veg-etation of braided channels of the Rhône River. J N Am Benthol Soc 13:439–467

Henry CP, Amoros C, Bornette G (1996) Species traits and recolonization processes after flood disturbances in riverine macrophytes. Vegetatio 122:13–27

Karaus U (2005) Ecology of lateral aquatic habitats along river corridors. PhD thesis, ETH, Zurich

Leslie DJ (2001) Effect of river management on colonially-nesting waterbirds in the Barmah-Millewa forest, south-eastern Australia. River Res Appl 17:21–36

Leslie DJ, Ward KA (2002) Murray River environmental flows 2000–2001. Ecol Manage Restor 3:221–223

Madsen JD, Chambers PA, James WF, Koch EW, Westlake DF (2001) The interaction between water movement, sediment dynamics and submersed macrophytes. Hydro-biologia 444:71–84

McClain ME (2002) The application of ecohydrological principles for better water resources management in South America. In: McClain ME (ed) The ecohydrology of South American rivers and wetlands. (Special Publication 6) International Associa-tion of Hydrological Sciences, Wallingford, pp 193–209

Nienhuis PH, Leuven RSEW (2001) River restoration and flood protection: controversy or synergism? Hydrobiologia 444:85–99

Paetzold A (2005) Life at the edge – aquatic and terrestrial interactions along river corridors. PhD thesis, ETH, Zurich

Paetzold A, Tockner K (2005) Quantifying the effects of riparian arthropod predation on aquatic insect emergence. J N Am Benthol Soc 24:395–402

Polis GA, Power ME, Huxel GR (2004) Food webs at the landscape level. University of Chicago Press, Chicago, 548 pp

Puckridge JT, Sheldon F, Walker KF, Boulton AJ (1998) Flow variability and the ecology of large rivers. Mar Freshwater Res 49:55–72

Quinn GP, Hillmann T, Cook R (2000) The response of macroinvertebrates to inundation in floodplain wetlands: a possible effect of river regulation? Regul River Res Manage 16:469–477

Reckendorfer W, Steel A (2004) Auswirkungen der hydrologischen Vernetzung zwischen Fluss und Au auf Hydrologie, Morphologie und Sedimente. Abh Zool Bot Ges 34:19–30

Reid MA, Quinn GP (2004) Hydrologic regime and macrophyte assemblages in temporary floodplain wetlands: implications for detecting responses to environmental water allocations. Wetlands 24:586–599

Roux AL, Bravard JP, Amoros C, Pautou G (1989) Ecological changes of the French upper Rhône River since 1750. In: Petts GE (ed) Historical change of large alluvial rivers: Western Europe. Wiley, Chichester, pp 323–350

Schiemer F, Reckendorfer W (2004) Das Donau Restaurierungsprojekt – Ökologische Auswirkungen. Abh Zool Bot Ges 34, 185 pp

Schiemer F, Baumgartner C, Tockner K (1999) Restoration of floodplain rivers: The Danube restoration project. Regul River Res Manage 15:231–244

Schiemer F, Keckeis H, Winkler G, Flore L (2001) Large rivers: the relevance of ecotonal structure and hydrological properties for the fish fauna. Arch Hydrobiol Suppl 135:487–508

Scholz O, Gawne B, Ebner E, Ellis I (2000) The effects of drying and re-flooding on nutrient availability in ephemeral deflation basin lakes in western New South Wales, Australia. River Res Appl 18:185–196

Siepe A (1994) Regeneration of floodplain biotopes on the upper Rhine – the „Polder Altenheim" case. Water Sci Technol 29:281–288

Tharme RE (2003) A global perspective on environmental flow assessment: emerging trends in the development and application of environmental flow methodologies for rivers. River Res Appl 19:397–441

Tockner K, Schiemer F, Ward JV (1998) Conservation by restoration: the management concept for a river-floodplain system on the Danube River in Austria. Aquat Conserv Mar Freshwater Ecosyst 8:71–86

Tockner K, Pennetzdorfer D, Reiner N, Schiemer F, Ward JV (1999) Hydrological connectivity, and the exchange of organic matter and nutrients in a dynamic river–floodplain system. Freshwater Biol 41:521–535

Tockner K, Malard F, Ward JV (2000) An extension of the flood pulse concept. Hydrol Process 14: 2861–2883

Tockner K, Paetzold A, Karaus U, Claret C, Zettel J (2005) Ecology of braided rivers. IAS Special Publication (in press)

Van den Brink FWB (1994) Impact of hydrology on floodplain lake ecosystems along the lower Rhine and Meuse. PhD thesis, University of Nijmegen, Nijmegen

Walker KF (1985) A review of the ecological effects of river regulation in Australia. Hydrobiologia 125:111–129

Ward JV, Tockner K, Schiemer F (1999) Biodiversity of floodplain river ecosystems: ecotones and connectivity. Regul River Res Manage 15:125–139

3 Sustainable Agriculture and Wetlands

Frank Rijsberman, Sanjini de Silva

3.1 Agriculture and Wetlands: Introduction

Agriculture and wetlands have not always had a very harmonious relationship. The need to expand agriculture to feed a growing population led in many places to a major conversion of wetlands into farmland. Worldwide, this led to the well known, often quoted estimate that 50 % of the world's wetlands have been lost to agriculture and urbanization in the twentieth century. For environmentalists in general and wetland specialists in particular, it is clear that this loss of wetlands should be stopped urgently – or reversed where possible. And indeed, in some parts of the world, some land reclaimed for agriculture has been given back to nature though not without sharp conflicts with farmers[1]. Since agriculture is by far the largest user of water and is often assumed to be a very inefficient, even wasteful, user, it is obvious to many in the environmental field that the expansion of future human water use will have to come from transfers of water out of agriculture (e.g. IUCN 2000). For agriculture, however, the primary objectives are the Millennium Development goals on the reduction of malnutrition and poverty. While it is recognized that reducing malnutrition has at least as much to do with access to affordable food as it has to do with increasing production, there is an increasing focus on the 900 million poor people in rural areas who depend on water for productive purposes (i.e. food and livelihoods). This still emphasizes the need to make water available to the currently water-insecure people in developing

[1] The idea, for example, some ten years ago, to turn small areas of reclaimed agricultural land in the estuary of the Western Scheldt into tidal wetlands to compensate for losses due to dredging in the Western Scheldt shipping lanes led to strong farmer protests (Rijsberman 1996). Around the same time, a LIFE project in Sweden (LIFE96ENV/S/346) created 65 small wetlands with a total area of 74 ha in Southern Sweden (lars.jacobson@lund.se). In Australia, Wetlandcare Australia (www.wetlandcare.com.au) encourages landholders to manage their marginal farmland as wetlands.

Ecological Studies, Vol. 190
J.T.A. Verhoeven, B. Beltman, R. Bobbink, and D.F. Whigham (Eds.)
Wetlands and Natural Resource Management

countries. Consequently, rather than moving water out of agriculture, agronomists see a need for more water (e.g. van Hofwegen and Svendsen 2000). This sets the scene in which the future relation between wetlands and agriculture has to develop.

Among wetland specialists the very high value of wetland ecosystems to society, in terms of biodiversity but also in terms of goods and services produced, is accepted as being obvious though difficult to quantify given their enormous heterogeneity and complexity, other than some specific case studies (e.g. Barbier et al. 1991; Costanza et al. 1997). Among agronomists it is not very long ago – and it partly persists today – that wetland systems were qualified first and foremost as wastelands, badlands, or sources of disease. In other words, systems that needed to be reclaimed or drained – often for agriculture – to become useful to society. Given these widely divergent perspectives, it is not surprising that the two communities have found it hard to communicate effectively. Nevertheless, this is exactly what is required to halt the ongoing destruction of wetlands in large parts of the developing world.

There are some positive signs. Until recently there was little evidence of positive interest in wetlands from the agricultural side and vice versa. The Bonn International Water Conference was the first major water event that put the balance of water used for food and nature on the agenda (Rijsberman and Molden 2001). "Eco-agriculture" is now gradually becoming a mainstream subject of interest for agricultural research (McNeely and Scherr 2001) and agriculture was on the agenda for the first time for the contracting parties of the RAMSAR Convention during COP8 in Valencia in 2002 (for acronyms, see Table 8.1). The 17th Session of the Global Biodiversity Forum, held in 2002 ahead of RAMSAR COP8, included a workshop on agriculture and wetlands. At the recent meeting of the UN Commission on Sustainable Development (CSD12, April 2004), while the overall focus was dominated by water supply and sanitation-related discussions, there was also promising attention for the balance of water use for agriculture–environment interactions (SIWI/IWMI 2004).

This paper explores the scope for a better, shared understanding of key water-related issues among the people involved in wetlands and agriculture. The experience of the authors and the work they draw on focuses primarily on Asia and Africa, with excursions into the experience of OECD countries. The authors start by exploring some basic definitions – that often lead to confusion, if not controversy – on the functions of water to produce food and generate livelihoods and to maintain environmental values and services. They then analyze some key trends in water resources management and in development that are likely to affect wetlands. They briefly discuss the results of the Dialogue on Water, Food and Environment, explore key elements of a joint wetlands–agriculture research agenda, and then attempt to draw some conclusions on the future of sustainable wetlands and agriculture.

Table 3.1 Key to acronyms

Acronym	Subgroup	Full name
CA		Comprehensive Assessment of Water Management for Agriculture; program of the CGIAR, convened by IWMI
CGIAR		Consultative group on International Agricultural Research
CPWF		Challenge Program on Water and Food; program of the CGIAR, implemented by a consortium of partners led by IWMI
CSD		Committee on Sustainable Development of the United Nations
DWFE		Dialogue on Water, Food and Environment
FAO		Food and Agriculture Organization of the United Nations
GWP		Global Water Partnership
ICID		International Commission on Irrigation and Drainage
IFAP		International Federation of Agricultural Producers
IRRI		International Rice Research Institute
IUCN		International Union for the Conservation of Nature
IWMI		International Water Management Institute
MEA		Millennium Ecosystem Assessment
RAMSAR	Convention	Convention on the Wise Use of Wetlands
	COP	Conference of the Parties to the Convention
	STRP	Scientific and Technical Review Panel
SIDA		Swedish International Development Agency
SIWI		Stockholm International Water Institute
UNEP		United Nations Environment Program
WHO		World Health Organization of the United Nations
WWF		World Wide Fund for Nature

3.2 Water for Food, Water for Environment

In our opinion, some ideas that are strongly held by either the environmental or the agricultural community, but not shared by the other side, are hampering effective communication among the two communities. This section explores several key (mis-)conceptions and proposes a new conceptual framework to overcome these.

3.2.1 "Ecosystems Produce the Water Used by Agriculture"

Many in the environmental community support the viewpoint that ecosystems produce water for human use. In the World Water Vision exercise an early classification as "water for people", "water for food" and "water for nature" was not accepted by the environmental community because it suggested that nature used water while the prevailing view was that nature produced water; the compromise accepted was "water and nature". This perspective is also shown clearly in the influential conceptual framework of the Millennium Ecosystem Assessment (MEA 2003, 2005) that analyzes the services provided by ecosystems as provisioning, supporting and regulating. It shows water as one of the key provisioning services of freshwater ecosystems. Thus, discussions on the balance of water resources to be allocated to different uses should not include allocations of water to nature, in this perspective, because nature does not use water – it produces water. In contrast, the traditional view held by many in agriculture – and still remarkably persisting in some parts of the community – is that water not captured for productive use by people is essentially "wasted".

Our view is that nature and agriculture are the two largest users of water by far, that there is real and increasing competition for water between these two major uses, and that resolution of the conflicts among the two requires the adoption of a shared world view that recognizes this. Compared with the competition between agriculture and nature, the competition for water with either urban or industrial uses is globally insignificant in volume and is also a lost cause. Locally, the agriculture–urban competition can dominate water-use discussions; and in the vicinity of major metropolitan areas, or highly urbanized countries such as the Netherlands, there can be serious conflicts. However, the value of a liter of water to a domestic user is so much higher than its value in agriculture or nature that domestic users will always be able to outbid agriculture and nature.

The root cause of much confusion is that the standard definition of water resources is highly biased towards water supply and sanitation. Only the rainfall that either runs off into streams and rivers (surface water) or recharges aquifers (groundwater) is counted as a renewable water resource in the standard definitions. The assessments of the world's water resources thus usually start with a resource base of some 40 000 km^3 $year^{-1}$(Shiklomanov 1998). This is only some 40 % of the total rainfall, however. In other words, more than half the hydrological cycle is not accounted for. The other 60 % of the rainfall infiltrates into the soil, is held there as soil moisture, and evapotranspires back into the atmosphere, either directly from the soil (evaporation) or through the vegetation (evapotranspiration), without entering streams or groundwater. While this is hardly a new discovery, virtually all world water assessments ignore the 60 % that Mahlin Falkenmark started calling "green water" and concentrate on the surface and groundwater or "blue water" (see Fig. 3.1; Cosgrove and Rijsberman 2000).

For most human uses, such as water supply and sanitation, this makes sense since the soil moisture cannot be captured or used and therefore can be safely ignored. Not so, obviously, for either agriculture or ecosystems. Both critically depend on this 60 % of the hydrological cycle, the "green water" or soil moisture. Figure 3.2 shows the parts of the world that rely on green versus blue water for their food production. Recent perspectives in agriculture indicate that green and blue water are highly related, and in fact to a large extent interchangeable, in food production. The traditional distinction between rainfed and irrigated agriculture has by and large become obsolete. At IWMI, we now emphasize the continuum of approaches to increasingly manage water for food production that range from purely rainfed agriculture, to rainwater harvesting, rainfed agriculture with supplemental irrigation, small-scale or micro-irrigation, groundwater irrigation with small motor pumps, and ultimately to large-scale surface water irrigation systems (Blank et al. 2001; Kijne et al. 2003; Rijsberman 2004, 2006). Rather than referring to rainfed and irrigated agriculture as two separate systems, we prefer to refer to the continuum of approaches as a single system: "*water management for agriculture*".

In a system that, as a simplified example, consists of rainfed agriculture, undisturbed wetlands and natural ecosystems, increasing agricultural production implies a choice between: (a) extending the agricultural area by con-

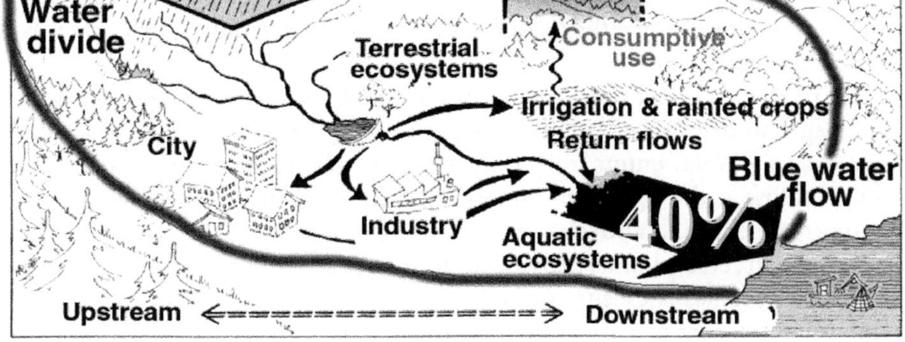

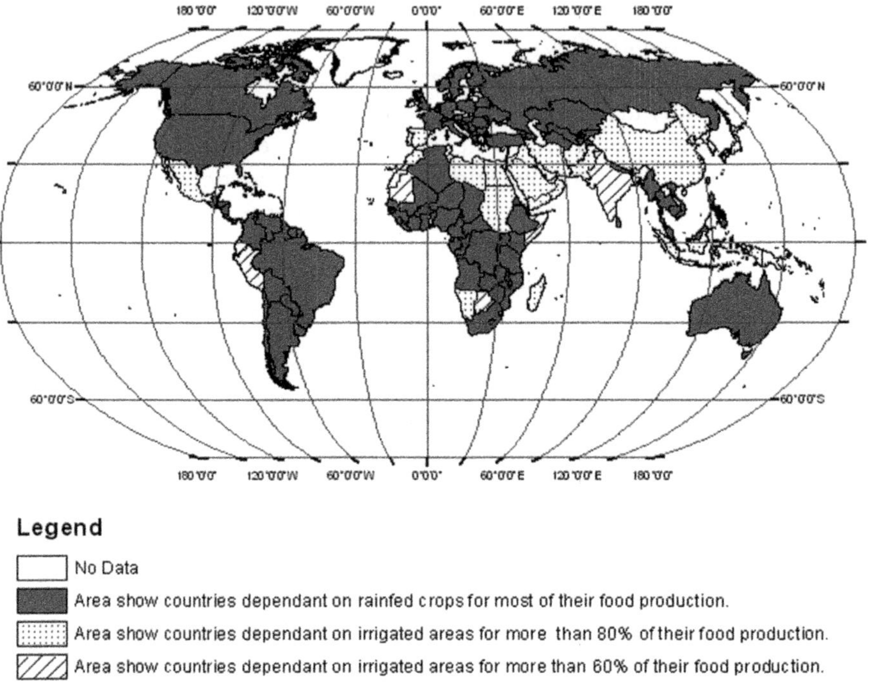

Fig. 3.2 Hatched areas (see legend) show countries depending for most of their food on rainfed crops (source: SIWI/IWMI 2004)

verting the natural area into rainfed agriculture and (b) irrigating the rainfed area by withdrawing water from the river – or the wetlands. The first option transfers the green water and associated land that used to support the natural area to the expanded rainfed area but does not affect the blue water use. The second option transfers blue water that used to support nature to irrigated agriculture. This demonstrates the importance of looking at all water resources, green and blue, and their functions in nature and agriculture, in a unified framework. A third option is to increase the productivity of the water used for either rainfed or irrigated agriculture, as argued later in this paper.

Ecosystems do not produce or provision any water resources and neither does agriculture. Water is "produced" in the hydrological cycle powered by solar energy. Water enters into ecosystems or agriculture as rainfall or surface water inflow at the upstream end, is partially used in these systems through evapotranspiration, and the excess water flows out downstream. Water resources, green and blue, provide a critical supporting service, just like soils, for both systems. They support vegetation and a host of other living organisms that are part of natural, semi-natural, domesticated, or agro-industrial landscapes. This does not diminish the importance of ecosystems that provide hugely important water regulating functions as "sponges" and "filters".

This view, then, allows us to talk about the consumptive use of water, green and blue, by agriculture and nature.

3.2.2 "Irrigated Agriculture Uses 70% of the World's Water"

Another well known, often quoted fact is that irrigation uses 70 % of the water in human use, with an efficiency of 40–50 %. On the face of it, that makes it understandable to jump to the conclusion that an increase of that efficiency to 80–90 % (possible with drip irrigation) would save half the water – and would be the obvious solution to the world's water problems.

More accurately expressed, irrigated agriculture is responsible for some 70 % of all blue water withdrawn for human use. At the turn of the century, some 10 % of all blue water, around 4000 km^3, was withdrawn for human use. In other words, a bit less than 3 % of all water (green and blue) was withdrawn for irrigated agriculture. In IWMI's estimates, irrigated agriculture was responsible for some 6 % of all evapotranspiration, while a further 16 % was used consumptively by rainfed agriculture (see Fig. 3.3). Fortunately, natural ecosystems, from forests to wetlands, still use far more water than agriculture does.

This is not to say that all is well, or that agriculture does not use water unsustainably; it does, in many locations around the world. The global numbers hide the local problems. Very large amounts of the blue water of the world occur in Canada, where they are not readily accessible for human use; and the areas where there is very significant overuse do affect some of the key irrigated areas, and breadbaskets, of the world (see e.g. Fig. 3.4).

It is also a misconception that irrigation efficiency can be increased (e.g. from 40 % to 80 %), which then would lead to major savings. As explained by

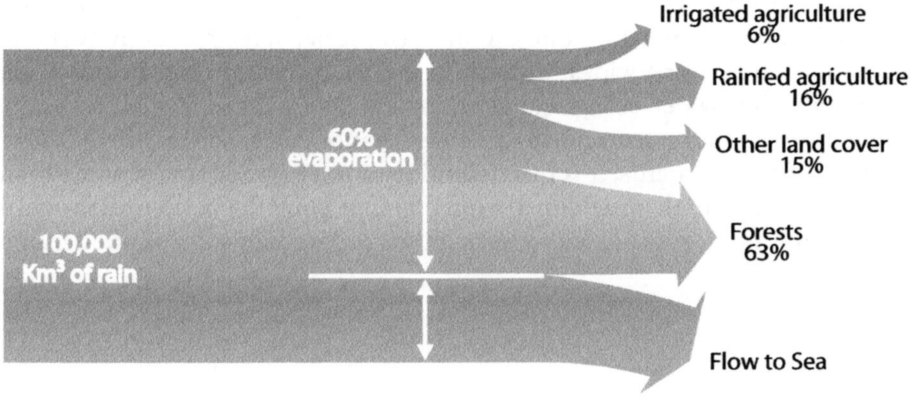

Fig. 3.3 Assessment of evapotranspiration from different land covers (source: IWMI)

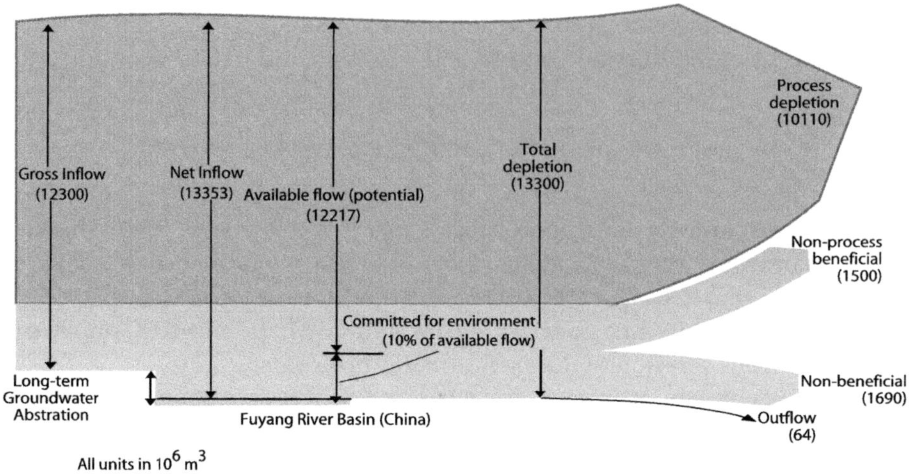

Fig. 3.4 Water allocation to different uses in the Fuyang river basin, China. Process depletion consists of crop evapotranspiration, domestic and industrial depletion, i.e. human uses (source: IWMI)

Seckler et al. (2003) "irrigation efficiency" is a very confusing term that has been defined in too many ways. Whether increased irrigation efficiency in a farmer's field does indeed save water depends very much on the fate of the return flow (the drainage water and recharge to the groundwater). If that return flow would be used downstream, then increasing the efficiency for the upstream farmer simply makes less water available to the downstream farmer. Water savings at the field scale do not always scale-up to savings at the basin scale. For this reason, IWMI prefers to focus on water productivity at different scales, i.e. the total value of goods and services produced per unit of water "depleted" (i.e. consumptively used, for evapotranspiration or for domestic and industrial purposes). At the scale of the farmer's field, water productivity can be measured in physical units of output (the crop per drop) or as the value in monetary units. At the basin scale, water productivity should be understood in the widest possible sense, i.e. including crop, livestock and fishery yields, wider ecosystem services and social impacts such as health, together with the systems of resource governance that ensure equitable distribution of these benefits. It is this definition of water productivity, at the basin scale, that we think provides a good starting point for a consideration of goods and services provided by agriculture and wetlands in a unified framework.

3.2.3 "Water Scarcity: Fact or Fiction?"

A review of water-scarcity indicators and global assessments (Rijsberman 2006) shows that the most widely used indicator, the Falkenmark Indicator (Falkenmark et al. 1989; Falkenmark 2000), is popular because it is easy to apply and understand; but it does not help much to explain the true nature of water scarcity. The more complex indicators are not widely applied because data are lacking to apply them and the definitions are not intuitive. It is clear that water is definitely physically scarce in the arid areas, Central and West Asia and North Africa, with projected availabilities of less than 1000 m^3 $year^{-1}$ per capita. This scarcity relates to water for food production, however, and not to water for domestic purposes that are insignificant at this scale. In most of the rest of the world, water scarcity at a national scale has as much to do with the development of the demand as the availability of the supply. Accounting for water for environmental requirements shows that abstracting water for domestic, food and industrial uses does already have a major impact on ecosystems in many parts of the world, even in areas not generally considered "water-scarce" (see Fig. 3.5).

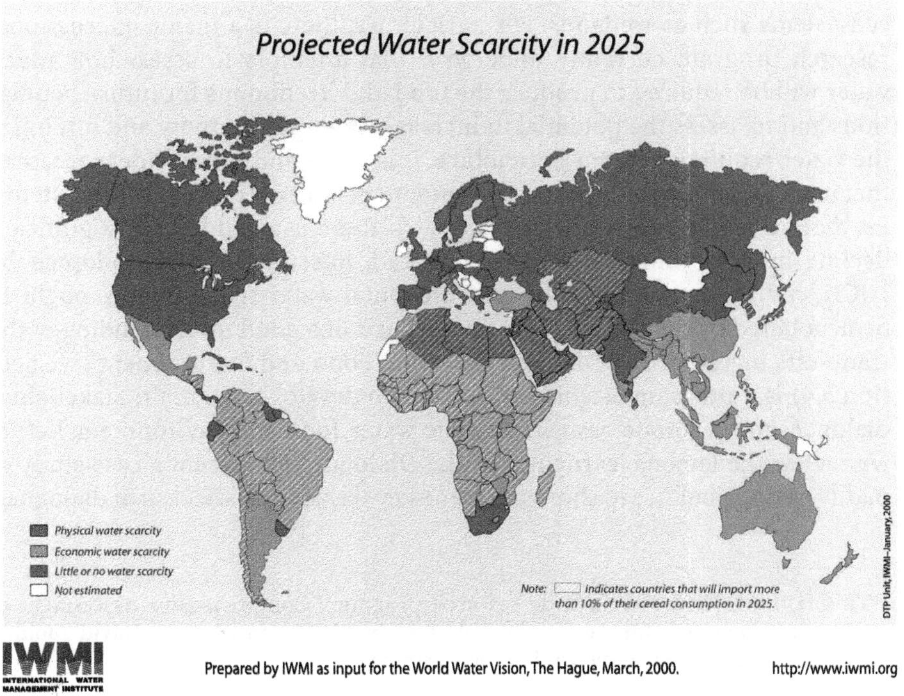

Fig. 3.5 A Global water scarcity projections of IWMI for 2025 when accounting for food production, domestic and industrial use (source: IWMI 2000).

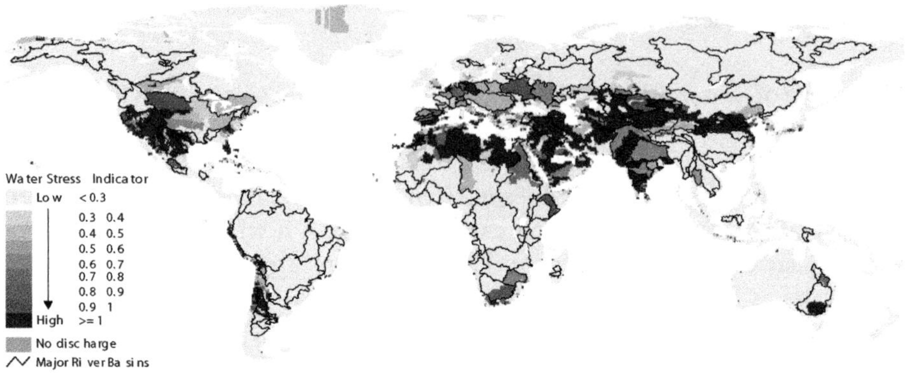

Fig. 3.5 B Areas where human withdrawals of blue water in 2000 already significantly affect environmental requirements (source: Smakthin et al. 2004b)

Rijsberman (2006) concludes that water will be a major constraint for agriculture in coming decades; particularly in Asia and Africa, this will require major institutional adjustments. He recommends a "soft path" to address water scarcity, focusing on increasing the overall water productivity (Rijsberman 2004), and taking into account the environmental water requirements of ecosystems such as wetlands. For agriculture, there is a major international research program currently underway[2] that attempts to assess how much water will be required to produce the food and livelihoods for future populations and to assess the potential to increase water productivity and minimize the water requirements for agriculture. It is more and more widely accepted that one cannot similarly assess the amount of water required by ecosystems, i.e. that there is no threshold below which there is a sudden and significant decline in environmental services. Instead, like the approach adopted by IUCN, environmental flows, or environmental water requirements, ought to be negotiated by the stakeholders on the basis of a good understanding of the trade-offs involved. The Dialogue on Water, Food and Environment (see Section 3.4) is a program established in 2001 to develop such multi-stakeholder dialogues to "negotiate" issues related to water, food and environment. Before we review the lessons learned from the Dialogue, we present a case study of paddy rice agriculture to showcase some key trends that affect such dialogues.

---

[2] This is the CGIAR system-wide research program "Comprehensive Assessment of Water management for Agriculture", hosted by IWMI and led by David Molden (d.molden@cgiar.org; www.iwmi.org/assessment).

3.3 Producing More Rice With Less Water

Rice is the most important cereal crop in the developing world and the staple food crop in some 33 countries. Rice growing is possibly the oldest form of intensive agriculture by man; it is cornerstone of the green revolution in Asia and a very large user of water. When people put pressure on agriculture to save water, then rice is the crop that comes to mind first. Can we grow more rice with less water?

Tuong and Bouman (2003) review the potential for increasing water productivity for rice cultivation in water-scarce environments. They distinguish between strategies that focus on: (a) increasing the yield per unit of water that is evapotranspired, (b) reducing the unproductive seepage, percolation and evaporation outflows, and (c) making more effective use of rainfall. According to Tuong and Bouman (2003), the modern "IRRI varieties" that were developed between 1960 and 1980 increased water productivity three-fold by reducing growth duration, but there has been limited further progress since. In particular, breeding plants for increased drought resistance has been hampered by the genetic complexity of the trait. However, breakthroughs in understanding of the molecular biology of plants, such as the successful sequencing of the rice genome, have led to optimism about future progress. Bennet (2003), for example, provides a thorough – and quite optimistic – overview of the potential for breeding plants for improved drought, salinity, waterlogging and submergence tolerance. These new rice plants would probably take another 10–15 years, however, to show up in farmers' fields. In the meantime, farmers in some parts of China have been successful in using a range of agronomic measures to reduce the unproductive outflows from their fields. They achieve this by implementing a range of measures, primarily a departure from continuously flooded fields to forms of alternate wet/dry irrigation systems. Another strategy that is pursued by IRRI scientists is to develop so-called "aerobic rice" – in essence rice that will be as productive as paddy rice, but will grow under upland conditions (i.e. no longer in flooded fields).

In other words, attempts to grow more rice with less water will lead to a reduction of the system of continuously flooded paddy fields, either though agronomic systems that focus on alternate wetting and drying (over cycles of several weeks) or to new rice varieties that no longer require flooded fields.

At the same time there is growing recognition of the relevance of paddy rice fields as an agro-ecosystem, as reviewed by Bambaradeniya and Amerasinghe (2004). Paddy rice fields, usually successors of shallow marshes, can be defined as temporary, seasonal wetland ecosystems managed for agriculture with a variable degree of intensity. There is a very considerable literature on the flora and fauna of rice fields but a conspicuous lacuna related to rice field biodiversity. Ecological studies contrasting traditional rainfed rice lands with more intensive irrigated systems have not been carried out so far. Bambaradeniya and Amerasinghe (2004) conclude that the long history of rice field cultivation has allowed an evolution of stable and balanced ecological relationships, but that the rapid changes introduced during the Green Revolution (increased fertilizer use, broad-spectrum biocides, short-duration varieties) are a threat to rice-field agro-ecology. In this light, it is worth considering carefully what the consequences will be of growing more rice with less water and its associated diminution of the volume and duration of freestanding water in the paddy fields.

3.4 Towards a Dialogue Among Agronomists and Environmentalists

Following the Second World Water Forum (The Hague, March 2000) and the World Water Vision process (Cosgrove and Rijsberman 2000), ten international organizations[3] agreed to the need for a sustained, action-oriented dialogue on the development of water management strategies and practices that enhance food *and* environmental sustainability. They jointly established the Dialogue on Water, Food and Environment (DWFE)[4] in 2001 for a first phase that ended with a workshop at the Stockholm Water Symposium in August 2004. The DWFE is an initiative that brought together agricultural and environmental communities to find ways of managing water to meet the needs of both. It was built on the principles of integrated water resources management but added the human dimension of conflict resolution, consensus building and social learning to the equation (see Fig. 3.6). It intended to demonstrate by example the tangible solutions to the problem and to build a recognized body of knowledge on critical issues. The body of knowledge was intended to provide answers to a set of issues (research questions) for which there was no consensus among the agricultural and environmental communities.

The DWFE started off with design workshops that led to a series of six working papers[5] containing guidelines for establishing dialogues at national, basin and local level, as well as proposed methodologies for several key issues. It subsequently started a process to identify and formulate dialogue processes in collaboration with various partner organizations. The partners met in a conference in Hanoi, Vietnam, in October 2002 to exchange experience

---

[3] FAO, IFAP, GWP, ICID, IUCN The World Conservation Organization, IWMI, UNEP, WHO, WWC and WWF.

[4] The Dialogue Secretariat is hosted by IWMI and coordinated by Domitille Vallee (d.vallee@cgiar.org).

[5] The working papers are available at: www.iwmi.org/dialogue.

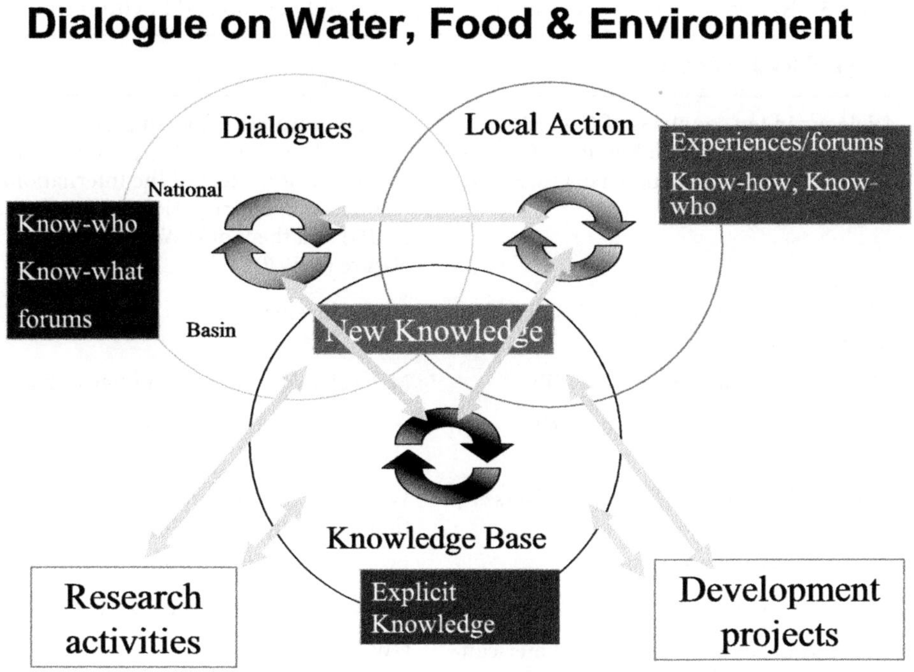

Fig. 3.6 An overview of the Dialogue on Water, Food and Environment (source: Bhatt and Vallee 2004)

(Wolter 2003). Since that time, about 30 dialogues have been initiated at country, basin and local level. Some of these have ended, while others are in progress or just started. A number of the consortium partners in the DWFE have also initiated (joint) activities which are intended to contribute to the water, food and environment knowledge base. One of these was an effort to look at "human health" as a cross-cutting issue for all water, food and environment dialogues (WHO 2004). Out of the dialogue process, the need for a more formal "learning framework" arose, both to help guide future dialogue-type activities and to capture the experience of current dialogue processes. A proposal to develop such a learning framework, and to apply it in a multi-year collaborative research program in southern Africa, was subsequently funded by SIDA. A first "learning framework meeting" was held in November 2003 in Nairobi and a second meeting in South Africa in July 2004. Eight national-level dialogue processes have been conducted or are in process (see Table 3.2). In addition, some 20 dialogue activities have been initiated at local or basin level (see Table 3.3).

As many DWFE activities are ongoing (and in some cases just beginning), it is not easy to draw specific lessons learnt. However, there are a number of general lessons learnt through the establishment of the process (Bhatt and

Table 3.2 National level dialogues in the Dialogue on Water, Food and Environment

National level dialogues	Organizers
Dialogue in ten countries of Central Europe: Bulgaria, Czech Republic, Estonia, Hungary, Latvia, Lithuania, Poland, Romania, Slovakia, Slovenia	Eastern Europe partnership of the Global Water Partnership (GWP) with regional committees of the International Commission on Irrigation and Drainage (ICID) and the World Wide Fund for Nature (WWF)
Water, food, environment (WFE) dialogue in Slovenia	GWP Slovenia
Malaysia national dialogue on WFE	Malaysian National Committee of ICID
Indonesia national dialogue on WFE	Indonesian national partnership of GWP
India: dialogue on the linking of rivers	WWF and Indian Participatory Irrigation Management (INPIM)
Indian National Consultation	Indian National Committee for ICID
Chinese National Consultation	Chinese National Committee for ICID
Dialogue debates in three Latin American countries: Peru, Chile, Uruguay	GWP

Vallee 2004). A survey amongst stakeholders concludes that at the global level – the level of the headquarters of the partner organizations – the DWFE was not very successful: there are still major disagreements among the ten partners. No real dialogue or organizational change was achieved; and this is blamed largely on the sectoral nature of these organizations as well as the rather short time that has passed since the start of the initiative. Even so, there was some success in recognition of the necessity to reach out across the sectoral divide, such as a joint UNEP/FAO workshop on environmental footprints of global food projections. At local, basin and national levels, there was more success reported. Some of the same partners seem to share a better understanding of conflicting issues at basin level through their national committees or regional offices than exists at their global headquarters.

Table 3.3 Basin and local level dialogues in the Dialogue on Water, Food and Environment

Basin and local dialogues	Organizers
India: Godavari basin dialogue	WWF
India: basin dialogues in Brahmani and Sabarmati	ICID
China: basin dialogues in Jiaodong and Qiantangjiang	ICID
Sri Lanka basin dialogues: Mahaweli, Malwathu Oya, Maha Oya, Menik Ganga	Lanka Jalani
Nepal: water food and environment management in East Rapti river basin	IWMI
Thailand: dialogue for integrated management of the Bang Pakong river basin – phase 1	Thai Ministry for Natural Resources and Environment with FAO, UNEP
Thailand: Songkhram river	Water and Nature Initiative (WANI) of IUCN
Vietnam: Perfume river	WANI/IUCN
India: developing an action plan for the revival of the dying river – Rajula block of Amreli District	Utthan Development Action Planning Team
El Salvador: Basin El Imposible/Barra de Santiago	WANI/IUCN
Honduras: diagnostic for dialogue in the Choluteca	GWP Central America
Mexico: Lerma Chapala basin	IMTA/SAGARPA/GTEPAI
Brazil: São Francisco river basin	Rios Vivos Network
Argentina: dialogue process in the Guarani aquifer	Foro Ecologista de Parana
Zambia: dialogue in the Kafue basin	WWF with FAO
Zambia: contribution of local action groups to the WWF dialogue in the Kafue plains	AREZ and WWF
Tanzania: Pangani basin	WANI/IUCN
South Africa: Pongolo floodplain system assessment	IWMI and University of Natal
Burkina Faso: Volta basin	FNPP program, FAO
Morocco: Soussa–Massa integrated water management	SIWM project supported by USAID

3.4.1 Water, Food and Environment Issues in Attapeu Province, Lao PDR

An example of the importance of health as a cross-cutting issue is shown in a joint FAO/Bangkok and IUCN participatory assessment of the role and nutritional value of aquatic resources in the livelihoods of the rural poor in Attapeu province, Lao PDR (FAO/IUCN 2003). Three communities participated in activities designed to assess the local availability and use of aquatic resources, their importance in local livelihoods, and the overall health and nutritional status of the villagers. It was found that fish and other aquatic animals make up the main animal protein sources in peoples' diets. Local health and nutrition conditions are quite poor, however, and evidence of malnutrition was apparent in significant numbers within the communities, particularly pregnant women and young children.

Given the prevalence of malnutrition within these communities, the role of aquatic resources is crucial in the coping strategies of the rural poor in periods of low rice availability. Strategies for food security and poverty alleviation should therefore pay special attention to sustainable aquatic resources management. However, provincial authorities are encouraging farmers to expand their rice cultivation. Land not used for cultivation is still viewed as "underused" and available for development. The areas earmarked for conversion into paddy are low-lying areas, including a wide range of aquatic habitats that play an important role in supporting aquatic biodiversity. While the assessment demonstrated the extensive indigenous knowledge of aquatic resources and well established coping strategies that are based on these resources, the assessment also exposed local practices and taboos related to the nutrition of pregnant women, young mothers and infants that are seriously deleterious to the health and wellbeing of these groups (FAO/IUCN 2003).

3.5 Research on Sustainable Agriculture and Wetlands

Until recently the agriculture–wetland interface was not on the agenda of agricultural research organizations such as IWMI. The environmental dimension of irrigated agriculture was traditionally defined as waterlogging, salinization and soil degradation. Since then IWMI has become a member of IUCN, an observer in the STRP of Ramsar and an active participant in the development of eco-agriculture as a research priority. The following overview of ongoing and recently completed research of IWMI and partners is intended as illustrative of the wetland–agriculture interface issues that are now on the agriculture research agenda.

- conceptual studies to define a systematic and semi-qualitative method of evaluation to classify the potential of using a "working wetland" (a managed wetland in which a rational compromise has been made between its ecological conditions and the level of human utilization; McCartney et al. 2005);
- simulation of reference hydrologic conditions of coastal lagoons prior to the development of irrigation systems upstream, to assess and monitor the impact of drainage water on coastal lagoons (Smakthin et al. 2004a);
- assessment of nutrient loadings from agriculture systems on coastal lagoons in Sri Lanka in general and the Bundala RAMSAR site in particular (Matsuno et al. 1998; Amerasinghe et al. 2002; Piyankarage et al. 2003, 2005);

- baseline biodiversity assessment before construction of a major irrigation system (the 5000-ha Kirindi Oya left-bank system) and subsequent monitoring of changes in biodiversity during construction[6];
- assessment of the importance of fisheries in reservoirs (Renwick 2001), as well as the impact of irrigation systems on fisheries in paddy fields, in floodplains and the river itself in Laos and Sri Lanka (Lorenzen et al. 2004; Nguyen-Khoa and Smith 2004; Nguyen-Khoa et al. 2005a, b; Smith et al. 2005);
- assessment of rural people's livelihood strategies in coastal zones in southern Sri Lanka and the role of fisheries and agriculture in these (Clemett et al. 2003; Senaratne Sellamuttu and Clemett 2003);
- trade-off analyses of wetlands-based livelihoods in the Limpopo basin to balance agricultural production with wise use of wetlands – recently initiated research[7];
- analysis of environmental water requirements at a global scale (Smakthin et al. 2004b) as well as (parts of) larger and smaller river basins in the Mekong and Sri Lanka;
- mapping wetlands and their consumptive use of water using remote sensing techniques developed for irrigated agriculture (Mohamed et al. 2004).

It is also significant that the CGIAR Challenge Program on Water and Food, the world's largest research program in tropical agriculture that deals with water, food and environment, has Aquatic Ecosystems and Fisheries as one of its five research themes[8].

3.6 Conclusions: Towards Sustainable Agriculture and Wetlands?

There is increased mutual interest from both the environmental and the agricultural communities in wetlands and agriculture. This presents an opportunity to attempt to provide a better basis for an effective dialogue than the two communities have enjoyed in the past. In order for the two communities to understand each other's language and perspective, several basic concepts are proposed here as "initial prerequisites", as follows:

- balancing water for wetlands and agriculture needs to take into account the full hydrological cycle, both the blue water (renewable water resources) and the green water (soil moisture);

---

[6] A joint IWMI/IUCN project, implemented with financial support of the Royal Netherlands Embassy and in collaboration with the Mahaweli Development Authority.

[7] Projects supported by both the CGIAR Challenge Program on Water and Food (www.waterforfood.org) and GEF.

[8] Led by V. Sugunan of the WorldFish Centre (v.sugunan@cgiar.org) and co-organizer of INTECOL Symposium 34.

- water resources provide a critical supporting service for vegetation and other living organisms in both wetlands and agriculture; the evapotranspiration of the vegetation in both systems is a measure of the water use (and water requirement) of both wetlands and agriculture; wetlands provide an important water resources regulation service; neither wetlands nor agriculture provide or provision water;
- water productivity, broadly defined at the basin scale (i.e. including crop, livestock and fishery yields, wider ecosystem services and social impacts such as health, together with the systems of resource governance that ensure equitable distribution of these benefits), provides a good framework to balance the needs of wetlands and agriculture at this scale level;
- water scarcity will be an increasingly important constraint for agriculture in Asia and Africa; this will lead to increasing competition for the resource between agriculture and wetlands – not domestic or urban use; sustainable food production and conception patterns as well as sustainable use of wetlands will only be possible if there is a considerably increased and effective dialogue among the environmental and agricultural communities;
- the Dialogue on Water, Food and Environment has provided an initial exploration of the potential and constraints to increased wetland–agriculture dialogue; its guidelines and learning framework (as well as the experience that is currently being developed with basin and sub-basin dialogues in the field) is worth taking into account;
- there is a considerably increased interest in the wetland–agriculture interface from the side of agricultural research organizations, such as IWMI and its counterparts in the CGIAR, that provides a significant opportunity for future collaboration among the environmental and agricultural research communities.

References

Amerasinghe FP, Matsuno Y, Silva RI de, Piyankarage SC, Bambaradeniya CNB, Mallawatanthri A (2002) Impact of irrigation water on water quality, fish and avifauna of three coastal lagoons in southern Sri Lanka. In: Ali A, et al. (eds) The Asian wetlands: bringing partnerships into good wetland practices. Proc Asian Wetland Symp 2001:27–30

Bambaradeniya C, Amerasinghe FP (2004) Biodiversity associated with the rice field agro-ecosystem in Asian countries: a brief review. (IWMI Working Paper 63) IWMI, Colombo

Barbier EB, Adams WM, Kimmage E (1991) Economic valuation of wetland benefits: the Hadejia–Jama'are floodplain, Nigeria. (LEEC Discussion Paper) IIED, London

Barbier EB, Acreman M, Knowler D (1992) Economic valuation of wetlands: a guide for policy makers and planners. Ramsar Convention Bureau, Gland

Bennet J (2003) Opportunities for increasing water productivity of CGIAR crops through plant breeding and molecular biology. In: Kijne JW, Barker R, Molden DJ

(eds) Water productivity in agriculture: limits and opportunities for improvement. CABI/IWMI, Wallingford, pp 103–126
Bhatt Y, Vallee D (2004) The global dialogue on water, food and environment: lessons learnt and looking ahead. IWMI, Colombo
Blank HG, Mutero CM, Murray-Rust H (eds) (2001) The changing face of irrigation in kenya: opportunities for anticipating change in Eastern and Southern Africa. IWMI, Colombo
Clemett A, Senaratna Sellamuttu S, Banda R (2003) Can coastal zone management in Sri Lanka sustain local livelihoods and ecosystems? (Policy Process Analysis Paper 2) Stockholm Environment Institute, Wallingford
Cosgrove WJ, Rijsberman FR (2000) World water vision: making water everybody's business. Earthscan, London
Costanza R, d'Arge R, Groot R de, Farber S, Grasso M, Hannon B, Limburg K, Naeem S, O'Neill R, Paruelo J, Raskin R, Sutton P, Belt M van den (1997) The value of the world's ecosystem services and natural capital. Nature 387:253–260
Falkenmark M (2000) Competing freshwater and ecological services in the river basin perspective: an expanded conceptual framework. Water Int 25:172–177
Falkenmark M, Lundquist J, Widstrand C (1989) Macro-scale water scarcity requires micro-scale approaches: aspects of vulnerability in semi-arid development. Nat Resour Forum 13:258–267
FAO/IUCN (2003) The role and nutritional value of aquatic resources in the livelihoods of rural people: a participatory assessment in Attapeu Province, LAO PDR. A contribution to the dialogue on water, food and environment. FAO, Bangkok
Hofwegen P van, Svendsen M (2000) A vision for water for food and rural development. Second World Water Forum, http://www.watervision.org
IUCN (2000) Vision for water and nature: a world strategy for conservation and sustainable management of water resources in the 21st century. IUCN, Gland
IWMI (2000) World water supply and demand in 2025. In: Rijsberman FR (ed) World water scenario analyses. World Water Council, Marseille
Kijne JW, Barker R, Molden DJ (eds) (2003) Water productivity in agriculture: limits and opportunities for improvement. CABI/IWMI, Wallingford
Lorenzen K, Smith LED, Burton M, Nguyen Khoa S, Garaway C (2004) Guidance manual on the management of irrigation development impacts on fisheries. (DFID project R7793) Imperial College, London
Matsuno Y, Hoek W van der, Ranawake R (1998) Irrigation water management and the Bundala National Park: proceedings of the workshop on water quality of the Bundala lagoons. IWMI, Colombo
McCartney M, Masiyandima M, Houghton-Carr H (2005) Working wetlands: classifying wetland potential for agriculture (research report). IWMI, Colombo (in press)
McNeely JA, Scherr S (2001) Common ground common future: how eco-agriculture can help feed the world and save wild biodiversity. IUCN, Gland
MEA (2003) Millennium Ecosystem Assessment conceptual framework document. MEA, www.millenniumassessment.org
MEA (2005) Millennium Ecosystem Assessment synthesis report. MEA, www.millenniumassessment.org
Mohamed YA, Bastiaansen WGM, Savenije HHG (2004) Spatial variability of evaporation and moisture storage in the swamps of the upper Nile studied by remote sensing techniques. J Hydrol 289:145–164
Nguyen-Khoa S, Smith, LED (2004) Irrigation and fisheries: irreconcilable conflicts or potential synergies? Irrig Drain 53:415–427
Nguyen Khoa S, Smith LED, Lorenzen K (2005a) Impacts of irrigation on inland fisheries: appraisals in laos and sri lanka research report comprehensive assessment programme. IWMI, Colombo (in press)

Nguyen Khoa S, Smith LED, Lorenzen K (2005b) Adaptive, participatory and integrated assessment of the impacts of irrigation on fisheries. Evaluation of the approach in Sri Lanka. (IWMI Working Paper) IWMI, Colombo (in press)

Piyankarage SC, Mallawatantri AP, Matsuno Y, Pathiratne AS (2003) Estimation of nitrogen and phosporus fluxes to Embilikala and Malala lagoons in Southern Sri Lanka. In: Ratnasiri J (ed) Assessment of material fluxes to the coastal zone in South Asia and their impacts. National Committee of IGBP, Colombo, pp 9–20

Piyankarage SC, Mallawatantri AP, Matsuno Y, Pathiratne AS (2005) Human impacts and the status of water quality in the Bundala Ramsar wetland lagoon system in southern Sri Lanka. Wetlands Ecol Manage (in press)

Renwick ME (2001) Valuing water in irrigated agriculture and reservoir fisheries: A multiple-use irrigation system in Sri Lanka. (IWMI Research Report 51) IWMI, Colombo, 34 pp

Rijsberman FR (1996) Feuchtgebietwiederherstellung in der Scheldt-Mündung: Kommunikation unter Landwirten, Umweltschützern und anderen Belteiligten. In: Trilateral Conference (ed) Sustainable agriculture in the Morava and Dyje floodplains. Stupava, Slovak Republic

Rijsberman FR (2004) Sanitation and access to clean water. In: Lomborg B (ed) Global crises, global solutions. Cambridge University Press, Cambridge, pp 498–527

Rijsberman FR (2006) Water scarcity: fact or fiction? Agric Water Manage 80:5–22

Rijsberman FR, Molden DJ (2001) Balancing water uses: water for food and water for nature. (Thematic Background Paper) Int Conf Freshwater 2001:43–56

Seckler D, Molden DJ, Sakhtivadivel R (2003) The concept of efficiency in water resources management and policy. In: Kijne JW, Barker R, Molden DJ (eds) Water productivity in agriculture: limits and opportunities for improvement. CABI/IWMI, Wallingford

Senaratna Sellamuttu S, Clemett A (2003) Avoiding conflict between fishers and farmers: a study of potential social conflict over scarce water resources in Kalametiya. In: Proceedings of Fourth National Symposium on Poverty Research in Sri Lanka. Sri Lanka Association for the Advancement of Science, Colombo

Shiklomanov IA (1998) World water resources: an appraisal for the 21st century. (IHP Report) UNESCO, Paris

SIWI/IWMI (2004) Water – More nutrition per drop: towards sustainable food production and consumption patterns in a rapidly changing world. (Background document for CSD12) SIWI, Stockholm

Smakthin VU, Pyankarage SC, Stanzel P, Boelee E (2004a) Simulating the hydrology of small coastal ecosystems in conditions of limited data. (IWMI Research Report 77) IWMI, Colombo

Smakthin VU, Revenga C, Doll P (2004b) Taking into account environmental water requirements in global-scale water resources assessments. (Comprehensive assessment of water management in agriculture research report 2) IWMI, Colombo

Smith LED, Lorenzen K, Nguyen Khoa S (2005) Livelihood functions of inland fisheries: policy implications in developing countries. Water Policy (in press)

Tuong TP, Bouman BAM (2003) Rice production in water-scarce environments. In: Kijne JW, Barker R, Molden DJ (eds) Water productivity in agriculture: limits and opportunities for improvement. CABI/IWMI, Wallingford, pp 53–68

WHO (2004) Health as a cross-cutting issue in dialogues on water for food and the environment. (Report of an international workshop; WHO/SDE/WSH/04.02) WHO, Geneva

Wolter H (2003) Overcoming the world water crisis: achieving water, food and environmental security. In: DWFE (ed) Proceedings of an international conference on the dialogue on water, food and environment. DWFE/ Ministry of Agriculture and Rural Development of Vietnam, Hanoi, www.iwmi.org/dialogue

4 Sustainable Water Management by Using Wetlands in Catchments with Intensive Land Use

Chengqing Yin, Baoqing Shan, Zhanpo Mao

4.1 Semi-Natural Wetlands Created by Humans Before the Industrial Age

Wetlands play important roles in regional hydrology. In those catchments with intensive land use, they perform a key function in hydrological regulation and mass cycling. In recent decades, there have been many studies on the ecological values of natural wetlands and many protection measures have been proposed. The ecological functions of semi-natural wetlands are, however, often ignored. People have used these semi-natural wetlands for many centuries in the development of civilization and the most important is water use. In this chapter, the ecological functions of the multipond systems are taken as an example and the wise use of these semi-natural wetlands is discussed.

In China, India and some other countries, there are many semi-natural wetlands that have been developed by man during a long era of civilization. Today, we can still learn much from the ancient management practices involved, as we face mounting problems of local and regional water management. Semi-natural wetlands occur in many landscapes and they can occupy a significantly large area. Examples of semi-natural wetlands are irrigation ponds, water gardens, reed beds and mulberry-coupled fishponds. Similar systems have also existed in central Europe for several centuries. Peasants in the Trebon Basin area of the Czech Republic created fishpond systems which have gradually been integrated into the landscape (Kvet et al. 2002). Management of fishponds has affected the surrounding primeval environment through dams, excavated hollows and artificial channels. In ancient times, semi-natural wetlands were usually closely related to food production, so that they were mostly distributed in the areas with extensive agricultural activities. At many places in China, for example, people still use them to produce rice, fish and other food products.

Ecological Studies, Vol. 190
J.T.A. Verhoeven, B. Beltman, R. Bobbink, and D.F. Whigham (Eds.)
Wetlands and Natural Resource Management

Pond systems played an important role in agricultural development by providing substantial amounts of water for irrigation (Agarwal and Narain 1997; Chen et al. 2000; Kakade et al. 2001; Li et al. 2000). In India and Sri Lanka, interconnected ponds are used to store rainwater to control flooding and prevent soil erosion during periods of heavy rainfall, and to supply water for irrigation in dry seasons. These measures provided an effective solution for local food production (Fox and Rockström 2000). In central Europe, the creation and management of ponds for water supply and fish production has developed in equilibrium with other socio-economic factors (Kvet et al. 2002).

The multipond system in China, the main focus of this chapter, is widely distributed in the southern part of China. It comprises a large artificial irrigation network system, composed of many tiny ponds, scattered in agricultural fields and connected by ditches and streams (Yin and Shan 2001). Typically, the total pond area occupies 6–10% of the total watershed and individual ponds vary in size from 0.05 ha to 1.0 ha, serving an area of 0.5–10.0 ha. In addition to providing water storage during periods of high runoff, the ditches that connect the ponds form a flow path to guide the water to the fields for purposes of irrigation and they connect to streams that carry excess water to downstream ponds during the growing season. In a landscape context, thousand of ponds are connected in a large network to provide water at a regional scale. Multipond systems were first developed in areas of igneous rocks where there was a scarcity of ground water. Accordingly, the pond systems were

Fig. 4.1 The multipond system, an overview picture

important for water storage and irrigation. In the hilly areas where the pond systems were developed, the location of ponds was determined mostly by landform and elevation (Fig 4.1).

4.2 Water Regulation by the Multipond Systems

One of the most important functions of semi-natural wetlands is to act as a buffer to regulate surface runoff. This development was necessary because the original forests, ecosystems that store large amounts of water, were cleared during agricultural development in history. Thus, the local hydrological cycling was changed and the ponds have replaced the forests as the element within the landscape that stores water. They reduce the threat of flood and drought and provide water for crop irrigation and natural vegetation uses. In this respect, wetlands function to compensate for the forest loss and contribute to a suitable water balance in the catchment.

4.2.1 Research Site Description

The case study was conducted in the Liuchahe watershed, along a stream flowing into Chaohu Lake, which is one of the five largest lakes by the Yangtze River. The Liuchahe watershed has an area of 691.6 ha, with a typical landscape of a multipond system in a low hill region (Fig. 4.2). The land-use percentages for the watershed are 45 % in rice cultivation, 26 % in dry land farming, 16 % in forest, 7 % in villages and 6 % in ponds. The multipond system is composed of 193 ponds, with a total area of 43 ha, which are connected by ditches. The mean pond depth is about 1.5 m. During the dry season some ponds are empty, while during the wet season the maximum depth of water amounts to 2.0 m. These ponds store surface runoff and agricultural drainage water, which is then used to irrigate rice and other crops. Each pond collects agricultural drainage and has an inlet and an outlet. The overflow from each pond goes to the next one in the chain, so that the ponds form a kind of cascade. In this manner, the runoff for the entire watershed is intercepted quantitatively in the multi-pond system during most rainfall events. The number of ponds that are integrated by ditches within each subwatershed ranges from five to ten, depending on the topography.

In this area, there is a triple-cropping system consisting of early rice, late rice and rape or wheat; and the farmland is located close to the ponds. Dry farmland is normally cultivated with wheat, cotton, peanut, soybean, sweet potato and vegetables year round. In this area, the mean annual temperature is 15.5 °C. Annual potential evaporation is 1484 mm and the average relative moisture is 77 %. Long-term average precipitation is 940 mm, of which 70 %

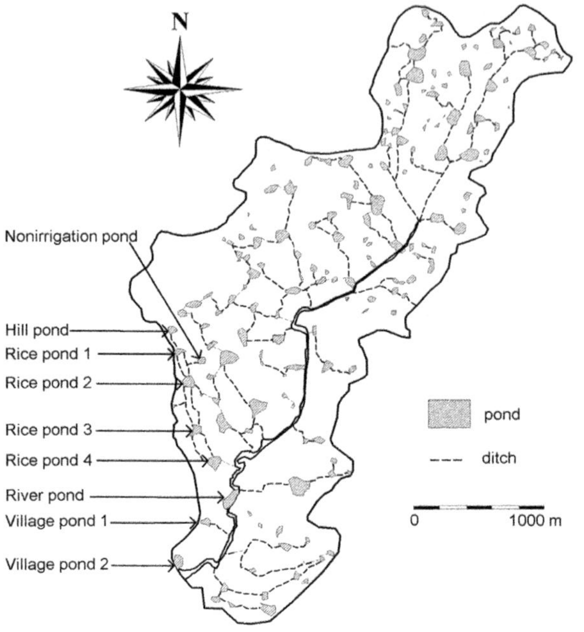

Fig. 4.2 The Liuchahe watershed with a multipond system near Chaohu Lake

falls from April to September. Each year there are three to five snowfall events.

Precipitation in the Liuchahe watershed is monitored at an official meteorological station. Three hydrological years (1990, 1995, 1998) were selected for comparing the flow pattern of surface runoff. The year 1990 was a normal year (annual precipitation 915 mm), while 1995 and 1998 were a dry and a wet year (annual precipitation 668 mm, 1174 mm), respectively. In most years, base flow only occurs at the lowest part of the Liuchahe River. The duration of water flow in the ditch and outflow from ponds depends on the rainfall intensity.

4.2.2 The Regulation Process for the Crop Water Supply by the Pond System

The ponds in the Liuchahe watershed have a storage capacity of 6.3×10^5 m^3, a volume that is equivalent to 91 mm of runoff depth. The volume of the multi-pond system is large enough to effectively influence water cycling on both a temporal and a spatial scale. In the normal year of 1990, there were 118 days with rainfall, mostly between February and August. The largest amount of monthly rainfall in this year was 113.4 mm in February and the least was only 19.1 mm in October. The crop water requirement was computed by multiplying the reference crop evapotranspiration with a crop coefficient, which was based on crop type and growth period to account for differences between grass and crop evapotranspiration (Kashyap and Panda 2001). Based on crop type and crop area in the Liuchahe watershed, the annual crop water requirement was calculated to be 573 mm in 1990 (Mao 2003), most of which was needed in May, June, July and October, accounting for 64 % of annual requirement. The highest monthly crop water requirement was 129 mm in July, while the lowest was 15 mm in January.

The annual water shortage calculated was 320 mm in the normal year 1990, without taking account of the ponds. The water shortage occurred from May to July and again in October. These periods of time accounted for 53 % of the annual shortage. The water shortage peak of 92 mm occurred in June and the lowest shortage was only 3 mm in January. The fluctuation in water shortage from January to May was much smaller than that of May to December and the largest difference amounted to 53 mm between May and June.

The annual surface runoff from all land use types in the normal year 1990 was calculated to be 355 mm. The maximum monthly runoff of 50 mm occurred in February with rainfall of 113 mm, while the lowest was 3 mm in October. The high surface runoff with low crop water requirement led to high water accumulation in the ponds, which then supplied water in the month of high crop water requirement and low precipitation. The storage increased gradually from January to May, because of low crop water requirement and high runoff production, and declined gradually from May to December due to high crop water requirement or low rainfall. The highest storage occurred in May, 96 mm and the lowest was zero in October. The annual runoff intercepted by the ponds reached 277 mm, accounting for 79 % of annual runoff and reducing the annual water shortage from 320 mm to 43 mm, which only occurred in October.

Figure 4.3 shows the influence of the multipond system on surface runoff in a normal year, 1990 (Mao 2003). The watershed would have a much higher runoff if there were no ponds, as shown in the line for surface runoff prior to pond regulation (SRPR). However, under the influence of ponds and ditches, the surface runoff became much lower. The majority of the runoff was stored in the wetlands as shown in the WS line. Rice and other subtropical crops have a high water demand; and drought often causes serious crop damage. The natural water supply usually cannot meet the crop water requirement, even in a wet year. With the pond influence, the actual water shortage became much smaller and crop damage was avoided (Fig. 4.4).

Further analysis showed that the runoff intercepted by the ponds was 277 mm in the normal year, 239 mm in the wet year and 195 mm in the dry year. The system influenced the hydrological process of runoff effectively and transformed most runoff events into intermittent flow. The multipond system was estimated to hold 90 % of runoff for a daily rainfall of 141 mm, with a peak discharge rate of 0.3 $m^3 s^{-1}$ on 10 July 1991. Simulations for crop water

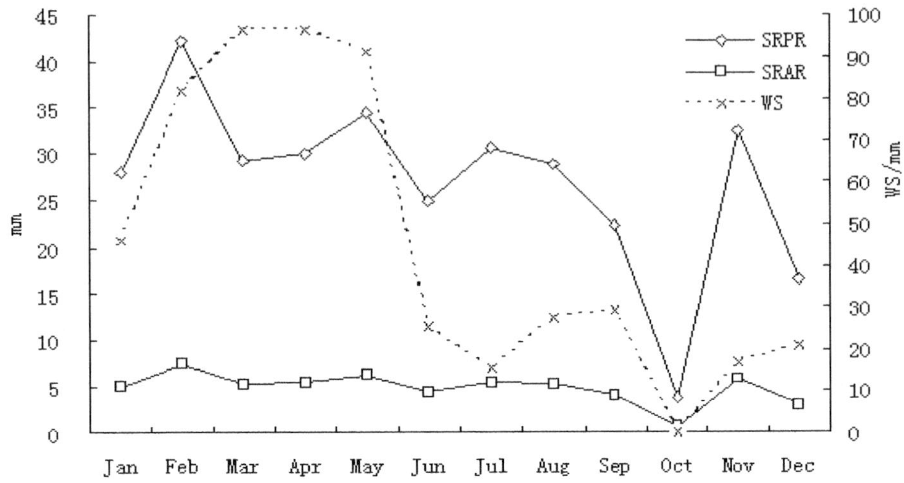

Fig. 4.3 The effect of the multipond system on surface runoff in a normal rainfall year, 1990. *SRPR* Surface runoff prior to pond regulation, *SRAR* surface runoff after pond regulation, *WS* water storage. Figure from Mao (2003)

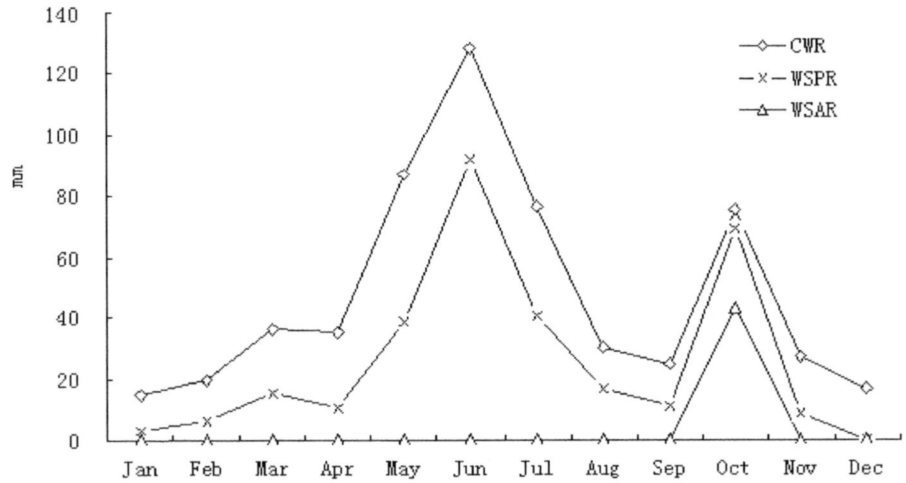

Fig. 4.4 The impact of the ponds on water shortage in a normal year, 1990, *CWR* Crop water requirement, *WSPR* water shortage without the pond regulation, *WSAR* water shortage at the regulation. Figure from Mao (2003)

Table 4.1 The influence of the multipond system for water resources in Liuchahe watershed. Under pond regulation, the actual water shortage was much less, in comparison with the calculated shortage if there were no ponds

Year	Rainfall (mm)	Crop water requirement (mm)	Water shortage (mm)	Surface runoff (mm)	Retained runoff (%)	Actual water shortage (mm)
1990	915	573	320	355	79	43
1991	1796	552	280	632	37	42
1995	658	523	333	236	83	138

requirement, water shortage, surface runoff and the influence of the multipond system during the three years are summarized in Table 4.1.

It can be concluded that the ponds effectively influenced water distribution and offset water shortage in normal years, minimized flooding in wet years and reduced crop drought damage in dry years. The annual surface runoff intercepted by the ponds reached 79 % in a normal year (1990), 47 % in a wet year (1991) and 84 % in a dry year (1995). In this way, the multipond system provides an effective alternative to attenuate the water shortage problem in the agricultural watershed. Crop water requirement consistently showed a pattern of high requirement in May, June and July; and the shortage in these months accounted for approximately 50 % of the annual water needs. Under the influence of the ponds, the actual water shortage was greatly reduced.

The influence of the ponds on flood control occurs in two ways. First, the multipond system reduces peak discharge and, second, it attenuates the water velocity within the pond system. In the simulation of rainfall on 10 July 1991, which was the day with the highest daily precipitation (141 mm) recorded in the past 20 years, the surface runoff from all land uses in the watershed was 224 402 m^3. Because the ponds were dry before this date, the runoff intercepted by the ponds was estimated to be 202 200 m^3. The results indicated that there was a 90 % reduction of runoff at the peak daily intensity. By regulating water in the watershed, the multipond system reduces the threats of flooding as well as drought.

4.3 Other Ecological Functions of Ancient Semi-Natural Wetlands in a Modern Scientific Context

Besides water regulation, the multipond systems have other ecological functions that stabilize adjacent ecosystems, i.e., sediment retention, nutrient retention and biodiversity conservation.

4.3.1 Sediment Retention Within the Watershed

Soil erosion is a serious problem in watersheds that have a high percentage of crops. The multipond system minimizes the impact of sediment loading by directing and storing water during runoff events. Surface runoff from agricultural fields flows into the adjacent ditches first and then passes through a series of ponds. The ponds greatly reduce the flow rate, resulting in significant sediment retention. On 1 May 1998, in a period with continuous runoff, the velocity of water when flowing from a ditch into a pond changed from 24.5 cm s^{-1} to 1.4 cm s^{-1}. The decrease in the water velocity made the suspended solids sink to the pond bottom and reduced the erosion of sediments from the ditch walls (Yin and Shan 2001). The multipond system tends also to trap sediments because of the huge water storage volume. When storm runoffs passed through the ditch–pond system, large amounts of solid particles were found to settle in the different ponds. Annual sediment accumulation was about 7.9×10^6 kg $year^{-1}$ and formed a 3-cm sediment layer. Ponds in the watershed had different accumulation rates of sediments as a result of the different export loading of ambient land uses (Table 4.2).

Ponds located in topographically different positions within the watershed were found to behave differently. The lowest accumulation rate was found in hill ponds, because the surrounding forested areas, that are higher in the watershed, had the lowest erosion rate. Non-irrigation ponds located in topographically lower areas had the highest accumulation rate. The suspended solids concentration entering ponds near villages was the highest and the accumulation rate in these ponds was the second highest. The area of the ponds in rice paddy areas was far greater than that of other types of ponds, so that the total amount of sediments accumulated in these rice ponds was the largest, although they had relatively lower accumulation rates. Through this

Table 4.2 Annual mean thickness and mean accumulation of sediments in different types of ponds in the Liuchahe multipond system

Type	Area (ha)	Thickness accretion (cm $year^{-1}$)	Content of nutrient, P (g kg^{-1})	Content of sediment (g cm^{-3})	Sediment accumulation (kg $year^{-1}$)
Hill pond	3.8	1.2	0.39	1.03	0.47×10^6
Non-irrigation pond	1.2	5.8	0.53	1.01	0.70×10^6
Rice pond	36.3	1.6	0.47	0.97	5.63×10^6
River pond	1.1	2.5	0.49	1.21	0.33×10^6
Village pond	1.9	4.0	0.76	1.10	0.81×10^6
Total	44.3	3.0[a]	0.53[a]	1.06[a]	7.9×10^6

[a] Mean value

process of sedimentation in ponds, over 80 % of the eroded soil particles can be retained in the watershed. The multipond system is thus an effective way to control sediment erosion from the watersheds.

4.3.2 Nutrient Retention and Recycling

The majority of lakes in East China are eutrophic. The diffuse pollution from rural areas contributes more than half of the phosphorus and nitrogen load to these lakes. Multipond systems have a large retention capacity to store nutrients. In the Liuchahe watershed, the concentrations of nutrients in runoff vary widely for villages, non-irrigated farmland and rice fields. The runoff water in the watershed is collected in the adjacent ditches and ponds, where the nutrients become part of the cycling and removing processes in the multipond system. As a result of these processes, the nutrient concentrations are reduced significantly. After the filtering, sedimentation and retention processes occurring while the water flows through the system, the nutrient concentrations at the watershed outlet are much lower than those at the source.

Because rice needs large quantities of water, irrigation of rice fields is the main use of pond water. In the Liuchahe watershed, annual irrigation schemes depend upon the intensity and frequency of rainfall. Generally, in one growing season, irrigation occurs four to eight times. After rainfall, the ponds receive runoff water and nutrients from nearby watersheds as well as upstream ditches. Whenever it is needed, the farmers drain or pump the pond water with nutrients to their lands for irrigation. Because of these flows and movements, it is suggested that the multipond systems provide intensive recycling centers among the agricultural fields.

The retention and removal mechanisms of nutrients in the multipond systems comprise sedimentation, adsorption, recycling through irrigation and uptake by aquatic plants (Yan et al. 1998; Yin and Shan 2001). In the five years of our studies, the reduction in nutrient output (P) from this catchment by the multipond systems was 97 % in 1987, 98 % in 1988, 99 % in 1994, 97 % in 1995 and 94 % in 1999. The reduced nutrients were retained in the multipond system or recycled back to the fields.

4.3.3 Landscape Complexity and Biological Diversity

The multipond system includes many small ponds that are scattered throughout the farmland and have strong interactions with the adjacent fields. Within the watershed, the areas occupied by ponds and ditches can be expressed as a landscape ecological index. For ponds the index is 0.28 ha^{-1}, for ditches it is 33.8 m ha^{-1} and for ecotones it is 45.5 m ha^{-1}. These values suggest that the multipond system is characterized by a very high habitat heterogeneity at the landscape scale, resulting in a high degree of connectivity between landscape elements.

The multipond system also provides critical habitat for aquatic plants and animals. In the ponds and ditches, species such as *Alternanthera philoxeroides*, *Phragmites australis*, *Potamogeton crispus* and *Myriophyllum verticillatum* are dominant. In regions with a high population pressure and an intensive agricultural activity, these semi-natural wetlands play important roles in maintaining biodiversity. However, in the function of biodiversity conservation, the semi-natural wetlands cannot provide a service similar to that of natural wetlands because of the priority given to human benefits in management decisions.

4.4 Wetlands and Human Activities in Harmony

The multipond system is a natural resource management system dating back to ancient times. In its first developmental stage, it was composed of a few ponds near the houses, assigned for the supply of drinking water for humans and livestock. Later on, more and more ponds were built near the croplands for irrigation. An ancient agricultural article describes the system as: "keeping nine parts of lands free of harm at the cost of converting one part into ponds". The harm means frequent droughts and floods in the course of history. In more recent times, many ditches were dug to guide the water flow from the land into the ponds or from one pond to the next in a sequence, so that the water would not wash away the fertile soil from the land. By practice and improvement during thousands of years, the multipond systems became part of Chinese agricultural culture, as a sustainable practice and way of land development. This system has become distributed over one-quarter of the territory of the People's Republic of China (Fig. 4.5). In many places, it appears as a landscape of thousands of ponds distributed over thousands of hectares of fields. In the Taoyuan County of Taiwan, the ancient pond system was used to adjust water from the reservoirs, to recycle agricultural discharge and to extend the irrigation capacity via the connection of over 8000 ponds for 23 833 ha of rice fields (Chen et al. 2000). As a semi-natural wetland, man had to manage it and external maintenance was necessary. The most important labor on it was to dredge the ponds once every three or four years and return the accumulated sediments back to the lands. In this way, the soil and nutrients were also recycled between the ponds and fields and the volume of the ponds could be kept to its original size.

As an ancient irrigation system in China, the multipond system played an important role in China's agricultural development. The system stabilized agricultural communities by maintaining high production yields through irrigation; and it integrated water resources into watershed management. It

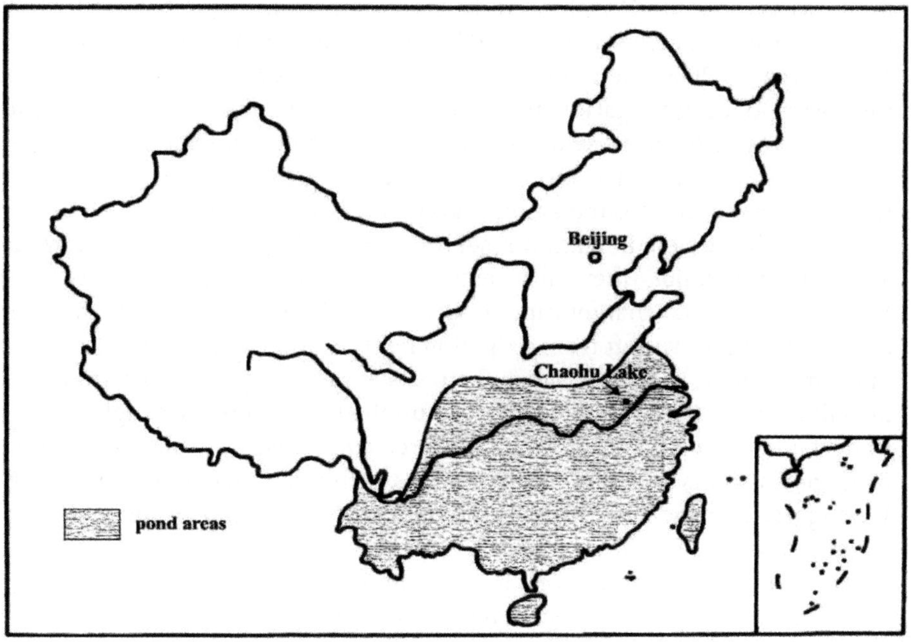

Fig. 4.5 The distribution area of multipond systems in China

remains important today, in the context of solving water-shortage problems, especially in areas with seasonally uneven precipitation distribution (Bouwer 2000; Johnson et al. 2001). The low cost of construction and energy consumption makes it an attractive agricultural system suitable for other developing countries with an increasing demand for food production and eutrophication control.

4.5 Protection of Semi-Natural Wetlands Together with Natural Wetlands

In recent years, the people in some regions begin to ignore the importance of the multipond systems, due to the development of modern reservoirs and irrigation channels. The water storage capacity of many ponds decreased because of sediment accumulation and a cessation in the removal of sediments through management. In the Liuchahe watershed, a pond originally had a typical depth of 2.0 m, but now most ponds have been reduced to a depth of 1.5 m. Throughout history, the ponds have been dredged by farmers and the pond mud has been applied to the surrounding land as fertilizer, so that ponds kept their volume. Now, farmers have stopped such actions in many areas. The number of ponds in the country was 8.3×10^6 in the 1950s and the irrigation area was 133×10^6 ha, accounting for 39% of the whole irrigation area. By 2000, the number of ponds was reduced to only 6.3×10^6 and the corresponding irrigation area was 80×10^6 ha (Hang 2001). Some ponds were destroyed to increase arable land and housing; and some ponds were filled with sediments that were not excavated to fertilize arable land. The distribution area of multipond systems gradually became reduced.

Although the new irrigation reservoirs have a large storage capacity and are multifunctional, they also have many disadvantages (Chiras 1994; Vitousek et al. 1997) that multipond systems could mitigate, such as the interception of surface runoff from irrigation fields, at low cost. As the transfer distance from pond to field is short, the water loss during transportation could be reduced; and the water temperature supplied by the ponds is high, which is important for healthy crop growth. Moreover, irrigation water from ponds contains nutrients and they can be recycled within the agricultural ecosystem. We suggest that the new reservoirs should be combined with the multipond system to form a modern irrigation system to improve agricultural production and at the same time protect ecosystem integrity.

A degraded multipond system has less ability to attenuate storm peak precipitation flows and to retain transported pollutants. In the 1990s, the traditional functions of the ponds were noticed again, because of the frequent occurrence of droughts and floods in southern China. With our continuous efforts, the Chinese central and provincial governments began to realize this problem. How to construct and manage the multipond system under the implementation of the "household responsibility system" is an important question that could seriously affect agricultural sustainable development and water environment protection. In the Province of Anhui, the government has now announced some polices to encourage farmers to manage old ponds and dig new ones. However, avoiding the degradation of the multipond system in most regions of China remains a future challenge.

References

Agarwal A, Narain S (1997) Dying wisdom: the decline and revival of traditional water harvesting systems in India. Ecologist 27:112–116

Bouwer H (2000) Integrated water management: emerging issues and challenges. Agric Water Manage 45:217–228

Chen PH, Leung KC, Wang JT (2000) Investigation of a ponding irrigation system to recycle agriculture wastewater. Environ Int 26:63–68

Chiras DD (1994) Environmental science: actions for a sustainable future. Benjamin/Cummings, New York, pp 222–224

Fox P, Rockström J (2000) Water-harvesting for supplementary irrigation of cereal crops to overcome intra-seasonal dry spells in the Sahel. Phys Chem Earth B 25:289–296

Hang HQ (2001) The effects of the ponds rising and falling (in Chinese). Anc Mod Agric China 4:12–17

Johnson N, Revenga C, Echeverria J (2001) Managing water for people and nature. Science 292:1071–1072

Kakade BK, Neelam GS, Petare KJ, Doreswamy C, Reddy GS (2001) Revival of rivulets through farm pond based watershed development. Balf Development Research Foundation, Pune

Kashyap PS, Panda RK (2001) Evaluation of evapotranspiration estimation methods and development of crop-coefficients for potato crop in a sub-humid region. Agric Water Manage 50:9–25

Kvet J, Jenik J, Soukupova L (2002) Freshwater wetlands and their sustainable future. Parthenon, New York

Li FR, Cook S, Geballe GT, Burch WR (2000) Rainwater harvesting agriculture: an integrated system for water management on rainfed land in China's semiarid areas. Ambio 29:477–483

Mao ZP (2003) Studies on the nutrients dynamics in multipond system, agricultural headwater and NPS models. PhD thesis, Chinese Academy of Sciences, Beijing

Vitousek PM, Mooney HA, Lubchenco J, Melillo JM (1997) Human domination of earth's ecosystems. Science 277:494–499

Yan WJ, Yin CQ, Tang HX (1998) Nutrient retention by multipond systems: mechanisms for the control of nonpoint source pollution. J Environ Qual 27:1009–1017

Yin CQ, Shan BQ (2001) Multipond systems: a sustainable way to control diffuse phosphorus pollution. Ambio 30:369–375

Section II

Wetland Science for Environmental Management

5 Constructed Wetlands for Wastewater Treatment

Jan Vymazal, Margaret Greenway, Karin Tonderski, Hans Brix, Ülo Mander

5.1 Introduction

Constructed wetlands (CWs) are engineered systems that have been designed and constructed to utilise the natural processes involving wetland vegetation, soils, and the associated microbial assemblages to assist in treating wastewaters. They are designed to take advantage of many of the same processes that occur in natural wetlands, but do so within a more controlled environment. CWs for wastewater treatment may be classified according to the life form of the dominating macrophyte into systems with free-floating, rooted emergent and submerged macrophytes (Brix and Schierup 1989). Most systems constructed around the world are planted with emergent macrophytes, but the design of these systems varies in terms of media and flow regime. The most common systems are designed with a free water surface (FWS CWs) or horizontal sub-surface flow (HF CWs), but vertical flow (VF CWs) systems are getting more popular at present (Brix 1994; Vymazal et al. 1998; Vymazal 2001). Constructed wetlands have traditionally been used for the treatment of domestic and municipal sewage from both separate and combined sewerage (Vymazal et al. 1998). However, since the late 1980s CWs have been used for many other types of wastewater, including agricultural wastewaters (cattle, swine, poultry, dairy), mine drainage, food processing wastewaters (winery, abattoir, fish, potato, vegetable, meat, cheese, milk, sugar production), heavy industry wastewaters (polymers, fertilisers, chemicals, oil refineries, pulp and paper mills), landfill leachate and runoff waters (urban, highway, field, airport, nursery, green house; Hammer 1989; Cooper and Findlater 1990; Kadlec and Knight 1996; Vymazal et al. 1998; Nehring and Brauning 2002; Dias and Vymazal 2003; Mander and Jenssen 2003). As many of these wastewaters are difficult to treat in a single-stage system, hybrid systems have been introduced which consist of various types of constructed wetlands staged in series.

Ecological Studies, Vol. 190
J.T.A. Verhoeven, B. Beltman, R. Bobbink, and D.F. Whigham (Eds.)
Wetlands and Natural Resource Management

The first experiments aimed at the possibility of wastewater treatment by wetland plants were undertaken by Dr. Käthe Seidel in Germany in the early 1950s at the Max Planck Institute in Plön (Seidel 1955, 1961, 1965a, 1976). In the early 1960s, Seidel intensified her trials to grow macrophytes – she planted macrophytes in the shallow embankment of tray-like ditches and created artificial trays and ditches grown with macrophytes. To go beyond the anaerobic septic tank systems, she integrated a stage of primary sludge filtration in vertically percolated sandy soils planted with *Phragmites australis*. So the system consisted of an infiltration bed through which the sewage flowed vertically and an elimination bed with a horizontal flow (Seidel 1965b). This system was the basis for hybrid systems which were revived at the end of the twentieth century. However, Seidel's concept to apply macrophytes to sewage treatment was difficult to understand for sewage engineers and therefore, it was no surprise that the first full-scale FWS CWs were built outside Germany, i.e. in The Netherlands.

5.2 Free Water Surface Constructed Wetlands

FWS CWs [also called surface flow (SF) CWs] typically consist of a sequence of shallow basins and a water control structure that maintains water depth. When rooted macrophytes are used, 20–40 cm of soil is needed to support the roots of vegetation if the beds are sealed. FWS CWs can use emergent, submergent, free-floating, and floating-leaved macrophytes. The latter two types, however, will not be covered in this paper (see Crites and Tchobanoglous 1998; Vymazal 2001).

FWS CWs function as land-intensive treatment systems. Inflow water containing particulate and dissolved pollutants slows and spreads through a large area of shallow water with emergent or submerged vegetation. Settable organics are rapidly removed by quiescent conditions, deposition, and filtration. Attached and suspended microbial growth is responsible for the removal of soluble organics. FWS CWs are very effective in removing suspended solids via filtration and sedimentation (Kadlec and Knight 1996). Nitrogen is most effectively removed in FWS CWs by nitrification/denitrification. Ammonium is oxidised by nitrifying bacteria in aerobic zones, and nitrate is converted to free nitrogen or nitrous oxide in the anoxic zones by denitrifying bacteria. FWS systems provide the sustainable removal of phosphorus, but at relatively slow rates because of limited contact between the water column and the soil (Kadlec et al. 2000).

In spite of many prejudices among civil engineers about odour nuisance, attraction of flies and poor performance in cold periods, the IJsselmeer Polder Authority in Flevoland in The Netherlands constructed its first FWS CW in 1967 (De Jong 1976; Veenstra 1998). A star-shape layout was chosen in order to obtain optimum utilisation of the available area which, however, complicated macrophyte harvesting and maintenance in general (De Jong 1976). Therefore, longitudinal channels were added to facilitate mechanical biomass harvesting and system maintenance. The new wetland design included channels 3 m wide and 200 m long, separated by parallel stretches 3 m wide, resulting in an increase in land requirement from 5 m^2 to 10 m^2 per population equivalent (PE). The system was called planted sewage farm (or Lelystad process; Greiner and De Jong 1984; Veenstra 1998). In 1968, a FWS CW was created in Hungary near Keszthely in order to preserve the water quality of Lake Balaton and to treat the wastewater of the town (Lakatos 1998). The FWS system in the Nyirbogdány Petrochemical Plant (Hungary) was created in the 1970s and has an area of 15 500 m^2 (Lakatos 1998; Lakatos et al. 1996).

In contrast to North America where they became quite common, FWS CWs had not spread significantly throughout Europe at the end of the twentieth century, while HF CWs drew much more attention (Vymazal et al. 1998). However, FWS CWs are now in operation in many European countries, e.g. Sweden, Poland, Estonia, United Kingdom and Belgium, often as part of hybrid systems. Recently, such systems have drawn more attention in Australia, while in Sweden they were constructed with nitrogen removal as a primary target. In the United Kingdom, FWS CWs are successfully used to treat highway runoff (Cooper et al. 1996). They are efficient in removing organics and suspended solids but nutrient removal is low, usually about 50 % when sewage is treated (Table 5.1). However, removal efficiency expressed as a percentage is affected by the fact that many FWS CWs are used for tertiary treatment and it is well known that the lower inflow concentration is the reason for the lower percent removal.

Table 5.1 Treatment efficiency (%) of FWS wetlands (Data from 85 systems in Australia, Canada, China, Netherlands, New Zealand, Poland, Sweden, and USA). Concentrations are given in mg l^{-1} and loadings are given in kg ha^{-1} day^{-1} for BOD_5, TSS and kg ha^{-1} $year^{-1}$ nutrients (N, P). Modified from Vymazal (2001)

	Concentration			Loading			
	Inflow	Outflow	Efficiency	Inflow	Outflow	Removal	Efficiency
BOD_5	34	9.5	70.3	12.0	4.3	7.7	68.4
TSS	53	14.4	72.9	16.2	4.7	11.5	71.0
Total N	14.3	8.4	51.8	4660	2190	2470	52.9
Total P	4.2	2.15	48.8	2680	1360	1320	49.1

5.2.1 Free Water Surface Wetlands for Treatment of Wastewater and Non-Point Source Pollution in Sweden

In Sweden, as well as in other European countries, much effort has been undertaken to reduce the load of nitrogen to coastal waters. FWS CWs are considered cost-efficient to achieve nitrogen removal in both wastewater and water draining from agricultural areas. A number of wetlands have for example been constructed in the 1990s with the main aim to remove nitrogen from wastewater treated in conventional wastewater treatment plants (WWTP). Design and operation conditions vary considerably between the wetlands, but satisfactory results for nitrogen removal have been achieved within the first year of operation (Andersson et al. 2000; Linde and Alsbro 2000). More than 2350 ha of wetlands were created in Sweden in the agricultural landscape between 1996 and 2002 with different kinds of subsidies; and in Denmark about 3200 ha were created by 2004. Many of those wetlands were created or restored with multiple aims, e.g. biodiversity, irrigation and nitrogen removal. In the past few years, the question has been raised whether wetlands can also function as traps for phosphorus in rivers discharging into sensitive lakes and coastal areas. Experiences from other regions suggest that wetlands act as net removers of total phosphorus in water. In this section, we review the Swedish experiences of the use of FWS wetlands as sinks for nitrogen and phosphorus, and see how that relates to literature data.

Four large wetlands treating wastewater were carefully monitored for 3–8 years (Table 5.2). Two of those wetlands, Magle and Ekeby, have predominantly open water with submerged plants and filamentous algae (Andersson et al. 2004). The other two are dominated by emergent macrophytes inter-

Table 5.2 Characteristics of seven FWS wetlands receiving water from point (wastewater treatment plants) or non-point sources (agricultural catchments)

Wetland	Period	Size (ha)	Hydraulic load (mm day^{-1})	Load (kg ha^{-1} $year^{-1}$)	
				Total N	Total P
Point sources					
Magle	1995–2001	20	57	4 200	33
Ekeby	1999–2001	28	155	6 300	77
Oxelösund	1994–2001	23	21	1 700	30
Alhagen	1999–2001	28	17	1 600	17
Non-point sources					
Råbytorp	1993–2002	0.75	383	14 270	175
Slogstorp	1997–2002	0.65	1784	57 344	433
Genarp	1998–2002	1.0	206	4 225	92

spersed with open water areas. The second data set is from three well monitored wetlands for agricultural runoff in South Sweden, where both water flow and water quality have been measured continuously. Those wetlands are all small open water ponds, dominated by submerged plants, or devoid of higher plants in the case of Råbytorp.

With respect to nitrogen removal, wetlands receiving wastewater had much higher relative removal efficiencies, varying from 24 % to 69 % of the load. In contrast, due to the high and variable load, wetlands receiving agricultural runoff removed a maximum of 9 % of the load and with a much higher difference in the specific removal per area of each wetland (Fig. 5.1). Looking more in detail at the wastewater-loaded wetlands, there was a clear difference in efficiency between the wetlands receiving water rich in ammonium and the other two. Because of design features, the detention time in wetland Alhagen was almost double that in wetland Oxelösund, despite the similar hydraulic load. Wetland Alhagen removed almost as much nitrogen per area as Magle wetland, and reached a record 70 % relative removal for the operational period. Oxelösund removed only about 40 % of the same load. Apart from one quarter in 2000, the wetland provided more than 50 % removal of nitrogen with little seasonal changes (Andersson et al. 2004), in

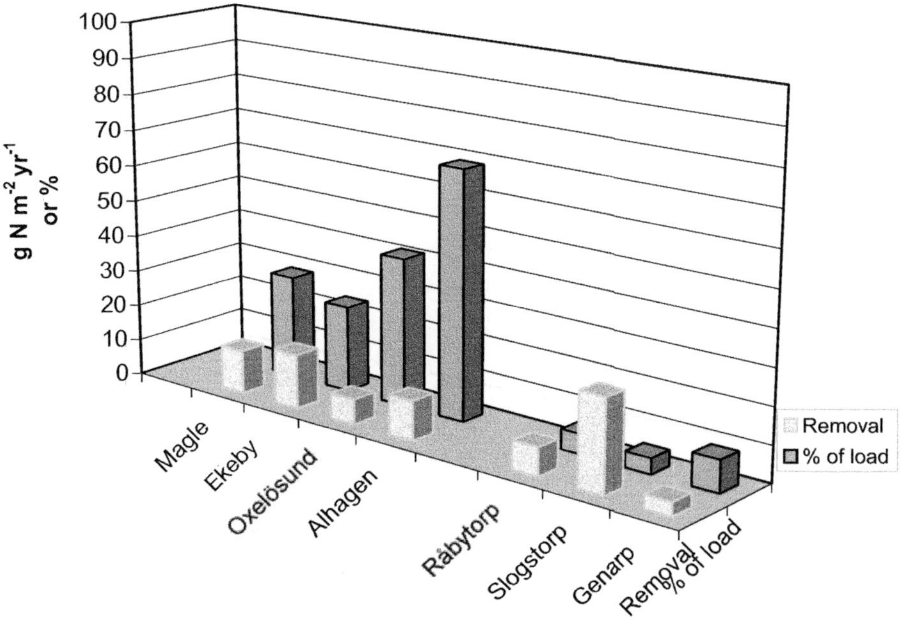

Fig. 5.1 Mean annual nitrogen area-specific and relative removal in wetlands receiving wastewater (Magle–Alhagen) and agricultural runoff (Råbytorp–Genarp) in South Sweden

contrast to what was observed in other wetlands (e.g. Kadlec 1999; Andersson et al. 2000; Kallner and Wittgren 2001). When adopting the first-order reaction rate equation (assuming plug flow conditions) to the annual mean values, the k_{A20} value (rate constant calibrated for 20 °C) for nitrogen removal in Alhagen was 50 % higher than in Oxelösund – 1.1 m^2 year^{-1} versus 0.61 m^2 year^{-1}. A probable explanation for that is the existence of an overland flow area in Alhagen, contributing significantly to nitrification of the ammonium in the wastewater (Andersson et al. 2004). For the wetlands receiving agricultural runoff, there was a clear dependence of annual removal on the nitrogen load

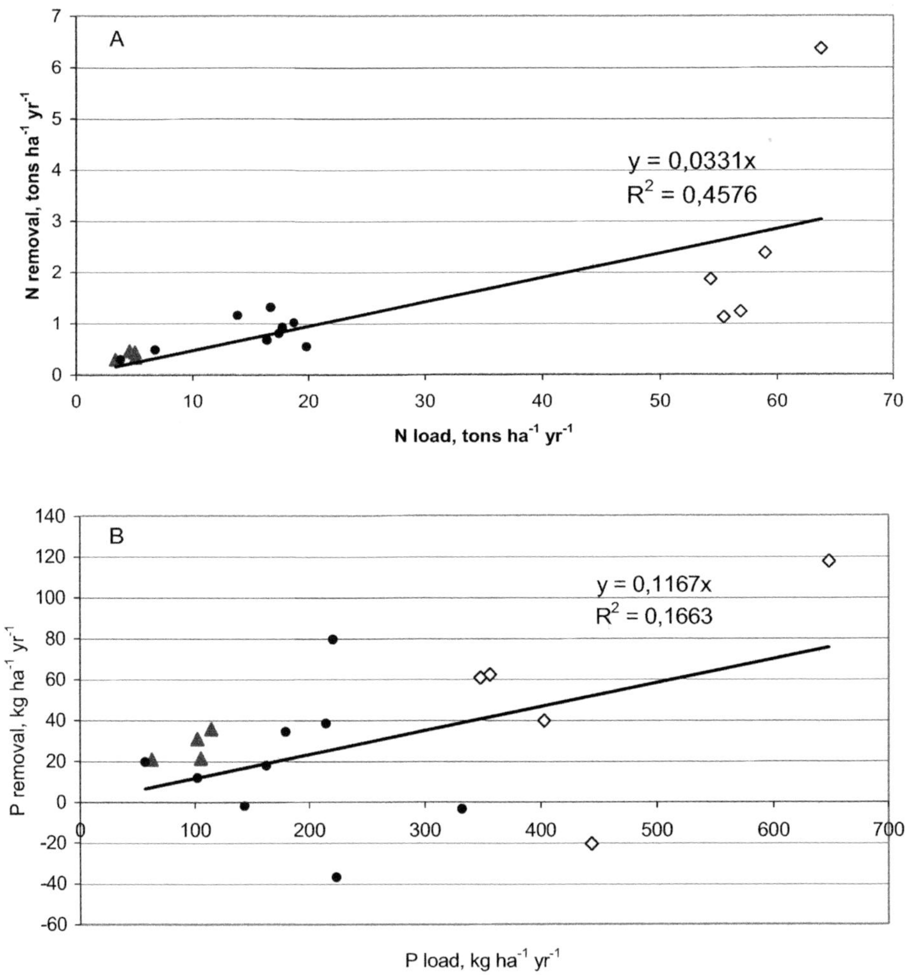

Fig. 5.2 Relationship between annual load and removal of total nitrogen (**A**) and total phosphorus (**B**) for the wetlands Genarp (*triangles*), Råbytorp (*circles*) and Slogstorp (*diamonds*) receiving agricultural runoff in South Sweden

with little variation, except for the first year in Slogstorp at high nitrogen load (Fig. 5.2A). This suggests a fairly predictable nitrogen removal in non-point source wetlands receiving nitrate-N.

The situation was quite different with respect to phosphorus, where the inter-annual variation was quite large (Fig. 5.2B). Sedimentation was the predominant removal mechanism, as a net build-up of sediments occurred in the two wetlands receiving high loads of phosphorus. Hence, sediment removal is a likely prerequisite to maintain the net phosphorus removal of FWS wetlands receiving water with similar quality from agricultural catchments.

In contrast, the wetlands receiving wastewater maintained a steady removal of phosphorus with little inter-annual variability (Andersson et al. 2004). The area specific removal amounted to 17–77 kg ha^{-1} $year^{-1}$, with the lowest removal observed in Magle wetland which received a filtered effluent containing a large proportion of PO_4^{3-}-P in the inflow water. This supports the suggestion that the most important mechanism for phosphorus removal in wetlands receiving effluent from WWTP with chemical precipitation is further precipitation and settling of phosphorus in the wetland (Andersson et al. 2000). Interestingly, the phosphorus removal in Magle was 1 g m^{-2} $year^{-1}$ (= 10 kg ha^{-1} $year^{-1}$), which has been suggested to be the loading threshold level below which the outflow phosphorus concentrations can remain very low (Richardson et al. 1997).

In conclusion, the experiences with FWS wetlands in Sweden suggest that a steady removal of nitrogen can be maintained both in wetlands receiving wastewater and in those receiving non-point source runoff. In wetlands receiving highly variable hydraulic loads and nitrogen concentrations, the areal removal is strongly correlated with nitrogen load (and concentration). For wastewater wetlands receiving ammonium-rich water, overland flow areas seem to be an important design component contributing to increased nitrification. A stable removal of phosphorus is achieved in wetlands receiving pretreated wastewater, independent of plant cover. Removal rates were in most cases considerably higher than can be explained by biological uptake and sediment accretion. This supports the suggestion that remaining precipitation chemicals play an important role in maintaining low phosphorus concentrations in the outlet of those wetlands. For wetlands receiving agricultural runoff, phosphorus removal was highly variable and could not be explained by simple flow and concentration variations.

5.2.2 The Role of Wetlands in Effluent Treatment and Potential Water Reuse in Subtropical and Arid Australia

Municipal wastewater provides two essential resources – nutrients and water. Both are essential for plant growth. Agriculture, horticulture, forestry, golf courses, parks, and gardens depend on these resources. In Australia, currently very little sewage effluent is reclaimed and reused. Typically treated effluent is discharged into rivers, estuaries, and oceans. Water for irrigation, industrial use, and human use is removed from rivers, lakes, reservoirs, and aquifers at a large environmental cost.

Now that water is recognised socially, economically, and politically as a critical resource requiring sustainable management, there needs to be a shift away from purifying water simply to be discharged directly back into rivers and oceans, unless it assists in restoring environmental flows. The cost of wastewater reclamation and the availability of water for reuse must be affordable to the community. We need innovative systems, designed to reclaim nutrients and water from wastewater for reuse, while also removing pathogens. Such systems must be environmentally sustainable, socially accepted, and cost-effective: constructed wetlands are a prime candidate.

Constructed wetland technology is used extensively in North America and Europe. However, despite several pilot projects in Australia in the 1980s (Greenway and Simpson 1996; Greenway and Woolley 1999, 2001; QDNR 2000), this wastewater treatment technology has not widely been adopted in Australia. Interest in constructed wetlands for the treatment of municipal wastewater diminished in the late 1990s. This may have been because of relatively poor phosphorus removal (Greenway and Woolley 1999; QDNR 2000), and government pressure to upgrade and augment sewage treatment plants to produce a very high-quality tertiary effluent, or concern that FWS CWs may be potential breeding sites for mosquitoes (NHMRC 1999; QDNR 2000). However, our recent study of mosquito larvae in four CWs in Queensland showed that predation of mosquito larvae by aquatic invertebrates controls the larvae and prevents the development of pupae (Greenway et al. 2003). Thus there are no health risks in terms of these wetlands being breeding grounds for mosquitoes, if designed and managed to maximise macroinvertebrate predators.

While most mosquitoes are opportunistic breeders, they will only deposit eggs if a suitable body of water is available. In aquatic ecosystems, mosquito larvae are an integral component of aquatic food webs. Mokany and Shine (2002) found that the presence of existing larvae was a strong attractant to further egg laying, indicating that the habitat is suitable for larval development. Thus a critical and significant issue for successful mosquito breeding is larval survival and whether adult mosquitoes emerge from pupae. If CWs are designed to function as wetland ecosystems with a diversity of aquatic organisms, then it is likely that the predator–prey mix would control mosquito breeding. Wetland plant diversity is important for determining macro-invertebrate associations (De Szalay and Resh 2000) and wildlife diversity (Knight et al. 2001) because of the creation of habitats and food resources. Wetzel (2001) noted that the most effective wetland ecosystems "are those that possess maximum biodiversity of higher aquatic plants and periphyton associated with the living and dead plant tissue".

A study of four municipal-treatment wetlands (Greenway et al. 2003) found that wetland with shallow marsh and deeper ponds had the greatest species richness of macrophytes (38 species) and macro-invertebrates (90 taxa) and the lowest occurrence of mosquito larvae (<1 % of dips). Wetlands dominated by dense monospecific stands of cattail (*Typha*) had fewer macro-invertebrate taxa and a higher proportion of mosquito larvae. The authors concluded that a marsh with a diversity of macrophytes appears optimal for macro-invertebrate biodiversity and the control of mosquito larvae by predation. The key to mosquito management is to ensure a well balanced ecosystem supporting a diversity of aquatic organisms.

Between 1992 and 1994, nine pilot wetlands were constructed in Queensland, Australia. Each wetland had a different configuration and received a different quality of effluent (Greenway and Woolley 1999; QDNR 2000). Performance efficiency (Table 5.3) varied widely between wetlands, but the water quality discharged was in general very good.

Constructed wetlands provide suitable conditions for pathogen removal, including filtration, adsorption and subsequent sedimentation, UV light, chemical oxidation, absorption, predation, attack by lytic bacteria and bacteriophages (viruses), and natural die-off. Background faecal coliform concentrations are generally $=10^{3}$ (often $<10^{2}$) colony-forming units (CFU) per 100 ml, but higher concentrations occur where a wetland supports duck and wading-bird populations. An approach to water reuse using CWs could eliminate disinfection by-products from treatment processes without undermining the control of microbial pathogens. Studies in Queensland (Bolton and Greenway 1999; QDNR 2000) have shown that CWs can remove 95 % of pathogen and indicator organisms. Despite high removal efficiencies, faecal coliform concentrations in the final effluent are often still between 100 CFU and 1000 CFU per 100 ml (Table 5.4).

All Queensland local governments are under considerable pressure to comply with new environmental-protection legislation, and the discharge of secondary-treated effluent into coastal and inland waters is being discouraged. In addition, the use of chlorine as a disinfectant has been discouraged. Effluent discharging into the waters of the Great Barrier Reef Marine Park, which extends for 2000 km along the Queensland Coast, is of particular concern because of the sensitivity of the marine ecosystems to nutrient enrichment and increased turbidity. Thus, effluent reuse including land use disposal via irrigation, is high on the agenda of the Queensland Government. Three examples of effluent reuse are given below.

1. Ingham Wetland, Hinchinbrook Shire Council, situated in tropical North Queensland. Effluent reuse includes: pumping the effluent to 10 ha of joint wetland adjacent to the sugar mill, scrubbing flue gases in the sugar mill, irrigating farms in the area, and supplementing water resources in a local creek.
2. Townsville Wetland, Townsville City Council, situated in the dry tropics. Effluent reuse includes: discharge to the adjacent Town Common to miti-

Table 5.3 Sewage concentrations (mg l^{-1}) into and out of five pilot surface-flow constructed wetlands in Queensland, Australia (modified from Greenway and Woolley 1999)

Location	Cairns		Ingham		Townsville		Emu Park		Blackall	
HRT	10 days		12 days		7 days		7 days		3 days	
Effluent HLR	500 m^3 ha^{-1} day^{-1}		400 m^3 ha^{-1} day^{-1}		NA		20 m^3 ha^{-1} day^{-1}		740 m^3 ha^{-1} day^{-1}	
(mg l^{-1})	In	Out	In	Out	In	Out	In	Out	In	Out
BOD_5	9	4	22	11	18	8	15	12	12	10
TSS	5	4	24	16	25	7	NA	NA	24	22
NH_4-N	0.3	0.2	8	5.4	11	2	2	0.1	9	8
NO_x-N	9	0.1	10	3	16	0.3	14	0.8	3	0.2
Total N	6.1	1.5	19.5	9.7	31.5	6.9	NA	NA	16.5	13.6
PO_4-P	7.5	6.8	6	5.4	6.2	6.5	6.2	5.5	2.1	4.2
Total P	7.8	6.9	6.8	5.4	6.8	7.1	NA	NA	3.3	3.1
Monitoring period	From 11/94 to 6/97		From 2/93 to 6/96		From 5/93 to 1/96		From 5/94 to 12/95		From 6/93 to 12/96	
Number of samples	70	30	8	70	65	65	6	6	6	7

HRT = hydraulic retention time, HLR = hydraulic loading rate

Table 5.4 Faecal coliform removal from pilot-scale sub-surface and surface-flow wetlands in Queensland

Location	Cairns[a]		Townsville[a]	Blackall[a]	Logan[b]	Oxley
Wetland type	Surface flow		Surface flow	Surface flow	Melaleuca	Sub-surface
Effluent	Secondary		Secondary	Secondary	Primary	Secondary
HRT (days)	16	11	7–10	4–5	2	7
Influent (CFU per 100 ml)	79 500	79 500	84 000	1600	1×10^7	36 000
Effluent (CFU per /100 ml)	1100	2900	300–700	300–900	<1000	110–195
Removal (%)	98.7	96.4	99.6	82	99.9	99.6

[a] Source: QDNR (2000)
[b] Source: Bolton and Greenway (1999)

gate drought situations and restore the natural wetland environment and wildlife populations. The Town Common is a popular tourist spot for bird watching and its permanent water ensures year-round visitors.

3. Blackall Wetland, Blackall Shire Council, situated in the arid western region (in the State's epicentre). Effluent reuse includes: irrigating commercial tree-lots of different species of flowering gums for export, developing a riverbank ecotourism wetland complex, using native plants and trees typical of western Queensland irrigated with effluent, maintaining water levels in the natural billabong and the duck pond, and irrigating community golf courses, parks, and gardens, in order to conserve the dwindling resources of the Great Artesian Basin.

5.3 Constructed Wetlands with Horizontal Sub-Surface Flow

The most widely used concept of constructed wetlands in Europe is that with horizontal sub-surface flow (HF CWs; Fig. 5.3). The design typically consists of a rectangular bed planted with common reed (*Phragmites australis*) and lined with an impermeable membrane. Mechanically pre-treated wastewater is fed in at the inlet and passes slowly through the soil medium under the surface of the bed in a more or less horizontal path until it reaches the outlet zone where it is collected before discharge via level control arrangement

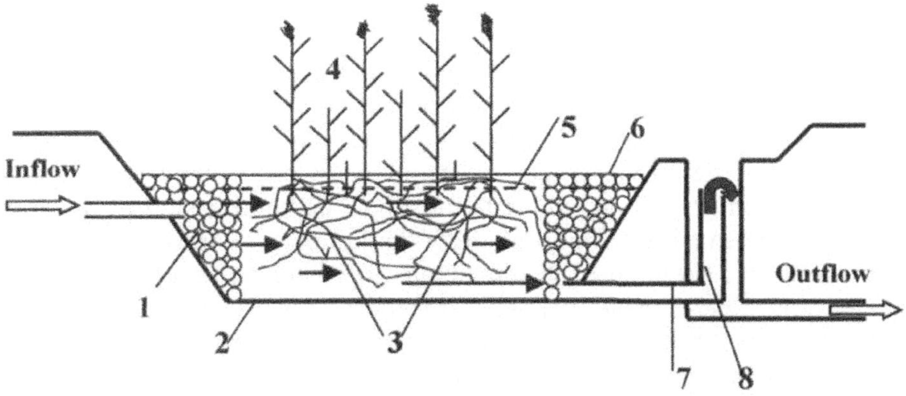

Fig. 5.3 Schematic representation of a constructed wetland with horizontal sub-surface flow. *1* Distribution zone filled with large stones, *2* impermeable liner, *3* filtration medium (gravel, crushed rock), *4* vegetation, *5* water level in the bed, *6* collection zone filled with large stones, *7* collection drainage pipe, *8* outlet structure for maintaining the water level in the bed. The *arrows* indicate only a general flow pattern (modified from Vymazal 2001)

at the outlet. During the passage of wastewater through the reed bed, the wastewater comes into contact with a network of aerobic, anoxic and anaerobic zones.

This concept was developed in the late 1960s in Germany by Käthe Seidel in Plön (Seidel 1965b). Seidel designed the system with filtration material with high hydraulic conductivity. However, Reinhold Kickuth from Göttingen University developed another system under the name "Root-Zone Method" (Kickuth 1969, 1977). Kickuth's system differed from Seidel's system in the use of cohesive soils with a high clay content. The first full-scale HF Kickuth's type CW for the treatment of municipal sewage was put in operation in 1974 in the community of Liebenburg–Othfresen (Kickuth 1977, 1978, 1981; Brix 1987a).

The problem in constructed beds where soil is used as medium is the low hydraulic conductivity, which results in surface flow and thereby a short retention time within the system. Kickuth (1977) also proposed a size of vegetated beds of only 2 m^2 PE^{-1}. This proved to be too small to achieve a satisfactory treatment effect. In 1983, German ideas were introduced in Denmark (Brix 1987b). Despite problems with surface flow soil-based systems, they exhibited a high treatment effect for most parameters if a reed bed area of 3–5 m^2 PE^{-1} was used (Brix 1998).

In 1985, first two HF CWs were built in the United Kingdom (here called "Reed Bed Treatment Systems") and by the end of 1986 more than 20 HF CWs were designed (Cooper and Boon 1987). The major change in the design was the use of very coarse filtration material, which ensured sub-surface flow similarly to the former design by Seidel (1965b). Also, a specific area of 5 m^2 PE^{-1} was used in the United Kingdom. In the late 1980s, the first HF CWs were built in many European countries; and they became the most widely used concept in Europe in the 1990s. The exchange of design and operational experience among nine European countries resulted in the European Design and Operations Guidelines for Reed Bed Treatment Systems which were presented at the Second Constructed Wetlands Conference at Cambridge (Cooper 1990).

A typical HF constructed wetland (Fig. 5.3) has a filtration bed depth of 0.6–0.8 m in order to allow roots of wetland plants, particularly common reed (*Phragmites australis*), to penetrate the whole bed and ensure oxygenation through oxygen release from roots. Roots and rhizomes of reeds and all other wetland plants are hollow and contain air-filled channels that are connected to the atmosphere for the purpose of transporting oxygen to the root system. The majority of this oxygen is used by the roots and rhizomes themselves for respiration, but as the roots are not completely gas-tight, some oxygen is lost to the rhizosphere (Brix 1994). According to the working principle of HF CWs, the amount of oxygen released from roots and rhizomes should be sufficient to meet the demand for aerobic degradation of oxygen-consuming substances in the wastewater as well as for nitrification of the ammonium. However, many studies have shown that the oxygen release from roots of different macrophytes is far less than the amount needed for aerobic degradation of the

Table 5.5 Treatment efficiency of vegetated beds of HF CWs – world-wide experience (data from Australia, Austria, Brazil, Canada, Czech Republic, Denmark, Germany, India, Mexico, New Zealand, Poland, Slovenia, Sweden, USA, UK). The number of systems is given in parentheses. Modified from Vymazal (2001). Concentrations are given in mg l^{-1} and loadings are given in kg ha^{-1} day^{-1} for BOD, COD, and TSS and given in kg ha^{-1} $year^{-1}$ for nutrients. Efficiency is given as a percentage. Note: Inflow is pre-treated wastewater, which means that the overall treatment effect of the entire systems is higher, especially for organics and suspended solids (common removal of pretreatment units is 30–40 % for organics and 40–50 % for TSS)

	Concentration			Loading			
	Inflow	Outflow	Efficiency	Inflow	Outflow	Removal	Efficiency
BOD_5	108	16.0	82.5 (164)	39.2	7.6	31.6	80.6 (131)
COD	284	72.0	74.6 (131)	120.0	34.6	85.4	71.2 (110)
TSS	107	18.1	83.1 (158)	53.6	11.6	42.0	78.4 (130)
Total P	8.74	5.15	41.1 (149)	1410	960	450	31.9(104)
Total N	46.6	26.9	42.3 (137)	6440	3940	2500	38.8 (113)
NH_4^+-N	38.9	20.1	48.3 (151)	3880	2550	1330	34.3 (90)
NO_3^--N	4.38	2.87	34.3 (79)	980	670	310	31.6 (66)

oxygen-consuming substances delivered with sewage (Brix 1990; Brix and Schierup 1990). HF systems are nevertheless very efficient in the removal of organics and suspended solids (Table 5.5) and usually fulfill the effluent quality criteria, which for small sources of pollution are restricted to BOD_5 and suspended solids (Brix 1994; Vymazal et al. 1998). However, the removal of nutrients is low mainly because of the system's inability to oxidise ammonium, the predominant form of nitrogen in domestic and municipal sewage, and low sorption capacity of the filtration materials used for phosphorus (Vymazal 1999). HF CWs are particularly frequent in Germany where the number of these systems may exceed 50 000 (Wissing, personal communication). Other countries with large numbers of HF systems are Austria (ca. 1000), United Kingdom (ca. 800), Denmark (ca 200), Italy (ca. 400), Czech Republic (ca 160), Poland, Portugal (ca. 120), Slovenia, France, Estonia, and Norway.

5.4 Constructed Wetlands with Vertical Sub-Surface Flow

Constructed wetlands with vertical flow were originally designed by Seidel (1965b) as pre-treatment units before wastewater treatment in horizontal flow beds. This design was used to treat the wastewater from a recreation site out-

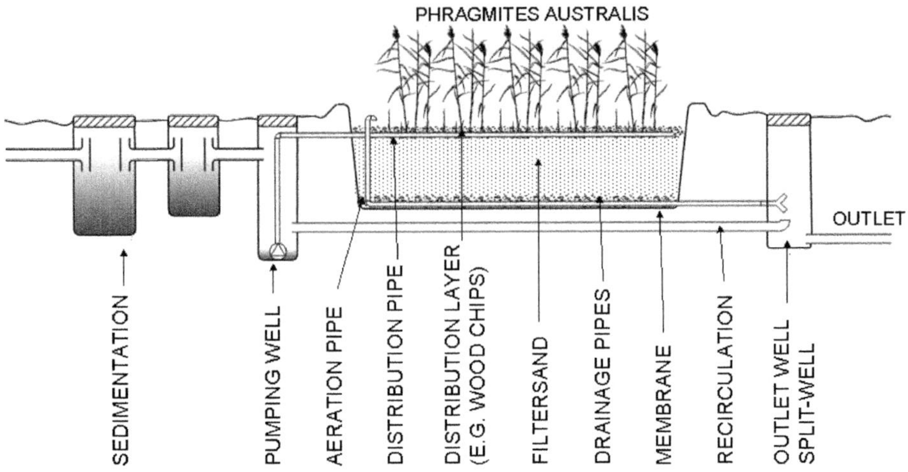

Fig. 5.4 Layout of a vertical flow constructed wetland system for a single household. Raw sewage is pre-treated in a 2-m^3 sedimentation tank. Settled sewage is pulse-loaded onto the surface of the bed by a level-controlled pump. Treated effluent is collected in a system of drainage pipes; and half of the effluent is recirculated back to the pumping well (or to the sedimentation tank)

side the municipal network in Lauwersoog in The Netherlands in 1975 (Greiner and De Jong 1984; Butijn and Greiner 1985). The system consisted of a preliminary settling/distribution ditch, four infiltration compartments and an effluent ditch.

At present, VF CWs usually comprise a flat bed of graded gravel topped with sand planted with macrophytes (Fig. 5.4). The size fraction of gravel is larger in the bottom (ca. 30–60 mm) and smaller in the top layer (ca. 6 mm). VF CWs are fed intermittently with a large batch, thus flooding the surface. Wastewater then gradually percolates down through the bed, which allows air to refill the bed. This kind of feeding provides good oxygen transfer and hence the ability to nitrify (Cooper et al. 1996).

VF CWs provide good removal of organics and suspended solids but they provide little room for denitrification, and therefore, ammonium-N is usually only converted to nitrate-N (Table 5.6). As a consequence, removal of total nitrogen is low, usually lower than in HF CWs. Removal of phosphorus is low unless special filtration material with high sorption capacity is used (Mæhlum and Jenssen 1998). Such a medium is, for example, Leca (light expanded clay aggregates) which is successfully used in Norway, Estonia, and Portugal. VF CWs require less land (2–3 m^2 PE^{-1}) as compared to HF systems (5–10 m^2 PE^{-1}) but require more maintenance and operation efforts because of the use of pumps, timers, and other electric and mechanical devices.

Table 5.6 Treatment efficiency of VF CWs (data from Australia, Austria, China, Denmark, France, Germany, Ireland, Poland, Nepal, Norway, Turkey, UK). For units and details see Table 5.5. Modified from Vymazal (2001)

	Concentration			Loading			
	Inflow	Outflow	Efficiency	Inflow	Outflow	Removal	Efficiency
BOD_5	145	27.2	81.6 (25)	93.2	8.6	87.5	85.3 (13)
COD	303	75.0	71.3 (16)	158.0	46.0	112.0	71.2 (14)
TSS	97	18.4	77.5 (14)	92.0	12.7	79.3	82.8 (14)
Total P	8.6	4.4	52.6 (16)	1390	700	690	55.2(10)
Total N	61	35	39.7 (18)	11 020	5370	5650	50.4 (11)
NH_4^+ -N	45.6	15.8	65.1 (22)	8490	2190	6300	62.2 (10)
NO_3^- -N	1.55	15.1	– (16)	260	3390		– (8)

5.4.1 Danish Experience with Vertical Flow Constructed Wetlands

During the past few years, investigations have been carried out in order to develop a constructed wetland system that will meet the most stringent treatment class for small wastewater-producers in rural areas of Denmark, i.e. 95 % removal of BOD, 90 % removal of total phosphorus, and 90 % nitrification (Arias et al. 2003a, b; Brix et al. 2003). An experimental VF CW system was established at a traditional municipal WWTP, so that the loading rate could be manipulated as desired; and a number of tests were conducted in order to establish the treatment capacity of vertical flow beds (Johansen et al. 2002). The studies also evaluated the effects of recycling nitrified effluent back to the sedimentation tank in order to enhance denitrification in the sedimentation tank and the overall treatment performance of the system (Brix et al. 2002; Marti et al. 2003). These studies determined the area-demand of the system and documented that the capacity of VF CWs to remove BOD and to nitrify is very high. The studies also proved that recycling nitrified effluent to the sedimentation tank enhanced the removal of nitrogen by denitrification and also improved and stabilised the overall treatment performance of the system.

Based on initial experiences from the experimental system, a full-scale system for a single house with four persons was constructed. The system consists of a 2-m^3 three-chamber sedimentation tank, a level-controlled pump and a 15-m^2 VF CW. Effluent from the system can be recirculated to the sedimentation tank to enhance removal of total nitrogen by denitrification (Brix 2003). These studies have produced the necessary background documentation for the development of official guidelines for the design and construction of a VF CW system use in rural areas. The guidelines are presently in the process of being printed by the Ministry of Environment and Energy. The main points of the guidelines are:

- The sewage must be pre-treated in a two- or three-chamber sedimentation tank (minimum size 2 m^3 for a single household with up to 5 PE).
- The necessary surface area of the VF CW is 3 m^2 per PE (minimum size for a single household is 15 m^2).
- The effective filter depth is 1.0 m. The filter medium is sand with a d_{10} between 0.25 mm and 1.2 mm, a d_{60} between 1 mm and 4 mm, and the uniformity coefficient ($U=d_{60}/d_{10}$) should be less than 3.5. The contents of clay and silt (particles less than 0.125 mm) must be less than 0.5 %.
- The filter bed must be enclosed by a tight membrane (minimum 0.5 mm thickness). The membrane must be protected by a geotextile on both sides.
- The bed is planted with common reed (*Phragmites australis*). The main function of the plants is to counteract clogging of the filter.
- The sewage is distributed evenly over the surface of the bed by a network of pressurised distribution pipes. The distribution pipes are insulated against frost by a 0.2-m layer of coarse wood chips or sea shells on the surface of the filter.
- The loading frequency of the bed is typically 16 to 24 pulses per day, when half of the effluent water is recycled within the system.
- The treated water is collected in an aerated system of drainage pipes placed in a 0.2-m layer of coarse gravel in the bottom of the filter.
- Half of the effluent water from the filter is recycled to the first chamber of the sedimentation tank or to the pumping well to enhance denitrification and to stabilise the treatment performance of the system.

Vertical flow systems constructed according to the guidelines will be able to fulfill the O and the SO treatment class, i.e. 95 % removal of BOD and 90 % nitrification. The VF systems are not able to fulfil the treatment classes requiring removal of phosphorus. The possibility of developing a separate unit containing a phosphorus-binding material has been studied extensively both in the laboratory, in experimental pilot-scale systems, and in a full-scale system; and a calcite material was identified with the right hydraulic properties and at the same time a high binding capacity for phosphorus (Brix et al. 2000; Arias et al. 2002; Arias et al. 2003a). However, unfortunately, the calcite material is no longer commercially available and therefore it has not been possible to include it in the guidelines. The possibility of precipitating phosphorus in the sedimentation tank by dosing with aluminium polychloride is presently being tested. Initial results are promising and suggest that the dosage needed only amounts to 30 l for a single-household system per year.

5.4.2 Constructed Wetlands with No Outflow

A novel CW system based on willows has been developed as a sewage disposal solution in rural areas in Denmark (Gregersen and Brix 2001; Brix and Gregersen 2002). The main attributes of the willow wastewater-cleaning facilities are that the systems have zero discharge of water (because of evapotranspiration) and part of the nutrients can be recycled via the willow biomass. Furthermore, the harvested biomass may be used as a source of bioenergy. Two sets of guidelines have been published: one guideline describes willow systems with no effluent (Ministry of Environment and Energy 2003a), and the other guideline describes willow systems that are not contained in a membrane-enclosed bed and thus allow some soil infiltration (Ministry of Environment and Energy 2003b). The system with infiltration is intended to be used on clayish soils, where infiltration is low.

The willow wastewater-cleaning facilities generally consist of ca. 1.5 m high-density polyethylene-lined basins filled with soil and planted with clones of willow (*Salix viminalis* L.). The surface area of the systems depends on the amount and quality of the sewage to be treated and the local annual rainfall. For a single household in Denmark, the area needed typically is between 120 m^2 and 300 m^2. Settled sewage is dispersed underground into the bed under pressure. The stems of the willows are harvested on a regular basis to stimulate the growth of the willows and to remove some nutrients and heavy metals.

Removal of water from the systems occurs by evaporation from the soil and plant surface, and by transpiration. The following factors are important for maximising the evaporative loss of water: high energy input (solar radiation), high air temperatures, low relative humidity in the air, exchange of air (wind), canopy resistance, stomata resistance, and leaf area index. Evaporation is enhanced by the "oasis" effect, which is the phenomenon where warmer, dry air in equilibrium with dry areas flows across vegetation with a high water availability (Rosenberg 1969). The vegetation experiences enhanced evaporation using sensible heat from the air as well as radiant energy, and air is cooled by this process. In addition, the so-called "clothesline" effect, where the vegetation height is greater than that of the surroundings (different roughness conditions), may increase evaporative water loss (Allen et al. 1998). This occurs where turbulent transport of sensible heat into the canopy and transport of vapour away from the canopy is increased by the "broadsiding" of wind horizontally into the taller vegetation. In addition, the internal boundary layer above the vegetation may not be in equilibrium with the new surface. Therefore, evapotranspiration from the isolated expanses, on a per unit area basis, may be significantly greater than the calculated potential evapotranspiration. Examples of the clothesline or oasis effects would be evapotranspiration from a single row of trees surrounded by short vegetation or by a dry non-cropped field, or evapotranspiration from a narrow strip of cattail (a hydrophytic vegetation) along a stream channel.

The main characteristics of the willow systems are:

- For a single household (5 PE) system, the sewage has to be pre-treated in a two- or three-chamber sedimentation tank with a minimum volume of 2 m^3 before discharge into the willow system.
- Closed willow systems are generally constructed with a width of 8 m, a depth of minimum 1.5 m, and with 45° slopes on the sides.
- The total annual water loss from the systems is assumed to be 2.5x the potential evapotranspiration at the location, as determined by climatic parameters.
- The necessary area of the systems is determined by the amount of wastewater, the "normal" precipitation, and the potential evapotranspiration at the location of the system.
- The bed is enclosed by a water-tight membrane and wastewater is distributed underground within the system by a level-controlled pump.
- A drainage pipe is placed in the bottom of the bed. The pipe can be used to empty water from the bed if salt accumulates after some years.
- One-third of the willows are harvested every year to keep the willows in a young and healthy state with high transpiration rates.

5.5 Hybrid Constructed Wetlands

Various types of CWs may be combined in order to achieve higher treatment effect, especially for nitrogen. However, hybrid systems comprise most frequently VF and HF systems arranged in a staged manner. There are now many fine examples of HF systems for secondary treatment; and they prove very satisfactory where the standard requires only BOD_5 and SS. However, there has been a growing interest in achieving fully nitrified effluents. Tertiary treatment HF systems produce well nitrified effluents (Cooper et al. 1996) but secondary treatment HF systems cannot do this because of their limited oxygen transfer capacity. VF systems, in contrast, do provide good conditions for nitrification but no denitrification occurs in these systems. Therefore, there has been a growing interest in hybrid systems (also called combined systems). In hybrid systems, the advantages and disadvantages of the HF and VF systems can be combined to complement each other. It is possible to produce an effluent low in BOD, which is fully nitrified and partly denitrified and hence has much lower total-nitrogen concentrations (Cooper 1999, 2001).

Many of these systems are derived from original hybrid systems developed by Seidel at the Max Planck Institute in Krefeld, Germany. The process (Fig. 5.5) is known as the Seidel system, the Krefeld system, or the Max Planck Institute Process (MPIP; Seidel 1965b, 1976, 1978). The design consists of two

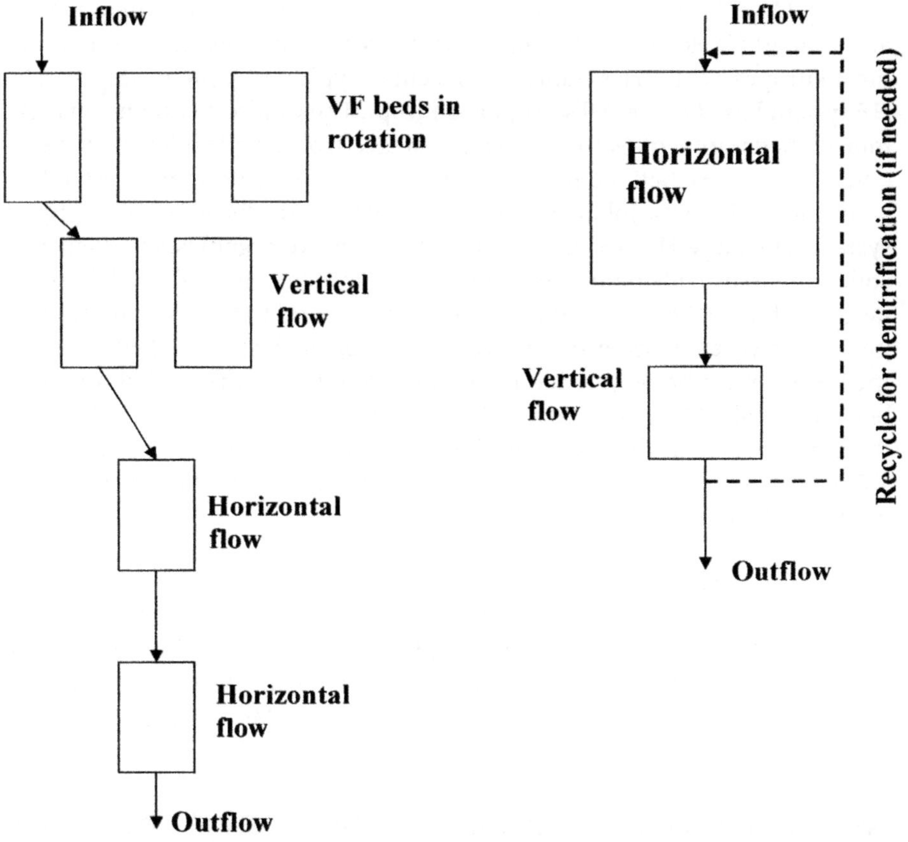

Fig. 5.5 Hybrid systems based on concepts by Seidel (*left*) and Brix and Johansen (*right*). Modified from Brix (1998) and Cooper (1999)

stages of several parallel VF beds followed by two or three HF beds in series. The VF stages are usually planted with *Phragmites australis*, whereas the HF stages contain a number of other emergent macrophytes, including *Iris*, *Schoenoplectus*, *Sparganium*, *Carex*, *Typha* and *Acorus*. The VF beds are loaded with pre-treated wastewater for 1–2 days and are then allowed to dry out for 4–8 days. The thin crust of solids that forms on top of the VF beds is mineralised during the rest period and achieves an equilibrium thickness (Brix 1994).

In the early 1980s, several hybrid systems of Seidel's type were built in France with the system at Saint Bohaire which was put in operation in 1982, being the best described (Boutin 1987; Vuillot and Boutin 1987). A similar system was built in 1987 in the United Kingdom at Oaklands Park (Burka and Lawrence 1990). In the1990s and early 2000s, VF-HF systems were built in many European countries, e.g. in Slovenia (Urbanc-Bercic and Bulc 1994),

Norway (Mæhlum and Stalnacke 1999), Austria (Mitterer-Reichmann 2002), and Ireland (O'Hogain 2003), and now this type is getting more attention in most European countries. Table 5.7 presents a typical treatment effect of a VF-HF system based on Seidel's original concept. The Colecott system consists of four VF beds (total of 64 m^2) at the first stage, two VF beds (60 m^2) at the second stage, one HF bed (60 m^2) at the third stage, and is designed for 60 PE.

In the mid-1990s, Johansen and Brix (1996) introduced a HF-VF hybrid system. The large HF bed is placed first, to remove organics and suspended solids and to provide denitrification. An intermittently loaded small VF bed is designed for further removal of organics and SS and to nitrify ammonium to nitrate. However, in order to remove total nitrogen, the nitrified effluent from the VF bed must be recycled to the sedimentation tank or the first horizontal bed. Brix et al. (2003) pointed out that special care must be taken not to affect the performance of the sedimentation tank or the nitrifying capacity of the VF bed by recycling too large a volume of wastewater. A similar system was built in Poland at Sobiechy (Ciupa 1996).

Recently introduced hybrid constructed wetlands also comprise more than two types of CWs and quite often include a FWS stage. For example, the system at Cecchi consists of a HF bed (480 m^2) followed by a FWS wetland (850 m^2). The system is heavily loaded – the organic load of the HF bed amounts to 1336 kg ha^{-1} day^{-1} – but the treatment effect is very high for organics, suspended solids, and nutrients (Table 5.8; Masi et al. 2002).

Table 5.7 Performance of the VF-HF Colecott hybrid system (data from O'Hogain 2003). Concentrations are given in mg l^{-1}, removal in %

Parameter	Inflow	VF1out	VF2out	HFout	Removal
COD	462	210	66	47	89.4
BOD_5	269	171	43	23	91.4
TSS	53	28	3	1	98.1
NH_4^+-N	45	28	16	7	84.4
NO_3^- -N	0.1	4.7	3.8	2.7	
NO_2^- -N	0.1	0.2	0.1	0.1	
PO_4^{3-}	18.0	16.0	15.0	11.0	38.9

Table 5.8 Casa Vincicola Cecchi, Italy – treatment performance for the period 13.2.01 – 11.3.03 (based on Masi et al. 2002). Values are given in mg l^{-1} except pH

	BOD_5	COD	TSS	Total N	Total P	pH
Inflow	1 833	3 906	213	18.9	4.7	6.1
HF out	49.4	161.0	13.3	4.8	1.5	6.9
FWS out	25.4	84.0	23.4	3.5	1.3	7.4

5.6 Trace Gas Fluxes from Constructed Wetlands for Wastewater Treatment

Wetlands are considered as globally significant sources of greenhouse gases like methane (CH_4; Bartlett and Harris 1992; IPCC 2001) and nitrous oxide (N_2O; Martikainen et al.1993). Numerous studies consider emissions and sequestration of carbon dioxide (CO_2) in wetlands (Mitsch and Gosselink 1993). Depending on meteorological and hydrological conditions, wetlands can be sources or sinks of carbon (Whiting and Chanton 2001). Constructed wetlands for wastewater treatment have also been considered as sources of CH_4 (Kadlec and Knight 1996; Tanner et al. 1997; Mander et al. 2003) and N_2O (Fey et al. 1999; Tanner et al. 2002; Mander et al. 2003). Denitrification, as the microbial reduction of NO_3^--N to NO_2^--N and further to gaseous forms NO, N_2O, and N_2 (Knowles 1982), has been found in numerous studies as a significant process in nitrogen removal in CWs (Spieles and Mitsch 2000). However, in well aerated, yet moist conditions (soil water-filled pore space at 40–60 %), N_2O can be emitted during nitrification (Mosier 1998). Both denitrification and methane formation depend on the oxygen status of the soil or sediment. In this relation, the spatial and temporal variability of fluxes of both N_2O and CH_4 is extremely high. CH_4 is produced in anoxic soils and sediments, while well drained soils act as a sink for atmospheric CH_4 due to methane oxidation, through ammonium oxidisers or methanotrophs (Hanson et al. 1994). Due to increasing human impact on the global environment, N_2O is increasing in the atmosphere at a rate of about 0.3 % per year. It has an atmospheric lifetime of about 120 years, a global warming potential of 296 relative to CO_2 over a 100-year time-horizon, and is responsible for about 6 % of anticipated warming (IPCC 2001).

N_2O, CH_4, and CO_2 emission rates were quantified from two sub-surface flow CWs (Kodijärve HF CW, Kõo hybrid CW) for municipal wastewater treatment in Estonia, using the closed chamber method. The cumulative emission of all studied gases from CWs varied over 16.3–21.9 kg, 230–295 kg, and 9100–9700 kg ha^{-1} year^{-1} for N_2O-N, CH_4-C, and CO_2-C, respectively. The emission level in Kõo always exceeded the corresponding values in Kodijärve, which was probably due to the relatively high loading of the vertical flow system (only two beds of 64 m^2 for about 300 PE).

In Kodijärve, the nominal loading is only 20–40 PE per 312.5 ha. When properly functioning, however, the vertical flow system can have a relatively small area, although this seems to enhance N_2O emission. Regarding CH_4 flux, it is crucial to avoid clogging both vertical flow and horizontal flow filters: clogging might help to generate a higher N_2 flux and correspondingly lower N_2O flux; however, it significantly increases methane emissions.

The Global Warming Potential (GWP) of the studied systems was calculated, converting the fluxes of N_2O and CH_4 into CO_2 equivalents (CO_2 eq;

IPCC 2001). In Kodijärve, the average N_2O flux from both beds was quite similar: 3.85±7.10 t CO_2 eq ha^{-1} $year^{-1}$ in bed 1 and 3.85±3.85 t CO_2 eq ha^{-1} $year^{-1}$ in bed 2. Methane flux rates, however, showed significant differences, ranging from 0.69±2.05 t CO_2 eq ha^{-1} $year^{-1}$ in bed 1 to 12.4±26.3 t CO_2 eq ha^{-1} $year^{-1}$ in bed 2. In Kõo, the highest GWP of N_2O was found in the vertical flow beds (12.4±19.5 t CO_2 eq ha^{-1} $year^{-1}$), while the horizontal flow bed showed a high methane flux (9.13±17.9 t CO_2 eq ha^{-1} $year^{-1}$). Both Kodijärve HF CW and Köo hybrid CW emitted remarkable amounts of CO_2-C, N_2O-N, and CH_4-C (Fig. 5.6). The cumulated emission of N_2O and CH_4 from riparian alder forest in Porijõgi was significantly lower than from the CWs (0.4±1.0 t and 0.1±0.30 t CO_2 eq ha^{-1} $year^{-1}$, respectively), whereas the CO_2-C flux was high but spatially variable (3.5±3.7 t ha^{-1} $year^{-1}$). When comparing the greenhouse potential of CH_4 and N_2O over a long time-scale (100–500 years), one can speculate that, due to the short adjustment time for CH_4 in the atmosphere (8.4 years; IPCC 2001), the radiative forcing of CH_4 will decline relative to CO_2 (Whiting and Chanton 2001; Brix et al. 2001). N_2O with its atmospheric lifespan of about 120 years and GWP value of 296 has, however, potentially high impacts that cannot be predicted at this time. Therefore, further investigations should concentrate on the factors that regulate N_2O and N_2 emission rates from constructed wetlands.

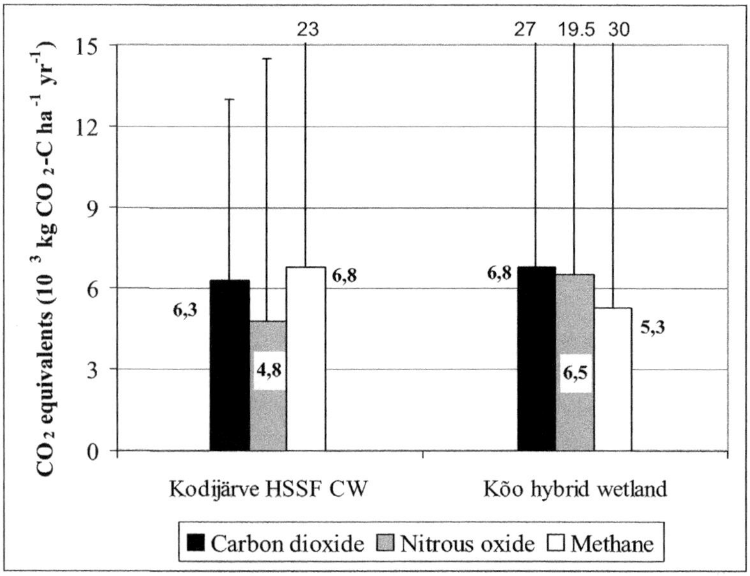

Fig. 5.6 Cumulated flux rates of major greenhouse gases from Kodijärve and Kõo constructed wetlands from October 2001 to November 2003, presented as CO_2 equivalent values (kg CO_2-C ha^{-1} $year^{-1}$; mean ±SD). The conversion of the flux rates into CO_2 equivalents is given with 296 for N_2O and 23 for CH_4 in a time horizon of 100 years (IPCC 2001)

In CWs, we found a remarkable variability in the average emission rates of N_2O-N, CH_4-C, and CO_2-C, ranging between 1 μg and 2600 μg m^{-2} h^{-1}, –1.7 μg and 87 200 μg m^{-2} h^{-1}, and –6.1 mg and 1050 mg m^{-2} h^{-1}, respectively. Release of all gases studied was significantly higher during the warmer period; however, there was no significant correlation between the N_2O flux and soil/water temperature. Similar to the purification performance, gaseous emissions in spring and early summer were significantly lower than in autumn. The most intensive flux of N_2O and CH_4 was observed in chambers installed above the inflow pipes of horizontal flow beds. The vertical flow wetland did emit significantly more N_2O than the horizontal flow beds. Although the emission of N_2O and CH_4 from constructed wetlands was found to be relatively high, their global influence is not significant. Even if all the global domestic wastewater were treated by wetlands, their share in the global trace gas emission budget would be less than 1 %.

5.7 Conclusion

Constructed wetlands have been used for wastewater treatment for nearly 40 years and have become a widely accepted technology available to deal with both point and non-point sources of water pollution. They offer a land-intensive, low-energy, and low-operational-requirements alternative to conventional treatment systems, especially for small communities and remote locations. Constructed wetlands also prove to be affordable tools for wastewater reclamation, especially in arid and semi-arid areas. Although the emission of N_2O and CH_4 from constructed wetlands was found to be relatively high, their global influence is probably not significant.

References

Allen RG, Pereira LS, Raes D, Smith M (1998) Crop evapotranspiration. Guidelines for computing crop water requirements. (FAO irrigation and drainage paper 56) FAO, Rome

Andersson JL, Wittgren HB, Kallner S, Ridderstolpe P, Hägermark I (2000) Wetland Oxelösund, Sweden – the first five years of operation. In: Jenssen P, Mander U (eds) Natural wetlands for wastewater treatment in cold climate. (Advances in ecological sciences) WIT, Southampton, pp 9–27

Andersson JL, Kallner Bastviken S, Tonderski KS (2004) Free water surface wetlands for wastewater treatment in Sweden – nitrogen and phosphorus removal. Proc Int Conf Wetland Syst Water Pollut Control 9:39–46

Arias CA, Brix H, Johansen NH (2002) Phosphorus removal from municipal wastewater in an experimental two-stage vertical flow constructed wetland system equipped with a calcite filter. Proc Int Conf Wetland Syst Water Pollut Control 8:952–960

Arias CA, Brix H, Johansen NH (2003a) Phosphorus removal from municipal wastewater in an experimental two-stage vertical flow constructed wetland system equipped with a calcite filter. Water Sci Technol 48:51–58

Arias CA, Cabello A, Brix H, Johansen NH (2003b) Removal of indicator bacteria from municipal wastewater in an experimental two-stage vertical flow constructed wetland system. Water Sci Technol 48:35–41

Bartlett KB, Harris RC (1992) Review and assessment of methane emissions from wetlands. Chemosphere 26:261–320

Bolton KGE, Greenway M (1999) Pollutant removal capacity of a constructed Melaleuca wetland receiving primary settled effluent. Water Sci Technol 39:199–206

Boutin C (1987) Domestic wastewater treatment in tanks planted with rooted macrophytes: case study, description of the system, design criteria, and efficiency. Water Sci Technol 19:29–40

Brix H (1987a) The applicability of the wastewater treatment plant in Othfresen as scientific documentation of the root-zone method. Water Sci Technol 19:19–24

Brix H (1987b) Treatment of wastewater in the rhizosphere of wetland plants – the root-zone method. Water Sci Technol 19:107–118

Brix H (1990) Gas exchange through the soil-atmosphere interphase and through dead culms of *Phragmites australis* in a constructed reed bed receiving domestic sewage. Water Res 24:259–266

Brix H (1994) Constructed wetlands for municipal wastewater treatment in Europe. In: Mitsch WJ (ed) Global wetlands: old world and new. Elsevier Science, Amsterdam, pp 325–333

Brix H (1998) Denmark. In: Vymazal J, Brix H, Cooper PF, Green MB, Haberl R (eds) Constructed wetlands for wastewater treatment in Europe. Backhuys, Leiden, pp 123–152

Brix H (2003) Danish experiences with wastewater treatment in constructed wetlands. In: Dias V, Vymazal J (eds) The use of aquatic macrophytes for wastewater treatment in constructed wetlands. ICN/INAG, Lisbon, pp 327–361

Brix H, Gregersen P (2002) Water balance of willow dominated constructed wetlands. Proc Int Conf Wetland Syst Water Pollut Control 8:669–670

Brix H, Schierup H-H (1989) The use of macrophytes in water pollution control. Ambio 18:100–107

Brix H, Schierup H-H (1990) Soil oxygenation in constructed reed beds: the role of macrophyte and soil-atmosphere interface oxygen transport. In: Cooper PF, Findlater BC (eds) Constructed wetlands in water pollution control. Pergamon, Oxford, pp 53–66

Brix H, Arias CA, Del Bubba M (2000) How can phosphorus removal be sustained in subsurface-flow constructed wetlands? Proc Int Conf Wetland Syst Water Pollut Control 7:65–74

Brix H, Sorrell BK, Lorenzen B (2001) Are *Phragmites*-dominated wetlands a net source or net sink of greenhouse gasses? Aquat Bot 69:313–324

Brix H, Arias CA, Johansen NH (2002) BOD and nitrogen removal from municipal wastewater in an experimental two-stage vertical flow constructed wetland system with recycling. Proc Int Conf Wetland Syst Water Pollut Control 8:400–410

Brix H, Arias C, Johansen NH (2003) Experiments in a two-stage constructed wetland system: nitrification capacity and effects of recycling on nitrogen removal. In: Vymazal J (ed) Wetlands: nutrients, metals and mass cycling. Backhuys, Leiden, pp 237–258

Burka U, Lawrence P (1990) A new community approach to waste treatment with higher water plants. In: Cooper PF, Findlater BC (eds) Constructed wetlands in water pollution control. Pergamon, Oxford, pp 359–371

Butijn GD, Greiner RW (1985) Afwalwaterzuivering met behulp van begroeide infiltratievelden (Wastewater treatment with vegetated percolation fields). In: Aart PJM van der (ed) Wetlands for the purification of wastewater. University of Utrecht, Utrecht, pp 64–89

Ciupa R (1996) The experience in the operation of constructed wetlands in north-eastern Poland. Proc Int Conf Wetland Syst Water Pollut Control 5:6

Cooper PF (ed) (1990) European design and operations guidelines for reed bed treatment systems. (WRc Report UI 17) European Community/European Water Pollution Control Association Emergent Hydrophyte Treatment System Expert Contact Group, Swindon

Cooper PF (1999) A review of the design and performance of vertical flow and hybrid reed bed treatment systems. Water Sci Technol 40:1–9

Cooper PF (2001) Nitrification and denitrification in hybrid constructed wetlands systems. In: Vymazal J (ed) Transformations on nutrients in natural and constructed wetlands. Backhuys, Leiden, pp 257–270

Cooper PF, Boon AG (1987) The use of *Phragmites* for wastewater treatment by the root zone method: the UK approach. In: Reddy KR, Smith WH (eds) Aquatic plants for water treatment and resource recovery. Magnolia, Orlando, pp 153–174

Cooper PF, Findlater BC (eds) (1990) Constructed wetlands in water pollution control. Pergamon, Oxford

Cooper PF, Job GD, Green MB, Shutes RBE (1996) Reed beds and constructed wetlands for wastewater treatment. WRc Publications, Marlow

Crites RW, Tchobanoglous G (1998) Small and decentralized wastewater management systems. McGraw Hill, Boston

De Jong J (1976) The purification of wastewater with the aid of rush or reed ponds. In: Tourbier J, Pierson RW (eds) Biological control of water pollution. Pennsylvania University Press, Philadelphia, pp 133–139

De Szalay FA, Resh VH (2000) Factors influencing macro-invertebrate colonisation of seasonal wetlands: responses to emergent plant cover. Freshwater Biol 45:295–308

Dias V, Vymazal J (eds) (2003) The use of aquatic macrophytes for wastewater treatment in constructed wetlands. Instituto da Conservação da Naturreza/Instituto Nacional da Água, Lisbon

Fey A, Benckiser G, Ottow JCG (1999) Emissions of nitrous oxide from a constructed wetland using a groundfilter and macrophytes in waste-water purification of a dairy farm. Biol Fertil Soils 29:354–359

Gregersen P, Brix H (2001) Zero-discharge of nutrients and water in a willow dominated constructed wetland. Water Sci Technol 44:407–412

Greenway M, Simpson JS (1996) Artificial wetlands for wastewater treatment, water reuse and wildlife in Queensland, Australia. Water Sci Technol 33:221–229

Greenway M, Woolley A (1999) Constructed wetlands in Queensland: performance efficiency and nutrient bioaccumulation. Ecol Eng 12:39-55

Greenway M, Woolley A (2001) Changes in plant biomass and nutrient removal in a constructed wetland, Cairns, Australia. Water Sci Technol 44:303–310

Greenway M, Dale P, Chapman H (2003) An assessment of mosquito breeding and control in 4 surface flow wetlands in tropical–subtropical Australia. Water Sci Technol 48:249–256

Greiner RW, Jong J de (1984) The use of marsh plants for the treatment of waste water in areas designated for recreation and tourism. (RIJP Report 225) RIJP, Lelystad

Hammer DA (ed) (1989) Constructed wetlands for wastewater treatment. Lewis, Chelsea, Mich.

Hanson GC, Groffman PM, Gold AJ (1994) Denitrification in riparian wetlands receiving high and low groundwater nitrate inputs. J Environ Qual 23:917–922

IPCC (2001) Atmospheric chemistry and greenhouse gases, chapter 4. In: Houghton JT, et al. (eds) Climate change: the scientific basis. Cambridge University Press, Cambridge, pp 239–287

Johansen NH, Brix H (1996) Design criteria for a two-stage constructed wetland. Proc Int Conf Wetland Syst Water Pollut Control 5:IX/3

Johansen NH, Brix H, Arias CA (2002) Design and characterization of a compact constructed wetland system removing BOD, nitrogen and phosphorus from single household sewage. Proc Int Conf Wetland Syst Water Pollut Control 8:47–61

Kadlec RH (1999) Chemical, physical and biological cycles in treatment wetlands. Water Sci Technol 40:37–44

Kadlec RH, Knight RL (1996) Treatment wetlands. CRC/Lewis, Boca Raton

Kadlec RH, Knight RH, Vymazal J, Brix H, Cooper PF, Haberl R (2000) Constructed wetlands for pollution control. Processes, performance, design and operation. (IWA Scientific and Technical Report 8) IWA, London

Kallner S, Wittgren HB (2001) Modeling nitrogen transformations in surface flow wastewater treatment wetlands in Sweden. Water Sci Technol 44:237–244

Kickuth R (1969) Höhere Wasserpflanzen und Gawässerreinhaltung. Schiftenreihe der Vereinigung. Dtsch Gewaessersch EV-VDG 19:3–14

Kickuth R (1977) Degradation and incorporation of nutrients from rural wastewaters by plant rhizosphere under limnic conditions. In: Utilization of manure by land spreading. Comm Eur Commun EUR 5672e:335–343

Kickuth R (1978) Elimination gelöster Laststoffe durch Röhrichtbestände. Arb Dtsch Fischereiverbandes 25:57–70

Kickuth R (1981) Abwasserreinigung in Mosaikmatritzen aus aeroben und anaeroben Teilbezirken. In: Moser F (ed) Grundlagen der Abwasserreinigung. Verlag Oldenburg, Munich, pp 630–650

Knight RL, Clarke RA, Bastian RK (2001) Surface flow (SF) treatment wetlands as a habitat for wildlife and humans. Water Sci Technol 44:27–37

Knowles R (1982) Denitrification. Microbiol Rev 46:43–70

Lakatos G (1998) Hungary. In: Vymazal J, Brix H, Cooper PF, Green MB, Haberl R (eds) Constructed wetlands for wastewater treatment in Europe. Backhuys, Leiden, pp 191–206

Lakatos G, Kiss MK, Kiss M, Juhász J (1996) Application of constructed wetlands for wastewater treatment in Hungary. Proc Int Conf Wetland Syst Water Pollut Control 5:IX/9

Linde L, Alsbro R (2000) Ekeby wetland – the largest constructed SF wetland in Sweden. Proc Int Conf Wetland Syst Water Pollut Control 7:1101–1110

Mæhlum T, Jenssen P (1998) Norway. In: Vymazal J, Brix H, Cooper PF, Green MB, Haberl R (eds) Constructed wetlands for wastewater treatment in Europe. Backhuys, Leiden, pp 207–216

Mæhlum T, Stålnacke P (1999) Removal efficiency of three cold-climate constructed wetlands treating domestic wastewater: effects of temperature, seasons, loading rates and input concentrations. Water Sci Technol 40:273–281

Mander Ü, Jenssen P (eds) (2003) Constructed wetlands for wastewater treatment on cold climates. WIT Press, Southampton

Mander Ü, Kuusemets V, Lõhmus K, Mauring T, Teiter S, Augustin J (2003) Nitrous oxide, dinitrogen, and methane emission in a subsurface flow constructed wetland. Water Sci Technol 48:135–142

Marti E, Arias CA, Brix, H, Johansen NH (2003) Recycling of treated effluents enhances reduction of total nitrogen in vertical flow constructed wetlands. Publ Inst Geogr Univer Tartuensis 94:150–155

Martikainen PJ, Nykänen H, Crill P, Silvola J (1993) Effect of a lowered water table on nitrous oxide fluxes from northern peatlands. Nature 366:51–53

Masi F, Conte G, Martinuzzi N, Pucci B (2002) Winery high organic content wastewaters treated by constructed wetlands in Mediterranean climate. Proc Int Conf Wetland Syst Water Pollut Control 8:274–282

Ministry of Environment and Energy (2003a) Guidelines for willow systems up to 30 PE (in Danish). Økologisk Byfornyelse og Spildevandsrensning 25

Ministry of Environment and Energy (2003b) Guidelines for willow systems with soil infiltration up to 30 PE (in Danish). Økologisk Byfornyelse og Spildevandsrensning 26

Mitsch WJ, Gosselink JG (1993) Wetlands. Van Nostrand Reinhold, New York

Mitterer-Reichmann GM (2002) Data evaluation of constructed wetlands for treatment of domestic wastewater. Proc Int Conf Wetland Syst Water Pollut Control 8:40–46

Mokany A, Shine R (2002) Competition between tadpoles and mosquitoes: the effects of larval density and tadpole size. Aust J Zool 40:549–563

Mosier AR (1998) Soil processes and global changes. Biol Fertil Soils 27:221–229

Nehring KW, Brauning SE (eds) (2002) Wetlands and remediation II. Battelle Memorial Institute, Columbus

NHMRC (1999) Draft guidelines for sewage systems: reclaimed water (Australia). NHMRC, Canberra

O'Hogain S (2003) The design, operation and performance of a municipal hybrid reed bed treatment system. Water Sci Technol 48:119–126

QDNR (2000) Guidelines for using freewater surface constructed wetlands to treat municipal sewage. Queensland Department of Natural Resources, Brisbane

Richardson CJ, Qian SS, Craft BC, Qualls RG (1997) Predictive models for phosphorus retention in wetlands. Wetlands Ecol Manage 4:159–175

Rosenberg NJ (1969) Advective contribution of energy utilised in evapotranspiration by alfalfa in the East Central Great Plains. Agric Meteorol 6:179–184

Seidel K (1955) Die Flechtbinse *Scirpus lacustris*. In: Oekologie, Morphologie und Entwicklung, ihre Stellung bei den Volkern und ihre wirtschaftliche Bedeutung. Schweizerbartsche Verlagsbuchnadlung, Stuttgart, pp. 37–52

Seidel K (1961) Zur Problematik der Keim- und Pflanzengewasser. Verh Int Verein Limnol 14:1035–1039

Seidel K (1965a) Phenol-Abbau in Wasser durch *Scirpus lacustris* L. wehrend einer versuchsdauer von 31 Monaten. Naturwissenschaften 52:398–406

Seidel K (1965b) Neue Wege zur Grundwasseranreicherung in Krefeld, Vol II: Hydrobotanische Reinigungsmehode. GWF Wasser/Abwasser 30:831–833

Seidel K (1976) Macrophytes and water purification. In: Tourbier J, Pierson RW (eds) Biological control of water pollution. Pennsylvania University Press, Philadelphia, pp 109–122

Seidel K (1978) Gewässerreinigung durch höhere Pflanzen. Z Garten Landschaft H1:9–17

Spieles DJ, Mitsch WJ (2000) The effects of season and hydrologic and chemical loading on nitrate retention in constructed wetlands: a comparison of low- and high-nutrient riverine systems. Ecol Eng 14:77–91

Tanner CC, Adams DD, Downes MT (1997) Methane emissions from constructed wetlands treating agricultural wastewaters. J Environ Qual 26:1056–1062

Tanner CC, Kadlec RH, Gibbs MM, Sukias JPS, Nguyen ML (2002) Nitrogen processing gradients in subsurface-flow wetlands – influence of wastewater characteristics. Ecol Eng 18:499–520

Urbanc-Bercic O, Bulc T (1994) Integrated constructed wetland for small communities. Proc Int Conf Wetland Syst Water Pollut Control 4:138–146

Veenstra S (1998) The Netherlands. In: Vymazal J, Brix H, Cooper PF, Green MB, Haberl R (eds) Constructed wetlands for wastewater treatment in Europe. Backhuys, Leiden, pp 289–314

Vuillot M, Boutin C (1987) Les systèmes rustiques d'épuration: aspects de l´expérience française; possibilités d'application aux pays en voie de dévelopment. Trib CEBEDEAU 518:21–31

Vymazal J (1999) Nitrogen removal in constructed wetlands with horizontal sub-surface flow-can we determine the key process? In: Vymazal J (ed) Nutrient cycling and retention in natural and constructed wetlands. Backhuys, Leiden, pp 1–17

Vymazal J (2001) Types of constructed wetlands for wastewater treatment: their potential for nutrient removal. In: Vymazal J (ed) Transformations on nutrients in natural and constructed wetlands. Backhuys, Leiden, pp 1–93

Vymazal J, Brix H, Cooper PF, Green M, Haberl R (eds) (1998) Constructed wetlands for wastewater treatment in Europe. Backhuys, Leiden

Wetzel RG (2001) Fundamental processes within natural and constructed wetland ecosystems: short-term versus long-term objectives. Water Sci Technol 44:1–8

Whiting GJ, Chanton JP (2001) Greenhouse carbon balance of wetlands: methane emission versus carbon sequestration. Tellus B53:521–525

6 Tools for Wetland Ecosystem Resource Management in East Africa: Focus on the Lake Victoria Papyrus Wetlands

Steven Loiselle, Andres Cózar, Anne van Dam, Frank Kansiime, Peter Kelderman, Matt Saunders, Silvio Simonit

6.1 Introduction

East Africa has several of the world's most important freshwater ecosystems, including the world's largest tropical lake, Lake Victoria, which covers over 68 000 square kilometres and is (or was) characterised by one of the richest ichthyofauna in the world (Greenwood 1965; Goldschmidt et al. 1993; Hecky 1993). For the populations in the Lake Victoria catchment, the lake and its extensive papyrus wetlands represent fundamental economic, social and nutritional resources. An integrated management of these ecosystems is crucial to maintaining or improving the living conditions of one of the world's fastest growing populations (3–4 % per annum; Okeyo-Owuor 1999). Past increases in population growth and economic development led to increased stress on both the aquatic and terrestrial resources, damaging the functional value of both. Fish remains the least expensive and most common (70 %) source of dietary protein in the region (Ogutu-Ohwayo et al. 1997). Drinking water is often drawn directly from the lake shores with little or no treatment. The lake and its wetlands provide a number of important secondary income services, both directly (tourism, transport, vegetation harvesting) and indirectly (climate control, storm water retention, groundwater replenishing).

During the past 50 years, many of the East African Great Lakes and Lake Victoria in particular underwent major changes, both in trophic structure and in water quality. Phytoplankton abundance and composition changed (Mugidde 1993), demonstrating an increasing eutrophication (Ochumba 1995), which led to anoxia and a loss of fish habitat (Kaufman 1992). The combined reduction of water quality and trophic complexity led to a simplification of the energy transfer within the ecosystem. The cascading effect in the

Ecological Studies, Vol. 190
J.T.A. Verhoeven, B. Beltman, R. Bobbink, and D.F. Whigham (Eds.)
Wetlands and Natural Resource Management

trophic chain continues to evolve (Balirwa et al. 2003) and the driving forces that led to the present state have yet to be completely identified and addressed. These include the increasing nutrient load, overfishing, climate change and a reduction in the functional capacity of the Lake's wetland borders (Ogutu-Ohwayo 1990; Hecky 1993; Lugn'ayia et al. 2001).

In the present condition of the Lake Victoria ecosystems, the role of the extensive papyrus wetlands along the lake borders is of vital importance in the protection and eventual restoration of the lake ecosystem. To help resolve this important problem, the European Commission is sponsoring an international research project to consolidate and reinforce the understanding of the role of wetlands in maintaining the resource quality in Lake Victoria (ICA4-CT-2001-10036). The project, entitled ECOTOOLS, focuses on the lake wetland ecotomes and their functional values with respect to the Lake and the region, in particular in the face of changing land uses in the larger Lake Victoria basin. Seven European and East African research institutions are working together in the four-year project which concludes in 2005. The study of the role of wetland vegetation in removing nutrients from urban and agricultural sources is being developed by the Makerere University Institute of Environment and Natural Resources, Uganda, together with the UNESCO-IHE Institute for Water Education, The Netherlands. The analysis of the impact of wetlands on the lake water quality and productivity is being conducted by the University of Siena, Italy, in collaboration with KMFRI, Kenya and the Water Resources Department, Uganda. The analysis of the role of wetlands in regulating carbon and water fluxes to the atmosphere is being made by the Trinity College Dublin, Ireland. The role of the wetlands in providing habitat for the ecotome fish population is being developed through historical data analysis and new analysis by the University RomaTre, Italy, in collaboration with local partners. The socio-economic component of the project aims at developing an integrated model of land–water interactions that includes the results from the project research on nutrients, habitat and carbon and water exchange. These modelling activities are being co-ordinated by the Environment Department at the University of York (UK), the Kenya Institute for Public Policy Research and Analysis and the Ugandan Ministry of Water Lands and Environment.

In the present chapter, we focus on the analysis and modelling of the fundamental functions that the Lake Victoria wetlands provide with respect to the lake and its environs. The main focus is directed toward the regulation of carbon and water fluxes by wetland vegetation, the nutrient retention capacity of wetlands and wetland impacts on the inshore lake water and habitat quality. These aspects are studied using in situ and mesocosm studies and modelled to examine key aspects of their management. The results are being combined into an ecological–economic watershed model to examine the effects of anthropogenic nutrient loadings and the modification of wetland integrity on fisheries in Lake Victoria.

6.2 Wetlands and Inorganic Carbon Retention

Tropical wetlands are highly productive ecosystems, using solar energy to convert carbon dioxide into organic carbon and releasing water vapour into the surrounding atmosphere. Decaying vegetation then releases carbon into the atmosphere as carbon dioxide and into the lake water as dissolved organic carbon. The carbon and water cycle of the wetlands has important impacts on the local hydrology, inshore water quality (organic carbon) and the overall value of the wetland as a carbon sink or source. To understand the relationship between wetland management and the carbon/water cycles, Trinity College Dublin used an open pathway eddy covariance (EC) system to measure the exchange of carbon dioxide and water vapour scalars between the vegetation and the atmosphere in the Kirinya and Nakivubo wetlands.

The Kirinya (West) wetland near Jinja on the northern shore of Lake Victoria is a relatively pristine *Cyperus papyrus* L. ecosystem. The Nakivubo wetland approximately 5 km outside Kampala on the western shore of Lake Victoria is a disturbed papyrus wetland where coco-yam (*Colocasia esculenta*) cultivation is extensive. Eddy covariance data was collected in Kirinya from May 2003 to May 2004 and then in Nakivubo from June to October 2004. In addition to the EC measurements, an estimation of standing biomass and productivity were made at the Kirinya site over a six-month period.

The EC tower consisted of an open path IRGA (Table 6.1; LI-COR 7500) and a 3-D sonic anemometer (CSAT-3; Campbell Scientific; for methodological information, see Baldocchi et al. 1996; Moncrieff et al. 1997; Aubinet et al. 2000; Wilson et al. 2002). In addition, a range of meteorological instruments were installed, measuring wind speed and direction, net radiation, photosynthetic active radiation, temperature, relative humidity and soil heat flux. The data was collected and gap-filled to produce diurnal flux patterns of both CO_2 and H_2O (Fig. 6.1).

The wetland carbon flux showed a diurnal pattern of carbon exchange between the papyrus vegetation and the atmosphere. During night-time peri-

Table 6.1 Explanation of acronyms and abbreviations

Acronym/abbreviation	Full text
ARIMA	Autoregressive integrated moving average
DW	Dry weight
EC	Eddy covariance
IRGA	Infrared gas analyser
PAR	Photosynthetic active radiation
P, N	Phosphorus, nitrogen
SVD	Singular value decomposition
UVA, UVB	Ultraviolet radiation A(280–320 nm), B(320–400 nm)

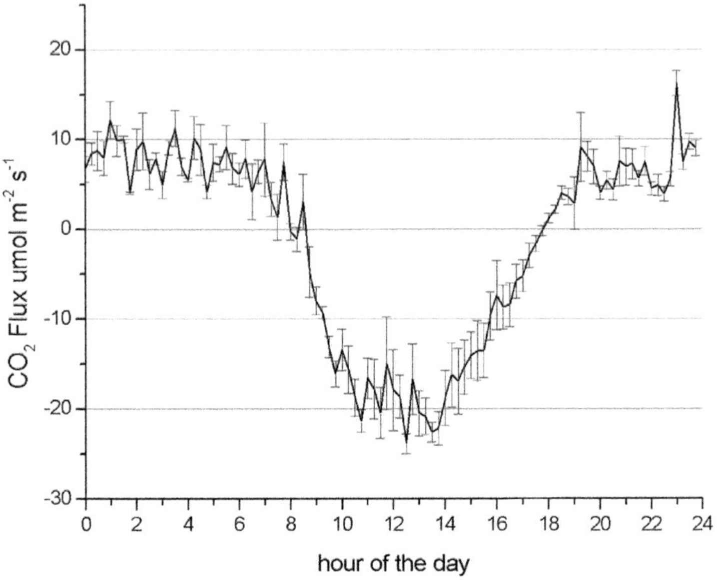

Fig. 6.1 Average diurnal CO_2 flux over a *Papyrus*-dominated wetland (negative values indicate carbon uptake) based on data collected from 22nd to 28th September 2003 (values indicate mean flux per time ±SE)

ods, carbon is released through the process of respiration, in both living and decomposing dead tissue; peak rates of release during this period range over 10–15 µmol CO_2 m^{-2} s^{-1}. After daybreak, as the intensity of photosynthetic active radiation (PAR) begins to increase, the plants start to photosynthesise and the system as a whole changes from a net source to a net sink of CO_2. Peak rates of carbon uptake occur slightly after midday and are associated with maximum values of PAR. Maximum rates of CO_2 uptake during this period range over 35–37 µmol CO_2 m^{-2} s^{-1}. As PAR levels decrease throughout the afternoon and into early evening, the rates of photosynthetic activity and therefore the rate of CO_2 uptake also tend to decrease and, as the sun sets, the system reverts back to being a net source of CO_2. Also during these periods, carbon is lost from the system through the decomposition and degradation of organic detrital material, due to the accumulation of senescing and dead plant material on the rhizome surface. Over time, this material moves through the rhizome mat and eventually into the water column below, either as suspended solid material or in dissolved organic form. The aerobic decomposition of this material on the surface of the plant rhizome is a contributing factor to both night-time and daytime flux dynamics.

The decomposition of this organic material through respiration is generally controlled by moisture content and temperature. Water levels in the lake vary slightly throughout the year and in relationship to inter-annual climate factors. Water levels are usually high enough to ensure that the papyrus rhizome system is always under water and largely anaerobic, thus preventing aerobic decomposition of the underlying peat deposits. Temperature patterns show stable diurnal trends, which are dependent on the solar energy input to the system. Detrital temperature at a depth of 5 cm remained constant at 23 °C throughout the day because it was shaded from sunlight. The overall balance of carbon exchange with the atmosphere is controlled by the relationship between the decomposition of the detrital organic matter and the productivity of the wetland vegetation. Over the experimental period, based on weekly diurnal flux patterns, the Kirinya wetland appears to sequester approximately 0.51 kg C m^{-2} $year^{-1}$. The disturbed wetland at Nakivubo show similar diurnal patterns of carbon flux to the Kirinya papyrus wetland. However, the uptake of carbon during the day is much lower, with peak uptake rates ranging over 20–23 µmol CO_2 m^{-2} s^{-1}.The data show similar rates of carbon release during the night, with values fluctuating over 5–15 µmol CO_2 m^{-2} s^{-1}. Based on the weekly flux dynamics, it is estimated that the Nakivubo wetland would sequester approximately 0.34 kg C m^{-2} $year^{-1}$.

In addition to the flux measurements, estimates were made of both the biomass and the net primary productivity (NPP) of the papyrus system. The biomass measurements involved measurements of increment in culm girth and the use of a culm girth/dry weight relationship. The productivity measurements were made over a six-month period. Table 6.2 provides a breakdown of the biomass components of papyrus at the Kirinya wetland and compares it with similar data taken from another papyrus wetland in Uganda. The Kirinya wetland shows very high rates of net primary productivity, approximately 8.08 kg dry weight (DW) m^{-2} yr^{-1} compared to values of 6.28 kg DW m^{-2} $year^{-1}$ estimated for papyrus vegetation on Lake Naivasha, Kenya by Jones and Muthuri (1997).

Table 6.2 Biomass and culm density estimations of *Papyrus* stands at two sites in Uganda. Values are mean kg DM m^{-2} ±SE

	Kirinya, Uganda	Mpigi, Uganda[a]
Shoot biomass live[b] (kg m^{-2})	2.26±0.3	2.1±0.27
Shoot biomass dead (kg m^{-2})	2.19±0.94	n.d.
Rhizome (kg m^{-2})	6.86±1.25	n.d.
Roots (kg m^{-2})	0.28±0.07	n.d.
Surface detritus[c] (kg m^{-2})	4.13±0.74	n.d.
Culm density (number m^{-2})	16±2.5	18.6±1.2

[a] Data taken from Jones and Muthuri (1997)

[b] Shoot biomass includes culm and umbel of individual plant unit

[c] Indicates dead and decaying rhizome, roots and detritus and the rhizome surface

Water vapour fluxes were also determined using the eddy covariance system and processed to produce diurnal fluxes. Water vapour exchange shows a typical diurnal flux pattern with an evapotranspiration peak during the day of approximately 12.5 mmol H_2O m^{-2} s^{-1}, which coincides with peak air temperatures. Based on the flux estimates over the measurement period, it is estimated that the average total daily flux of water from the system through canopy evapotranspiration is approximately 4.6 kg H_2O m^{-2} day^{-1}. The Nakivubo site shows a similar pattern of water vapour flux. There are lower losses of water during both daytime and night time measurements with peak rates of evapotranspiration around 8.9 mmol H_2O m^{-2} s^{-1}. Based on the data collected at Nakivubo, it was estimated that the total daily flux of water from the coco-yam system is approximately 3.24 kg H_2O m^{-2} day^{-1}.

The wetlands studied clearly play an important role in both carbon and water cycles. Pristine papyrus stands have the ability to sequester and store a significant amount of carbon. Papyrus plants also have the ability to efficiently use transpired water, which is important with regard to localised climate regulation. The degradation of these wetlands through coco-yam agriculture reduces the rates of carbon sequestration but does not necessarily alter the role of these wetlands from net sinks to net sources of carbon. The rate of water transpiration over the coco-yam stands does however alter the energy balance of these areas. Lower rates of water vapour flux to the atmosphere, in the form of latent heat, result in an increase in sensible heat flux. This may for example increase the temperature of the localised area, which may in the long term initiate changes in climatic conditions.

6.3 Wetlands and Nutrient Retention

Littoral wetlands in Lake Victoria provide a substantial nutrient and pathogen removal when receiving wastewater or runoff from population centres, industry and agricultural areas (Kansiime and Nalubega 1999). The growing threat of eutrophication to Lake Victoria with the coincident increasing anoxia of the lake bottom water layer and the prolonged stratification of the lake are producing conditions that threaten vital fishing activities. The Lake has reached this state due to its extended flushing times, the heavily populated watershed, the reduced retention capability of wetland areas and climate change. While much of the nutrient flux in the open waters is related to atmospheric deposition, inshore areas are impacted by nutrient loads from the upland watershed. In these vital areas, littoral wetlands retain phosphorus and nitrogen from stream and watershed runoff in relation to their dominant vegetation, extension and the dispersion of runoff waters (Kansiime et al. 2003). An ecosystems approach was used by the ECOTOOLS nutrient analysis teams to determine the wetland-related nutrient retention by: (1) examining the characteristics of the wastewater and runoff that enter into two important Ugandan wetlands, i.e. Nakivubo (Kampala) and Kirinya (Jinja), (2) identifying the wastewater flow patterns within the wetlands, (3) characterising the dominant wetland plant species, (4) identifying the in situ nutrient retention in two wetlands and (5) determining the specific retention capacity of the major wetland plant species (those which naturally occur and those which are planted for subsistence farming). The results from experimental microcosms using an effluent flow from a wastewater treatment in Kampala indicate that naturally occurring papyrus vegetation has a higher nutrient retention potential than the vegetation introduced in the converted wetland areas by subsistence farmers (coco yams).

In situ measurements in Kirinya and Nakivubo wetlands (receiving the wastewater and runoff from the cities of Jinja and Kampala, respectively), were carried out along transects cut across the wetlands. Results indicate that the retention of nutrients in the two wetlands varies with distance within the wetland and wetland vegetation. In general, there is a reduction in concentrations of nutrients as the wastewater flows from the inflow (transect 1) towards the swamp lake interface (transect 3 for Kirinya, transect 4 for Nakivubo; Figs. 6.2, 6.3). Natural papyrus vegetation growing in the Nakivubo wetland was found to have higher biomass production (6.8 kg DW m^{-2}) compared to yams (3.4 kg DW m^{-2}). The superior characteristics of papyrus (high biomass production, high nutrient removal capacity) compared to yams explain the role that littoral papyrus wetlands have in controlling the nutrient concentrations and eutrophication level of the inshore areas.

Some of the nutrients that are not retained within the wetland are released into the inshore areas of the Lake, where eutrophic or hyper-eutrophic conditions dominate. However, the retention and release of nitrogen and phosphorus follow different paths that are dependent on wetland conditions. Nitrogen

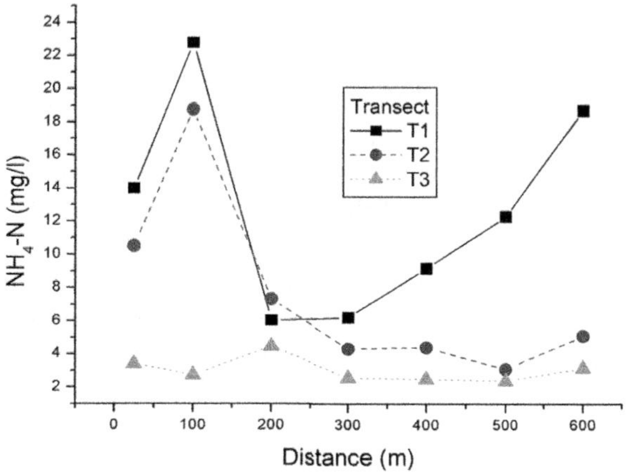

Fig. 6.2 Variation of NH_4-N in different transects T1, T2 and T3 in Kirinya wetland. Distances represent the distance from the western wetland border of the transect. Each transect represents a cross-section of the wetland

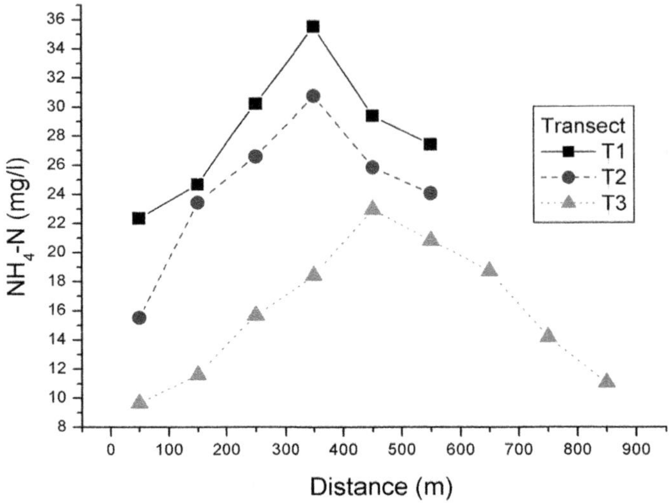

Fig. 6.3 Variation of NH_4-N along T2, T3 and T4 in the Nakivubo wetland in Kampala. Distances represent the distance from the western wetland border of the transect. Each transect represents a cross-section of the wetland

is entrapped physically and removed by plant uptake and eventually re-released into the water column in the form of organic nitrogen. The uptake is therefore a function of vegetation community, mat structure and the conditions of the interstitial areas of the mat where vegetation roots are present. Phosphorus (P), in contrast, is largely removed physically, entrapped in the wetland mat detritus and sediment. Plant uptake appears to play only a minor role in the P dynamics of the wetland (on the order of 10 %, according to literature in comparable wetlands). Laboratory experiments on sediment cores from Kirinya wetland showed marked P uptake capacities of sediment. The absorption was found to follow both Langmuir and Freundlich isotherm characteristics with uptake capacities of 1.7–6.8 mg P g^{-1} sediment. The highest P uptake capacities were found for sediments closest to the lake (furthest from the wetland inlet), where relatively lower wastewater P loadings are present. P removal rates did not seem to be affected by seasonal variations in the Kirinya wetlands. There was no significant variation between the rainy or dry seasons (Fig. 6.4a). However, a significant increase in P removal was observed in 2002/2003 in comparison to the transect measurements performed in 2000/2001. This increase was most likely due to an improvement of the distribution efficiency following the installation of a wastewater distribution pipe at the wetland inlet (Fig. 6.4b).

In the coming decades, the P uptake capacity of the Kirinya wetland will gradually decline under the present P wastewater loadings. Using the potential P uptake capacities and the estimated actual P removal rates determined from the sediment cores and measured in the Kirinya wetlands, it is possible to estimate that the wetland sediment will reach P saturation within the coming 25–50 years. Periods of exceptional rains may result in high flow

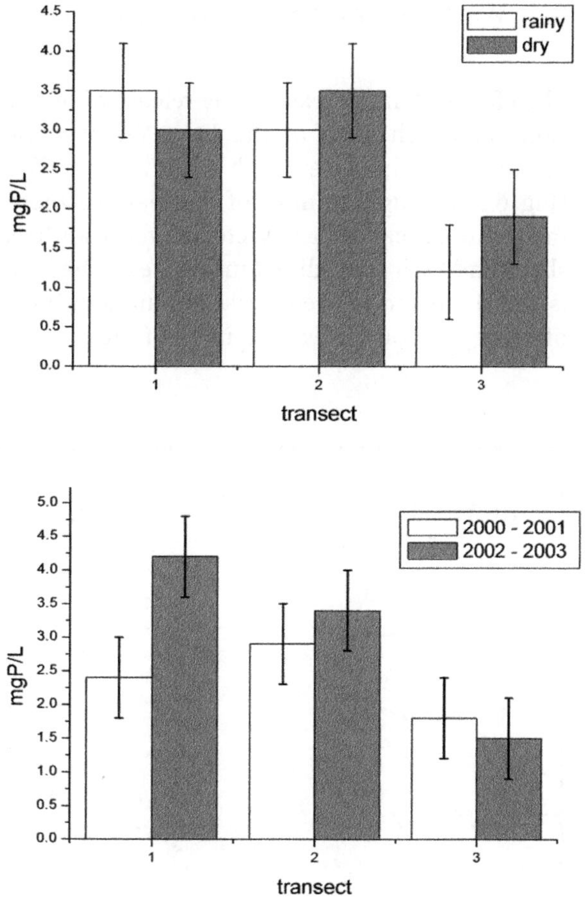

Fig. 6.4 Phosphorus removal efficiency in Kirinya wetland. No significant differences were observed between seasons (**A**) but a significant increase in retention was observed after the wastewater distribution was improved (**B**)

rates which may transport some of the nutrient rich sediment into the lake, in particular during first flush conditions. This is currently being investigated.

The capacity of the Lake's littoral wetlands to remove or retain incoming nutrients will depend on their extension, vegetation composition, physical structure, biomass and wastewater flow rate and distribution. All these parameters are influenced by the management of the watershed and wetland. Incorrect management will result directly in the degradation of the inshore waters of the Lake and eventually contribute to the eutrophication of the open lake waters. Impacts of management alternatives on retention capacities were examined using modelling approaches (see Section 6.5).

6.4 Wetlands and Eutrophication

The effects of an increase in the release of nutrients into the lake led to conditions in which most of the Lake Victoria waters can be characterised as eutrophic and hyper-eutrophic, with elevated concentrations of algal biomass (Fig. 6.5) throughout most of the year. Nowhere is this more apparent that in the inshore areas of Lake Victoria, where anthropogenic changes in the watersheds had their most direct influence on the Lake. This coastal strip of the lake is closely linked to the littoral population. Industrial and domestic activities often depend on the water quality of the inshore waters; and fishing activities

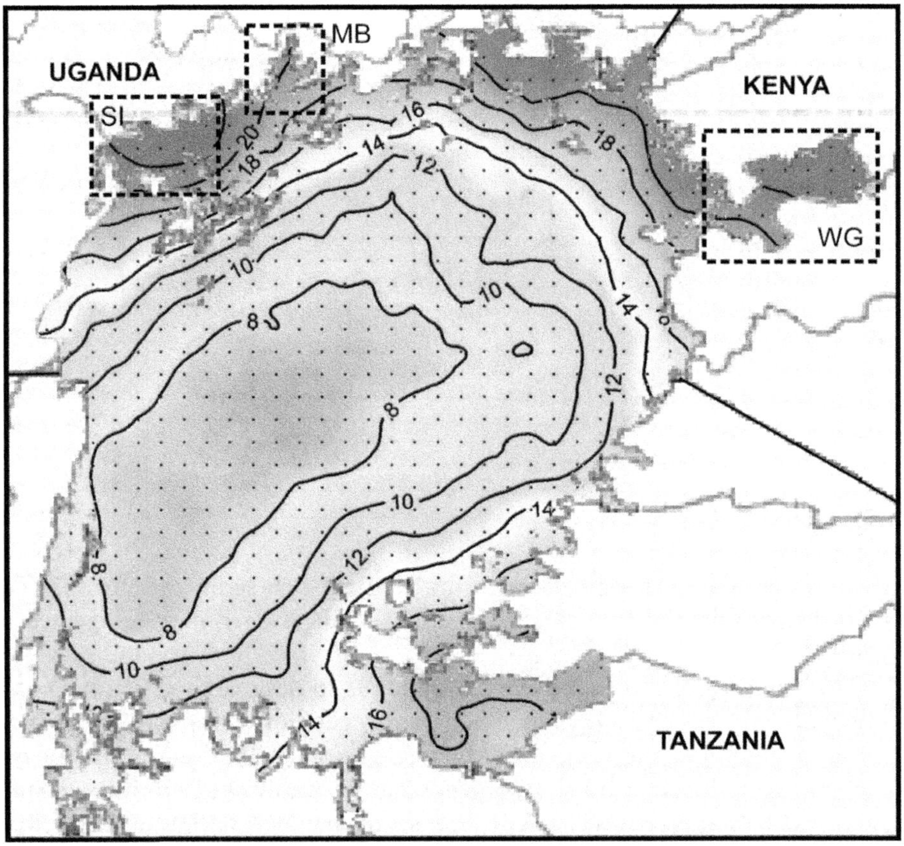

Fig. 6.5 Average SeaWiFS-derived chlorophyll-*a* concentrations (mg m^{-3}) in Lake Victoria during the period 1997–2004. Note the strong coast effect on the distribution of chlorophyll-*a*. Main inshore areas which co-vary independently from the offshore regions are delimited within *squares*: Sese Islands (*SI*), Murchinson Bay (*MB*) and Winam Gulf (*WG*; see Fig. 6.6). The highest population densities in the lake catchment are found in the MB and WG areas. The main rivers in the watershed area are shown

are based on the use of canoes and artisanal methods. The growing threats of eutrophication and pollution are producing conditions that threaten the use of these resources by the lake populations. Therefore, the characterisation of these coastal areas represents a primary task in ecosystem resource management.

To analyse the spatio-temporal variability in phytoplankton biomass of the lake, a seven-year series of SeaWiFS imagery (150 scenes) was analysed, using a combination of SVD and ARIMA methodologies. In addition to describing differences in phytoplankton concentrations between inshore and offshore areas, the analysis allowed us to identify differences in the temporal evolution of phytoplankton concentrations throughout the lake. The offshore region showed periodic patterns closely linked to the seasonal variability in wind speed and rainfall. This seasonal variability is associated to the latitudinal movement of the inter-tropical convergence zone, allied with the monsoon influence on the region. Results showed a wide interannual variability during the study period (1997–2004). Correlations between global climatic events such as El Niño and variations in phytoplankton concentrations also emerged from this analysis. Inshore areas were found to have consistently higher concentrations of phytoplankton and showed temporal patterns that were delayed with respect to offshore dynamics, as shown for the Sese Islands and Murchison Bay in Uganda and Winam Gulf in Kenya (Fig. 6.6). Apart from the climatic conditions, the degree of eutrophication and the phytoplankton dynamics of the inshore area depend directly on the catchment properties and in particular the wetland extension, population density and agricultural activities. Differences between major bays can be seen, with Murchison Bay displaying more variable concentrations than Winam Gulf. This is the result of a higher nutrient load (higher population density) with respect to bay area.

To better examine the relationship between wetlands and lake water quality in the inshore areas, the University of Siena team performed a series of sampling transects along the coastal areas of Uganda and Kenya, including the Sese Islands, Murchison Bay and Winam Gulf. During these campaigns at over 300 sites, measurements of temperature, pH, dissolved oxygen, nutrients, turbidity, vertical attenuation of visible and ultraviolet radiation and chlorophyll-*a* were measured. Overall physico-chemical characteristics (temperature, pH, DO) of the inshore waters were found to be strongly controlled by the presence of wetlands. Semi-enclosed coasts with large littoral wetlands were correlated with cooler, more acidic and poorly oxygenated waters. The spatial distribution of nutrients (phosphorus and nitrogen) and chlorophyll concentrations was found to be influenced by the presence of point loadings such as rivers (e.g. Katonga river, Sondu river) and major population centres such as Kampala (in Murchison Bay) and Kisumu (in Winam Gulf). Based on nutrient and chlorophyll data, it can be stated that the Lake Victoria inshore waters show clear signs of hyper-eutrophication. In

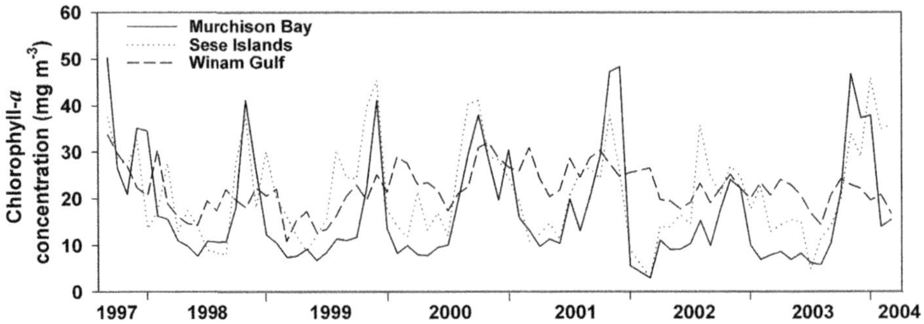

Fig. 6.6 Monthly evolution of the SeaWiFS-derived chlorophyll-*a* concentration in areas of Murchinson Bay, Sese Islands and Winam Gulf

several bays, the areas receiving inputs of urban warm waters or infested by water hyacinth showed an exacerbation of the effects of eutrophication. In such local areas, relatively stable thermal stratification and anoxic deep layers appeared in the water column despite its shallowness (<10 m). Total N (TN) to total P (TP) ratios indicated that approximately 44 % of the sampled sites showed P-deficiency (TN:TP ratio >50, using the criteria of Guildford and Hecky 2000) and 45 % of the observations were indicative of N- and P-deficiency. Only local singular areas were found to have conditions of N-deficiency (TN:TP ratio <20).

The spatial distribution of chlorophyll-*a* concentrations showed an irregular pattern and no clear relationship with nutrient concentrations was found. As phytoplankton growth is controlled not only by nutrients but also by light availability, this was studied to determine how organic matter released by the wetland controls optical conditions and phytoplankton growth in the inshore areas. In fact, highly coloured organic matter, released from the floating vegetation mats, can have a significant role in the reduction of ultraviolet and visible light. The optical characteristics of the water column in six bays (Table 6.3), each with a different quantity of littoral vegetation, was investigated and compared with measurements of chlorophyll-*a* concentrations and nutrients.

By examining the relationship between the main attenuation components in the water column (phytoplankton, inorganic particles, dissolved organic matter, the water itself), it was concluded that the dominating light attenuation component in the inshore areas was dissolved organic matter. In addition, a significant and logarithmic relationship was found between wetland area/open water area and the vertical attenuation of visible light, indicating that wetland extension indirectly influences the availability of solar radiation in the water column.

To determine if wetland-induced light attenuation was sufficient to limit phytoplankton growth, an analysis of the water column irradiance profile in

Table 6.3 Optical properties of Lake Victoria inshore waters in relation to wetland extension. The wetland extension is reported as total wetland area (m^2), divided by total open water area (m^2) and percent coastline (km) covered by wetlands with respect to total coastline (km)

	Wetland extension		DOM		KPAR		KUVB	
Site	Area/open water (m^2)	Coastline (%)	Average	SD	Average	SD	Average	SD
Katonga (June 2003)	90	82	10.5	6.0	3.3	1.4	35.4	18.0
Katonga (Nov. 2002)	90	82	8.6	4.1	3.3	0.9	23.6	7.7
Bunjako (Nov. 2002)	12	6	1.7	0.7	2.0	0.4	7.0	1.6
Nsonga (June 2003)	160	59	4.5	4.0	2.3	0.6	17.3	9.4!
Nabugabo coast (Nov. 2002)	<1	0	0.9	0.8	0.9	0.3	3.2	0.5
Lulagwe (June 2003)	<1	0	1.2	0.1	1.0	0.1	3.3	0.4
Tende (June 2003)	3	7	3.5	0.4	1.5	0.3	5.0	1.5
Murchinson (June 2003)			2.2	1.4	1.7	0.7	9.7	7.3
Winam (Mar. 2004)	5	13	2.3	0.5	2.8	1.6	8.8	10.5
Yala coast (Mar. 2004)	NA	NA	1.4	0.9	1.0	0.4	4.3	1.8

each shoreline site (wetland, non-wetland) was made. The availability of solar radiation was found to be a limiting condition in those inshore areas where wetlands were present. A negative and significant relationship was seen between the degree of light limitation and distance from the wetland. This indicates that wetland-released organic matter may limit primary production in the inshore areas.

Regarding the attenuation of damaging ultraviolet radiation and the possible relationship with photoinhibition and damage to the DNA of hatching fish, an analysis was made using the vertical profiles of UVA and UVB radiation in the water column of each inshore site. Using data for the photoinhibition of a tropical phytoplankton community (Villafañe et al. 1999), it was determined that UV-induced photoinhibition is possible in inshore areas where wetlands have been eliminated or are not present. Wetland generated dissolved organic matter strongly attenuates ultraviolet radiation to the point where only a small percentage of the water column is exposed to high irradiances.

This study allowed us to characterize the current state of the Lake Victoria inshore waters and to identify the main areas currently affected by eutrophication. The presence of wetlands is a significant factor determining the spatial heterogeneity in the inshore area. The combined effect of wetlands on limiting nutrient loads and reducing available light protects inshore areas from excessive phytoplankton growth and the related damage to the overall water quality and fisheries.

6.5 Ecological Modelling

Wetlands in Lake Victoria influence the energy and material fluxes of the lake and watershed ecosystems. To examine how management alternatives may influence nutrient retention, carbon release and water quality in the inshore areas, ecological models were developed and linked through an ecological–economic watershed model. Each model is focused on a specific part of the wetland functioning. The main elements are a nitrogen retention model, a phosphorus retention model and a carbon/biomass model. An ecological–economic watershed model links the nutrient retention capacity of the littoral wetlands to increases in the phytoplankton biomass and negative effects on Lake fisheries.

The nitrogen model examines the cycle of nitrogen within the floating wetland stands, which includes the wetland vegetation aboveground biomass, root and detritus compartments and the water compartment. The model aims to: (1) describe nitrogen transformations in the wetland, (2) understand the processes contributing to nitrogen retention capacity of the wetland and (3) evaluate the effects of potential management strategies on the nutrient absorption capacity of the wetland (e.g. harvesting of vegetation). The model has three layers: vegetation, water and sediment. The main state variables are: papyrus biomass (aboveground biomass plus rhizomes and roots), papyrus mat detritus (the dead organic matter in the papyrus mat), ammonium (NH_4^+) and nitrate (NO_3^-) in the water column and organic nitrogen in the sediment. The major transformation pathways are nitrification, uptake by plants, plant mortality, mineralization of detritus, settling of detritus to the sediment sludge, denitrification and remineralization of organic matter from the sludge (see Fig. 6.7).

Plant growth was modelled with a logistic equation, in which nitrogen growth was the balance of uptake by the plants and decay (mortality) of the plants. The maximum N uptake rate of the plants was limited by the concentrations of NH_4^+ and NO_3^- (Michaelis-Menten-type equation with half saturation constants). Nitrification, plant mortality, mineralization of detritus, denitrification and settling of detritus were modelled as first-order reactions (Table 6.4).

Several simplifying assumptions were made. Temperature and dissolved oxygen concentration were assumed to be constant. Both NH_4^+ and NO_3^- were taken up by papyrus. Ammonia volatilization was assumed to be unimportant because of the low pH in wetland and the fact that most of the water surface was covered by the papyrus mat. NH_4^+ and NO_3^- in the sediment were not modeled to keep the model simple. It was assumed that a net flux of ammonium from the sediment to the water occurred. For parameterization, a com-

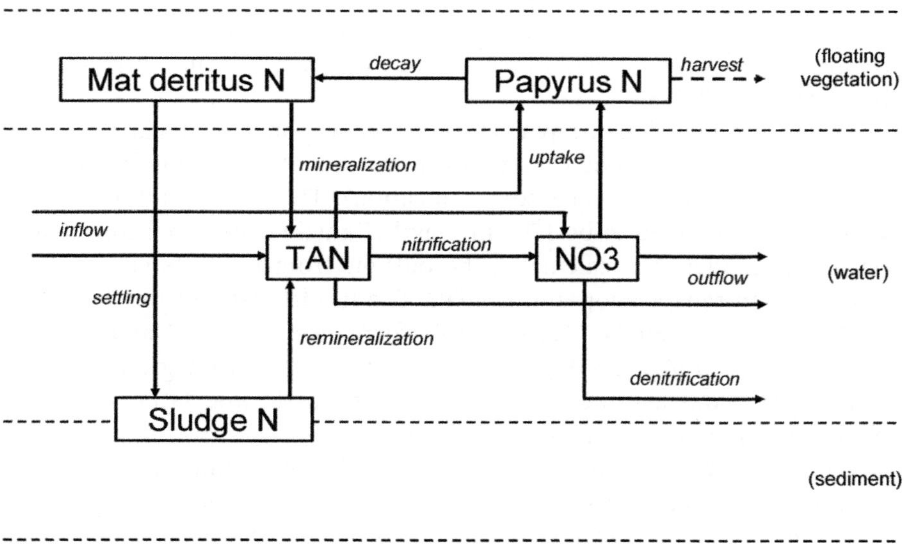

Fig. 6.7 Schematic diagram showing the flows of nitrogen in the model

Table 6.4. Parameter values, initial conditions and wetland dimensions in the model

Parameter	Value	Unit
K NH_4^+ (half saturation constant)	0.7	g m^{-3}
K NO_3 (half saturation constant)	0.1	g m^{-3}
Papyrus N content	1.5	% DM
Mat N content	1.0	% DM
Sludge N content	1.0	% DM
K denitrification (first order constant)	0.1	day^{-1}
K mineralization (first order constant)	$6.5e^{-5}$	day^{-1}
K nitrification (first order constant)	$5.0e^{-4}$	day^{-1}
K settling (first order constant)	$2.3e^{-4}$	day^{-1}
Max. papyrus biomass	8000	g DM m^{-2}
Max. NH_4^+ uptake rate of papyrus	0.05	day^{-1}
Max. NO_3 uptake rate of papyrus	0.05	day^{-1}
Initial conditions		
Concentration of NH_4^+ inflow	10	g m^{-3}
Concentration of NO3 inflow	0.5	g m^{-3}
Initial papyrus biomass	7800	g m^{-3}
Wetland dimensions		
Surface area	147 000	m^2
Average depth	0.65	m

bination of direct measurements in the Kirinya and Nakivubo wetlands in the period 2000–2003 and literature values on papyrus wetland ecology and nitrogen cycling in aquatic systems were used, including data published by Delincé (1992), Jamu (1999), Jiménez-Montealegre (2001) and Jørgensen and Bendoricchio (2001).

To run the model, a one-year rainfall and evaporation scenario was constructed using meteorological data from weather stations in Gaba, Kampala and Jinja (Kaggwa, personal communication). The average inflow into the wetland was assumed to be 1500 m^3 day^{-1} (Kansiime, personal communication). The model then calculated the outflow from the difference between inflow, evapotranspiration and rainfall. The model was run for three years (1095 days) using Stella 7.0.2 (High Performance Systems, Hanover, USA), with a time-step of 0.125 day, using Euler (rectangular) integration. For the simulation, it was assumed that the papyrus was in the stationary growth phase and therefore that biomass was more or less stable at a level of 7800 g DW m^{-2}. Calibration consisted of adjusting the maximum nutrient uptake rates of NH_4^+ and NO_3^- to achieve a stable papyrus biomass.

The maximum uptake rate at which this was achieved was 0.05 day^{-1} for NH_4^+ and NO_3^-. This resulted in an average nitrogen uptake by the papyrus in the model of about 0.13 g N m^{-2} day^{-1}. Because of the rapidly declining NO_3^- concentration (Fig. 6.8), this uptake mainly (95 %) consisted of NH_4^+. The regular fluctuations in simulated NH_4^+ concentrations around 5 g m^{-3} are caused by the yearly rainfall pattern.

Based on this analysis, the N retention of the wetland was estimated to be about 25 g m^{-2} $year^{-1}$. Because of the stability in plant and mat biomass, most of this retention was realised by the accumulation of organic N at the bottom of the wetland. Some of this organic sludge can be expected to be flushed out of the wetland during periods of heavy rainfall and high outflow, which would reduce this estimate of the N retention capacity of the wetland.

To analyse the effects of harvesting papyrus on N retention, three harvesting scenarios were simulated: 0 %, 10 % and 20 % harvesting of the total papyrus biomass once a year. Regarding the total biomass after harvesting, the 10 % regime showed a complete biomass recovery within a year. At 20 %, there was a decline in the overall biomass levels, indicating that this would be a non-sustainable level of harvesting. The model predicted an increase in NH_4^+ retention with harvesting. This was caused by the re-growth of the papyrus after harvesting, which leads to higher uptake of NH_4^+ from the water compared to the situation where the papyrus stand is stable at its maximum density. Nitrogen retention increased with harvesting because of the increased uptake of nitrogen (Fig. 6.9). The model showed an optimal N retention at around 10 % harvesting (26 g N m^{-2} $year^{-1}$ at 0 % harvesting, to

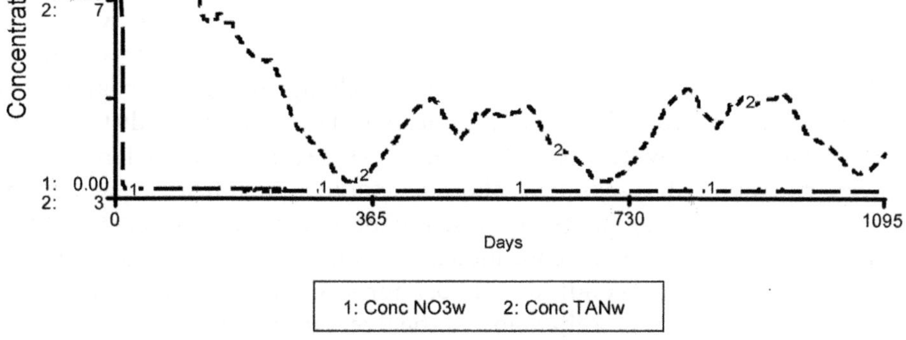

Fig. 6.8 Concentrations of total ammonia nitrogen (NH_4^+w) and nitrate (NO_3^-w) in the water of Kirinya wetland, as simulated by the model

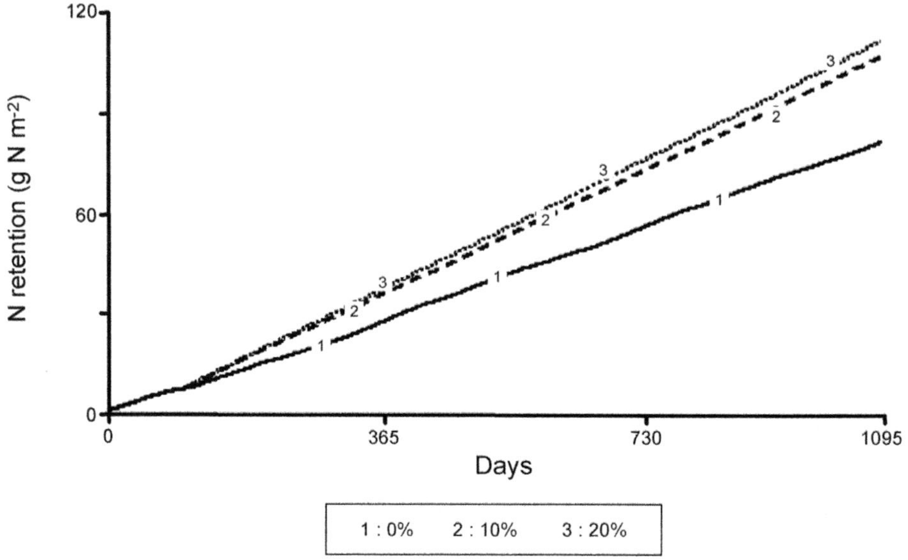

Fig. 6.9 Nitrogen retention of Kirinya wetland as simulated by the model when once every year 0 %, 10 % or 20 % of the aboveground biomass is harvested and exported from the system

34 g N at 10 % and 36 g N at 20 %), considering the productivity of the papyrus stands determined in Section 6.2.

Generally, the model results in biomass levels and nutrient concentrations that resemble actual measurements in the field. The model shows that, with sustainable harvesting, the nutrient and carbon uptake capabilities of the wetland can improve. However, the model does not quantify settling and outflow of detritus from the wetland, which should be further investigated. Furthermore, preferential pathways of water flow within the wetlands would reduce the effective surface area and therefore the N retention capabilities.

Phosphorus dynamics were modelled using a plug flow model with transect data from Kirinya wetland over the years 2000–2003, together with laboratory experiments of the sediment–water exchange using 24 wetland sediment cores. Data from the wetland indicated that, with proper distribution, the wetland can provide a 60 % P removal rate. At higher P contents in the overlying water, the sediment showed marked P uptake capacities of 1.7–6.8 mg P g^{-1} sediment. Results on sediment resuspension, simulating "first flush" effects are less clear. Trends indicate enhanced P release rates compared to stagnant sediment conditions. Phosphorus budget calculations on the Kirinya west wetland show that most of the P retention can be ascribed to sediment uptake. Plant uptake, in contrast, only plays a minor role in the P dynamics of the wetland.

The capability of the wetland to retain nutrients is linked to wetland hydrology and vegetation productivity. The latter depends on management practices and vegetation types. A model of carbon flows and productivity was constructed to study how variations in solar radiation, harvesting regimes and detritus accumulation may influence biomass growth. The conceptual model described the fluxes of carbon between each carbon pool in the papyrus wetlands (Fig. 6.10).

These models were then used to examine how wetland extension can effect nutrient concentrations in the lake and then eventually lake fisheries. Researchers focused on specific areas where wetland and watershed management are critical. Nutrient run-off was estimated using land use information and change on a watershed scale. The nutrient buffering capacity of the littoral wetlands was utilised to model the nutrients removed from the overall load from the watershed. The resulting nutrient load to the lake was then linked to phytoplankton biomass through a P limitation model (Jones and Bachmann 1976). Variations in water quality (phytoplankton concentrations) were modelled to impact fisheries through a modified Gordon–Schaefer model (Gordon 1954; Schaefer 1954, 1957; Simonit and Perrings 2004). The model parameters were econometrically estimated using water quality variables. We used annual data on fishing effort and CPUE for the Kenyan fisheries of Lake Victoria (Othina and Tweddle 1999) and a time series of water quality data estimated through an Ecosim (Walters et al. 1997) dynamic simulation using Ecopath (Christensen and Pauly 1992). Biological inputs and

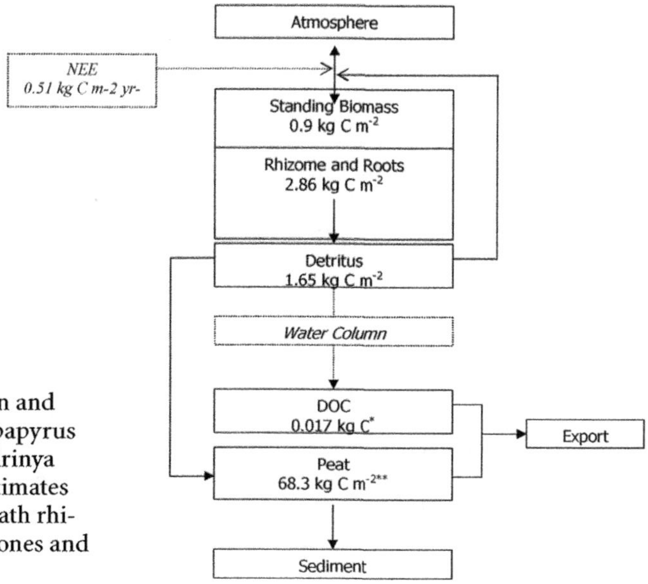

Fig. 6.10 The distribution and flux of carbon within a papyrus wetland, based on the Kirinya wetland, Uganda and estimates of 2 m water depth beneath rhizome. Data taken from Jones and Muthuri (1997)

diet composition data were taken from the Ecopath model for the Kenyan waters of Lake Victoria developed by Villanueva and Moreau (2002).

The model shows that, in mesotrophic and oligotrophic conditions, the growth of the fish stock is positively affected by increases in chlorophyll-*a* concentrations. However, when chlorophyll-*a* values increase above a critical value (17 mg m^{-3}), any further increase in algal biomass causes a negative impact on fisheries. The maximum sustainable yield at this critical phytoplankton biomass value is 195 368 t and the fish stock 227 126 t. As the lake conditions become more eutrophic, the fish yield and fish stock decrease as the environmental conditions deteriorate. Further increases in nutrient load to the Kenyan waters will continue to reduce fisheries values.

The yield was determined to be at its maximum at 1550 t $year^{-1}$. It should be noted that, even with no loading from the watershed, the yield will remain positive as primary productivity would be entirely supported by atmospheric deposition, which is assumed to be constant. The actual loading of phosphorus from the Kenyan basin is estimated around 1925 t P $year^{-1}$ (COWI 2002), which leads to the highly eutrophic conditions of the Kenyan waters, well above critical value determined in the model. In this case, any further increase in nutrient loading will reduce fishery stocks and yields.

The model was applied to the Yala wetland, the largest on the Kenyan Lake Victoria coast. This wetland has been modified by human development since the mid-1960s. It is presently undergoing partial reclamation and plans continue for further reclamation. To determine the retention effect of the Yala wetland on the watershed phosphorus load, we used a general mass balance model (Kadlec and Knight 1996) and an aerial removal rate of 12 m $year^{-1}$ for TP. It is also assumed that hydraulic flow through and out of the wetland is equivalent to the water inflow (Q_{in}). In this manner, we estimated nutrient outflow from the wetland to the main lake as a function of wetland area. Using the

Table 6.5 Impact of wetland reclamation on fish yields under different fishery management scenarios. Note: X and Y are respectively fish biomass and yield under maximum sustainable yield (msy), open access (oa) and profit maximising (*) regimes

	Steady state values (t)	Steady-state values after reclamation of Area II (t)	Difference (t)	Economic loss to the fishery in Lake Victoria (US$ $year^{-1}$)
Xmsy	213 601	210 969	–2632	–
Xoa	21 595	21 595	–	–
X*	171 408	167 182	–4226	–
Ymsy	172 794	168 561	–4233	2,152,692
Yoa	33 172	32 742	–430	218,676
Y*	166 052	161 300	–4752	2,416,630

wetland retention model and long-term averages for hydrological and P loading (COWI 2002), we estimated that the 9200 ha of the wetland that is proposed to be reclaimed may currently retain 34 t P $year^{-1}$, from a total wetland inflow of 102 t P $year^{-1}$. However, as our field experiments for the Kirinya wetland may suggest, the retention function for TP could be considerably higher. The proposed reclamation implies that TP loading into Lake Victoria from the Kenyan watershed would increase by 1.7 %. This scenario can be translated in its relative annual loss (at prices and exchange rates for year 2000) to the fishery sector by the decreased level of fish stock and fishery yields under three possible fishery management regimes: (1) open access, (2) maximum sustainable yield and (3) optimal harvesting (Table 6.5).

6.6 Discussion

The Lake Victoria wetlands, as most wetland systems in East Africa, are transitional zones between an increasingly developed watershed and an overexploited lake ecosystem which provide fundamental resources for the regional population. The sustainable utilisation of wetlands is a great hope for the future of African wetlands (Denny 1993). Presently, the wetlands are being used for agriculture, fish farming, tourism and as a source of domestic water supply. Is it possible to utilise these ecosystems and maintain the functions on which the Lake and wetland populations of 30 million persons depend? To do so, the functional capabilities and limits of these ecosystems must be understood and this information needs to be made available to those persons who manage and govern these resources.

Several studies have been made to assess the retention capacity of natural and artificial wetlands in tropical Africa. Kizito (1986), Kansiime et al. (1994) and Kansiime and van Bruggen (2001) have studied the role of Nakivubo wetland in reducing pollutant loads in Murchison Bay, Uganda. Kipkemboi et al. (2002) compared the biomass production of *Papyrus* and *Miscanthidium* in different wetlands under different nutrient conditions. Azza et al. (2000) carried out a comparative study on the permeability of the root mats of *Papyrus* and *Miscanthidium*; and the performance of *Papyrus* was recently examined in a constructed wetland. The present research, focusing on major wetlands in Uganda and Kenya, built upon this information base while focusing on the impacts of disturbance and management on wetland functions. Efforts were made to better define the nutrient and carbon cycles and examine the indirect role of wetlands on lake productivity. By modelling these functions and their impacts on the water quality of the lake, it was possible to examine how reclamation or encroachment in these systems can aggravate the already compromised capacity of the Lake to supply fundamental resources to the regional population.

The results presented here indicate that multiple use of these wetlands is possible. As a support for subsistence agriculture, it appears possible that a limited area of the wetland can be dedicated to agricultural products without severely reducing the functional and structural value of the ecosystem. Harvesting the wetland vegetation on a sustainable basis will both provide an income source for wetland populations and improve the nutrient retention and carbon sequestration of the wetland. Productivity studies show that papyrus wetlands are some of the most productive systems on the planet and, if managed correctly, will produce an important quantity of exportable biomass (carbon and nutrients). It is therefore possible to manage these ecosystems on a multi-use basis, providing that the structural and functional aspects of these systems are not compromised.

6.7 Conclusion

Lake Victoria is an ecological and economic resource of primary value. The wetlands that cover much of the lake shoreline provide a fundamental service in maintaining water quality, fisheries productivity and local meteorology. The management and maintenance of these papyrus wetlands are basic for any hope of controlling or reversing the eutrophication of the lake waters, in particular in the vital inshore area where fisheries are most important. In these areas, the presence of coastal wetlands strongly influences the optical and trophic characteristics of the inshore waters. Local water and energy exchanges are also strongly influenced by wetland conditions. The role of latent heat exchange in maintaining the micro-meteorological of the local watershed is influenced by the wetland vegetation and extension.

In conclusion, we have shown that uncontrolled reclamation or disturbance to these wetland ecosystems can have a significantly negative impact on the quality of Lake Victoria and its resources. In the future, when pressure on the lake ecosystem from development and population growth will no doubt increase, the littoral wetlands will become even more important, as will the understanding of their limits and functions. It is clear that any management of the lake resources will have to consider the littoral wetland and watershed activities in an integrated manner, considering the ecological and economic links that tie these systems together.

References

Aubinet M, Grelle A, Ibrom A, Rannil U, Moncrieff J, Foken T, Kowalski A, Martin P, Berbigier P, Bernhofer C, Clement R, Elbers J, Granier A, Grunwald T, Morgenstern K, Pilegaard K, Rebmann K, Snijders W, Valentini R, Vesala T (2000) Estimates of the annual net carbon and water exchange of forests: the EUROFLUX methodology. Adv Ecol Res 30:113–175

Azza NGT, Kansiime F, Nalubega M, Denny P (2000) Differential permeability of papyrus and *Miscanthidium* root mats in Nakivubo swamp, Uganda. Aquat Bot 67:169–178

Baldocchi D, Valentini R, Running S, Oechels W, Dahlman R (1996) Strategies for measuring and modelling carbon dioxide and water vapour fluxes over terrestrial ecosystems. Global Change Biol 2:159–168

Balirwa JS, Chapman CA, Chapman LJ, Cowx I G, Geheb K, Kaufman L, Lowe-McConnell RH, Seehausen O, Wanink JH, Welcomme RL, Witte F (2003) Biodiversity and fishery sustainability in the lake victoria basin: an unexpected marriage. BioScience 53:703–715

Christensen V, Pauly D (1992) Ecopath II – a software for balancing steady-state ecosystem models and calculating network characteristics. Ecol Model 61:169–185

COWI (2002) Integrated water quality/limnology study for Lake Victoria environmental management project, part II technical report. COWI, Nairobi

Delincé G (1992) The ecology of the fish pond ecosystem with special reference to Africa. Kluwer, Dordrecht

Denny P (1993) Wetlands of Africa. In: Whigham DF, Dykyjova D, Hejny S (eds) Wetlands of the world 1: inventory, ecology and management. Kluwer, Dordrecht, pp 1–128

Goldschmidt T, Witte F, Wanink J (1993) Cascading effects of the introduced Nile perch on detritivorous/phytoplanktivorous species in the sublittoral areas of Lake Victoria. Conserv Biol 7:686

Gordon HS (1954) The economic theory of a common property resource: the fishery. J Political Econ 62:124–142

Greenwood PH (1965) The ciclid fishes of Lake Nabugabo, Uganda. Bull Br Mus Nat Hist 12:313–357

Guildford SJ, Hecky RE (2000) Total nitrogen, total phosphorus and nutrient limitation in lakes and oceans: is there a common relationship? Limnol Oceanogr 45:1213–1223

Hecky RE (1993) The eutrophication of Lake Victoria. Verh Int Ver Limnol 25:39–48

Jamu DM (1999) Modeling organic matter and nitrogen dynamics in integrated aquaculture/agriculture systems: effects of cycling pathways on nitrogen retention and system productivity (Dissertation). Abstr Int B Sci Eng 60:420

Jiménez-Montealegre RA (2001) Nitrogen transformation and fluxes in fishponds. PhD thesis, Wageningen University, Wageningen

Jones JR, Bachmann RW (1976) Prediction of phosphorus and chlorophyll levels in lakes. J Water Pollut Control Fed 48:2176–2182

Jones M, Muthuri F (1997) Standing biomass and carbon distribution in a papyrus (*Cyperus papyrus* L) swamp on Lake Naivasha, Kenya. J Trop Ecol 13:347–356

Jørgensen SE, Bendoricchio G (2001) Fundamentals of ecological modeling. Elsevier, Amsterdam

Kadlec RH, Knight RL (1996) Treatment wetlands. CRC/Lewis, Boca Raton

Kansiime F, Bruggen JJA van (2001) Distribution and retention of faecal coliforms in the Nakivubo wetland in Kampala, Uganda. Water Sci Technol 44:199–206

Kansiime F, Nalubega M (1999) Wastewater treatment by a natural wetland: the Nakivubo swamp, Uganda (PhD thesis). Balkema, Rotterdam

Kansiime F, Nalubega M, Bugenyi FWB, Tukahirwa EM (1994) The role of Nakivubo swamp in maintaining the water quality of Inner Murchison Bay – Lake Victoria. J Trop Hydrobiol Fish 5:79–87

Kansiime F, Nalubega M, Bruggen JJA van, Denny P (2003) The effect of wastewater discharge on biomass production and nutrient content of *Cyperus papyrus* and *Miscanthidium violaceum* in the Nakivubo wetland, Kampala Uganda. Water Sci Technol 38:233–240

Kaufman L (1992) Catastrophic change in species rich freshwater ecosystems, the lessons of Lake Victoria. Bioscience 42:846–858

Kipkemboi J; Kansiime F, Denny P (2002) The response of *Cyperus papyrus* (L) and *Miscanthidium violaceum* (K Schum) Robyns to eutrophication in natural wetlands of Lake Victoria, Uganda. Afr J Aquat Sci 27:11–20

Kizito YS (1986) The evaluation of pollution levels in Nakivubo Channel, Kampala Uganda. MSc thesis, Makerere University, Makerere

Lung'ayia H, Sitoki L, Kenyanya M (2001) The nutrient enrichment of Lake Victoria (Kenyan waters). Hydrobiologia 458:75–82

Moncrieff J, Massheder J, Bruin H de, Elbers J, Friborg T, Heusinkveld B, Kabat P, Scott S, Soegaard H, Verhoef A (1997) A system to measure surface fluxes of momentum, sensible heat, water vapour and carbon dioxide. J Hydrol 1997:188–189, 589–611

Mugidde R (1993) The increase in phytoplankton primary productivity and biomass in Lake Victoria. Verh Int Ver Limnol 25:846–849

Ochumba PBO (1995) Limnological changes in Lake Victoria since the Nile perch introduction. In: Pitcher TJ, Hart PB (eds) The impact of species change in African lakes. Chapman & Hall, London, pp 33–44

Ogutu-Ohwayo R (1990) The decline of the native fishes of Lake Victoria and Kyoga (East Africa) and the impact of introduced species, especially the Nile perch, *Lates niloticus*, and the Nile tilapia, *Oreochromis niloticus*. Environ Biol Fish 27:81–96

Ogutu-Ohwayo R, Hecky RE, Cohen AS, Kaufman L (1997) Human impacts on the African great lakes. Environ Biol Fish 50:117–131

Okeyo-Owuor JB (1999) A review of biodiversity and socio-economics research in relation to fisheries in Lake Victoria. Socio-economics of the Nile perch fishery of Lake Victoria. (Project Report 5) IUCN-EARO, Nairobi

Othina A, Tweddle D (1999) The status of the artisanal fishery of Lake Victoria, Kenya, with notes on improvements to the catch data collection system. In: Cowx IG, Tweddle D (eds) Report on fourth FIDAWOG workshop. (Lake Victoria Fisheries research project phase II, LVFRP/TECH/99/07) FIDAWOG, Nairobi, pp 78–91

Schaefer MB (1954) Some aspects of the dynamics of populations important to the management of commercial marine fisheries. Bull Inter-Am Trop Tuna Comm 1:27–56

Schaefer MB (1957) Some consideration of population dynamics and economics in relation to the management of marine fisheries. J Fish Res Board Can 14:669–681

Simonit S, Perrings C (2004) Trophic state variables and prices in bioeconomic fishery models: indirect economic indicators of freshwater fish stocks. (Centre for Environment and Development Economics working paper, CEDE/04-01) University of York. http://wwwyorkacuk/res/cede/resources/CEDE_04_01pdf

Villafañe VE, Andrade M, Lairana AV, Zaratti F, Helbling EW (1999) Inhibition of phytoplankton photosynthesis by solar ultraviolet radiation: studies in Lake Titicaca, Bolivia. Freshwater Biol 42:215–224

Villanueva MC, Moreau J (2002) Recent trends in the Lake Victoria fisheries assessed by ECOPATH. In: IG Cowx (ed) Management and ecology of lake and reservoir fisheries. Fishing News, Oxford, pp 96–111

Walters CJ, Christensen V, Pauly D (1997) Structuring dynamic models of exploited ecosystems from trophic mass-balance assessments. Rev Fish Biol Fish 7:139–172

Wilson K, Goldstein A, Falge E, Aubinet M, Baldocchi D, Berbigier P, Bernhofer C, Ceulemans R, Dolman H, Field C, Grelle A, Ibrom A, Law B, Kowalski A, Meyers T, Moncrieff J, Monson R, Oechel W, Tenhunen J, Valentini R, Verma S (2002) Energy balance closure at FLUXNET sites. Agric For Meteorol 113:223–243

7 Predicting the Water Requirements of River Fisheries

Robin L. Welcomme, Christophe Bene, Cate A. Brown, Angela Arthington, Patrick Dugan, Jackie M. King, Vasu Sugunan

7.1 Introduction

The past decade has seen a steady increase in international attention given to the crisis facing the world's water resources and the urgent need for more efficient use of this increasingly scarce resource (Baron et al. 2002). In the developing world in particular, such improvements in water use efficiency have been widely regarded as essential for improved food security and poverty alleviation. However, to date most of the attention being given to improved water productivity has been restricted to irrigated plant crops, and less to the importance of natural freshwater ecosystems and fisheries in sustaining food production and generating employment and income. In addition to supporting food crop security, increasing demands are being placed on water for a wide range of other human uses, including industry, domestic supply power generation, navigation, flood control, and land reclamation for urban development and agriculture, which may not consume water directly but influence the patterns of flow and, through this, the form and ecological function of river systems.

Landscape changes such as deforestation or poor agricultural practice can change river hydrographs by influencing the speed of runoff and the capacity of infiltration and storage capacity of soils. Such changes do not substantially alter the quantity of water flowing through a system during the year or the timing of flood peaks, but are likely to affect the magnitude and duration of peak flows (i.e., shorter, larger flood peaks). Large dams that store water for power generation, and abstractions that remove large quantities of water from rivers, exert major control over flow, affecting the total amount of water in the system and the timing with which the remainder is discharged (World

Ecological Studies, Vol. 190
J.T.A. Verhoeven, B. Beltman, R. Bobbink, and D.F. Whigham (Eds.)
Wetlands and Natural Resource Management

Commission on Dams 2000). Such changes in discharge patterns affect the species composition, abundance, and viability of river biota and may damage the sustainability of river fisheries (Poff et al. 1997; Bunn and Arthington 2002). These fisheries are of critical importance for rural development and food security in many developing countries. Widely dispersed and easily accessible to marginal and/or isolated communities, they provide alternative sources of income and food when the others are lacking.

Efforts are now being expended globally to provide for environmental flows (EFs) that satisfy the needs for sustainability of fish and other aquatic organisms in rivers. These (EFs) may be defined as *that portion of the original flow of a river that is needed to maintain specific, valued features of its ecosystem* or *the quantity of water that must be maintained in a river system at all times to protect the species of interest for fisheries or for conservation of the environments on which they depend.* The process of evaluating these needs concentrates on: first, calculating the amount of water that can be withdrawn from a river as part of the process of granting licenses for water abstraction; second, assessing the impact of new dams and river control structures on future flow quantities and patterns; and third, calculating discharges from reservoirs for the express purposes of conserving the aquatic environment downstream. This paper is arranged in two parts. The first summarizes technical issues surrounding such assessments, briefly reviews the water requirements of river fish and fisheries, and examines some of the ways for predicting these requirements. The second part examines practical experiences in southern Africa by outlining ten general principles that can guide the environmental flow assessment process.

7.2 The Hydrological Regime and Fisheries in Rivers

Natural hydrographs depend on rainfall and the nature and area of the river basin for their seasonal variations in discharge. Variations in flow can be very rapid in small basins as they respond to local rainfall but, as river order increases, the flood curve becomes smoother as the varied flows of numerous tributaries in the larger catchment combine into the main channel flow. Most natural flow regimes consist of two components: (a) base flows, when water is confined to the main channel and remnant floodplain water bodies, and (b) the flood, when high flows overbank and flood the surrounding seasonal wetlands (floodplains), reconnecting isolated floodplain water bodies and the main channel. Human activities typically alter the natural hydrograph by reducing the absolute amount of water available to the river, by changing the timing of peak discharges, or by suppressing them partially or completely, so that the water is confined to the main river channel. Such control is not always complete, however, and occasional catastrophic floods do

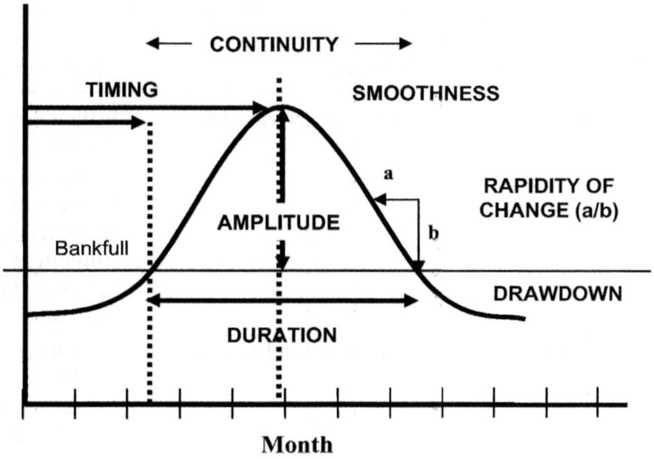

Fig. 7.1 Diagram of a typical unimodal hydrograph, identifying biologically active characteristics

still occur in exceptionally wet years when the control capacity of the system is exceeded.

Figure 7.1 is a diagrammatic representation of a hydrograph and shows various characteristics that are of significance for fish. Changes in these characteristics and their impact on fish in various ways, have been more fully described by Welcomme and Halls (2001, 2004) and Bunn and Arthington (2002) and are summarized in Table 7.1.

Flows can also be classified by the way they affect fish communities and the following are proposed as an illustration:

- *Population flows* influence biomass of the fish community through density-dependent interactions. They regulate the volume of water or flooded area in a system. They also change the relative areas of different types of habitat in the channel, as well as accessibility to floodplains, backwaters, and off-channel structures. The main hydrological features arising from population flows that influence fish populations are volume, depth, connectivity, and wetted or flooded area in a system.– *Critical flows* trigger events such as migration and reproduction. The main features of critical flows are flow velocity and timing coupled with season, temperature regimes, and sometimes day length and/or lunar phase.
- *Stress flows* (also known as channel-forming discharges) may be exceptional high- or low-water events that endanger fish, either because of excess velocity at high water or through desiccation at low water. These may be expressed as extreme flows occurring as isolated peaks in an irregular hydrograph, or as floods or droughts of very long return periods. Catastrophic events occur even in highly controlled rivers when rainfall exceeds the control capacity of the system. Such flows can have profound impacts on fish communities and on the riverine environment, frequently acting as "reset events".

Table 7.1 Effects of changes in characteristics of flow regimes on fauna and flora

Flow characteristic	Change	Effect on fauna and flora
Timing	Delay of arrival of peak flows	Influences physiological readiness of fish to mature, migrate and spawn; Synchronises maturation of drifting larvae and movement to floodplains and backwaters; Influences thermal coupling between flood and temperature of the water.
Continuity	Interruption to flood	Stranding of fish in temporary pools; Failure of eggs and larvae to colonize floodplain; Exposure of spawning substrates with stranding and desiccation of eggs.
Smoothness	Increased flashiness	Exposure of nests and spawning substrates with stranding and desiccation of eggs or larvae.
Rapidity of change	Overly rapid rise in level	Failure of floodplain vegetation; Submergence of nests and spawn in too great a depth.
	Overly rapid fall in level	Increased stranding mortality of fish in temporary floodplain waterbodies.
	Excessive flow in main channel	Can sweep drifting larvae past suitable habitat.
Amplitude	Decreased depth of flooding	Less space for reproduction, refuge and feeding of young and adult fish during flood; Fewer floodplain water bodies connected to main system.
	Decreased level in main channel during dry season	Smaller refuges for fish; Increased risk of anoxia; Greater mortality in main channel through competition and predation.
Duration	Reduced time when floodplain submerged	Less time for growth of fish; Less time for fish to remain in floodplain refuges.
	Increased time when dry phase in main channel	Greater exposure of fish to negative condition in main channel; Greater risk of desiccation of main channel, backwaters and floodplain waterbodies; Greater exposure to fishermen and predators.

- *Habitat flows* maintain the availability of habitat and environmental quality, including temperature, dissolved oxygen levels, sediment transport, and environmental support systems such as vegetation and food organisms. They also maintain the balance of ion and sediment concentrations in the river. They may operate directly on fish, as for instance in flows that maintain aeration in salmonid spawning redds, or indirectly via habitat, food competition, and availability. The key features are volume of water, flow velocity, connectivity, and wetted or flooded area in a system.

7.2.1 Fish Responses to River Flow

Two major motives drive the protection of flows in rivers: (a) preservation of the diversity of species in river faunas and floras, and (b) managing river biota, including fisheries for human benefit. The great diversity of river fish faunas, especially in large, tropical systems (see Oberdorff et al. 1995; Pusey et al. 1998), makes it almost impossible to consider the flow requirements of all species individually. Attempts to protect the fish fauna are often based on defining flows for a few species that are either prestigious because of their size, or are of major importance to fisheries or conservation.

An alternative approach groups fish into guilds according to their different flow requirements. Ecological guilds have been defined differently in various parts of the world. Regier et al. (1989) proposed an early classification based on the traditional South East Asian usage for tropical systems; and Bain et al. (1988) developed a classification of functional groupings for United States rivers. Aarts et al. (2004) summarize the guild classification for major European rivers and Welcomme et al. (2006) propose a global system for classifying guilds. Table 7.2 combines elements of these, together with some of Balon's (1975) reproductive guilds, to illustrate the way in which guilds respond to characteristic changes in river flows. The three main groups of fish and their sub-groups respond to changes to natural hydrographs in very different ways, which typically favour generalist, eurytopic species at the expense of the more specialist limnophilic and rheophilic ones.

7.2.2 What River?

Rivers undergo characteristic morphological changes in response to human activities (Table 7.3). These changes have been documented in many rivers, e.g., Fremling et al. (1989) for the Mississippi, Hesse et al. (1989) for the Missouri, Hohensinner et al. (2004) for the Danube, Brenner et al. (2004) for the Rhine, and Piskozub (1982) for the Vistula. Changes in form also involve changes in the type of fish that are present. Aarts et al. (2004) suggest that lateral interactions between rivers and their floodplains greatly diminish as the

Table 7.2 Response of the main behavioural guilds to changes in flow regimes in large rivers

Behavioural guild	Typical behaviour		Reaction to changes in hydrograph
	General	Specific	
Black fish – limnophilic species	Floodplain residents move little between floodplain pools, swamps, and inundated floodplain; Repeat breeders with specialized reproductive behaviour;	A: Tolerant of low dissolved oxygen tensions only	Tend to disappear when floodplain disconnected and desiccated through poldering and levee construction; May increase in number in shallow, isolated wetlands, rice-fields, and drainage ditches.
	Predominantly polyphils, nest builders, parental carers, or live bearers; Tolerant of low dissolved oxygen or anoxia (auxiliary breathing adaptations).	B: Tolerant of complete anoxia	Persist in residual floodplain water bodies; Principal component of rice field and ditch faunas.
White fish – rheophilic species	Long-distance migrants; One breeding season a year; Intolerant of low oxygen.	A: Main channel residents not entering floodplain; Predominantly psammophils, lithophils or pelagophils; Often have drifting eggs and larvae.	Tend to disappear when river dammed to prevent migration; When timing of flood inappropriate to their breeding seasonality; If flow excessive or too slow for the needs of drifting larvae.
		B: Use floodplain for breeding, nursery grounds, and feeding of juvenile and adult fish; Predominantly phytophils; Usually spawn at floodplain margin or on floodplain; sometimes have drifting eggs and larvae.	Tend to disappear when river dammed to prevent migration: Damaged when access to floodplain denied to developing fry and juveniles.

Grey fish – eurytopic species	Tolerant of low dissolved oxygen; Repeat breeders; Predominantly phytophils but some nesters or parental carers; Short-distance migrants often with local populations.	A: Occupy main channel, generally benthic	Able to adapt behaviourally to altered hydrograph; Generally increase in number as other species decline; Impacted negatively by flows that change depositional siltation processes and alter the nature of the bottom.
		B: Occupy riparian vegetation	Able to adapt behaviourally to altered hydrograph; Generally increase in number as other species decline; Impacted negatively by flows and management that changes riparian structure.
		C: Occupy larger and better oxygenated floodplain water bodies	Sensitive to isolation of floodplain water body but can colonize river if flow slowed sufficiently; Often form basic colonizers of reservoirs and dams.

Table 7.3 Characteristics of various developmental stages of a river, impacts on flood regimes, and form of large lowland rivers

Development stage	State of river channel	State of floodplain	Floodplain use	Fisheries
Unmodified	Hydrograph completely natural; Channel highly diverse, often freely meandering or anastomosing with islands; In-channel habitat structure diverse.	Natural flooding; Often forested at least as a riparian fringe; Plain interspersed with floodplain water bodies.	Wild-life hunting; Gathering seeds and fruits.	Fish assemblage intact; Capture fisheries on wild fish stocks.
Slightly modified	Hydrograph natural but may be influenced by changes in basin use; Channels diverse; Often freely meandering or anastomosing with islands; In-channel habitat structure diverse.	Natural flooding; Some forests, but many plains, savannah with floodplain grasses.	Seasonal cattle rearing; Draw-down agriculture; Floating rice culture.	Larger fish species may diminish in abundance; Capture fisheries on wild stocks; Some management of floodplain water bodies, brush park and drain-in fisheries.
Modified	Hydrograph modified, locally regulated with moderate local damming and leveeing but with some reaches still relatively unregulated; Some backwaters persist; Habitat structural diversity locally reduced.	Floodplain partially modified, deforested; Seasonal floods reduced, floodplain water bodies sometimes isolated; Local poldering and flood control structures.	Seasonal flood rice culture; Vegetable gardening; Dry season crops; Cattle rearing; Increasing human occupation.	Some rheophilic and limnophilic species locally threatened, larger fishes disappeared; Capture fisheries in main channel, and in rice fields; Some rice-fish culture, brush park fisheries; Control of water level in some floodplain water bodies; Floodplain species threatened.

Highly modified	Hydrograph considerable modified; Main stem and tributaries heavily dammed, sometimes in cascades; Fully regulated and channelized, often with revetted banks and dredged navigation channels; Backwaters eliminated; Main channel structural diversity low.	Floodplain dry or flooding completely controlled with extensive drainage and irrigation canals; Off channel water bodies largely eliminated or isolated; May be heavily poldered.	Intensive agriculture and human occupation; Irrigated rice culture and other dry-season crops.	Limnophilic and rheophilic species absent: assemblage dominated by eurytopic species and introduced species; Capture fisheries in river channel only: cage culture in river and irrigation channels; Small-scale pond farming and rice-fish farming on floodplain, stocked fisheries in surviving water bodies.

river becomes increasingly regulated and there is a shift from flood pulse-dominated natural rivers (Junk et al. 1989) to the longitudinal processes of the River Continuum (Vannote et al. 1980).

There are considerable differences in the degree of river modification between continents, based on fragmentation caused by dams alone (Revenga and Kura 2003; Nilsson et al. 2005). For example 23 % of systems are unfragmented globally, whereas only 11 % (five) rivers are undammed in Europe as against 45 % in North and Central America. When other forms of river modification, such as leveeing and channel-straightening, are taken into account the numbers of modified systems would be even greater. The process of modification is practically irreversible, for social and economic reasons. For example, control of flooding is accompanied by the occupation of the floodplain by housing and agriculture that, in its turn, creates a demand for further flood control structures and modification of the hydrological cycle.

7.2.3 Linkages Between Hydrological Regime and Fish Catch

Rivers, their floodplains, and riparian wetlands are intimately linked during the flood phases of the hydrological cycle (Junk et al. 1989; Junk and Wantzen 2003). The flooding phase can be a period of very high productivity. For example, the mean catch from Latin American floodplains is 28 kg ha^{-1} when measured at maximum extent of flooding. Similar figures from Africa are 60 kg ha^{-1} and from Asia around 100 kg ha^{-1} (Welcomme 2001). These figures demonstrate the potential impact of reductions in flooding due to flow regulation. Estimating the productivity of different types of floodplain habitat can possibly refine such estimates. For example, rice fields on floodplains are known to produce about 150 kg ha^{-1} (Gregory and Guttman 2002), so changes to the hydrographs that alter the ratio of rice paddy to other types of floodplain habitat will influence yields accordingly.

The close linkage between the floodplain and the river is shown by year-to-year variations in fish catch in the same river system as a response to changes in flood intensity. Such relationships have been recorded from numerous systems in both temperate and tropical systems (Table 7.4). Studies from the 1960s and 1970s, such as those of Krykhtin (1975), Welcomme (1979), Holcik and Kmet (1986) and Smolders et al. (2000) show that the catch in any year depended mostly on the strength of flooding two to five years previously, reflecting the time taken for the large fish forming the bulk of the catch to enter the fishery. More recent studies show the best correlations to be with the floods of the same year (Lae 1992; Halls 1998) or the previous year (Arthington et al. 2005), indicating that fish now enter the fishery in their first year of life in today's heavily exploited fisheries (Welcomme 1999; Albaret and Lae 2003).

Basin-specific relationships between flood intensity (or area of floodplain inundated) and catch are available for some systems (see Table 7.5) and can be

Table 7.4 Studies demonstrating correlations between flood intensity and fish catch

Climatic zone/ geographic region	River	Authors
Temperate		
Europe	Danube (Serbia)	Stankovic and Jankovic (1971)
	Danube (Slovakia)	Holcik and Kmet (1986)
Asia	Amur	Krykhtin (1975)
	Mahakam	Christensen (1993)
Tropical		
Asia	Mekong (Dai fishery)	Baran et al. (2001)
	Ganges/Brahmaputra delta, Bangladesh	De Graaf (2003)
Africa	Cross	Moses (1987)
	Niger	Welcomme (1979), Lae (1992)
	Kafue Flats, Kafue River	University of Michigan (1971)
	Shire	Welcomme (1985)
	Lake Kariba (river-driven reservoir)	Karenge and Kolding (1994), Kolding (1994)
	Lake Turkana (river-driven lake)	Kolding (1993)
South America	Orinoco	Novoa (1989)
	La Plata system	Quiros and Cutch (1989)
	Pilcomayo	Payne and Harvey (1989)
	Amazon	Lambert de Brito Ribeiro and Petrere (1990)
Australia	River-driven estuarine/ coastal zone	Loneragan and Bunn (1999)
Australia	Cooper Creek (arid zone river)	Arthington et al. (2005)

further refined as new data become available. Such relationships can be used as rapid predictors of the effects of changes in the flood season components of flow regimes. In some systems, correlations between fish production and flow have also been established with low-water (dry season) phases (University of Michigan 1971; Quiros and Cutch 1989; Arthington et al. 2005) and these types of relationship may be further developed to cover the whole hydrograph. Care has to be exercised, however, as in some cases fish populations of controlled rivers can be maintained by recruitment from tributaries that still have natural flood regimes. Flood–catch relationships such as these are based on rivers having intensive food fisheries that provide a cheap, long-

Table 7.5 Sample criteria for visual scoring and simulation models in upstream and lowland rivers

Criteria	Upstream (rhithronic)	Downstream (potamonic)
Common criteria	Flow velocity Flow timing Bank structure Emergent vegetation Riparian vegetation Longitudinal connectivity	Flow velocity (rheophilic species) Flow timing Bank structure Emergent vegetation Riparian vegetation Longitudinal connectivity
Unique criteria	Wetted area Pool – riffle area Substrate structure and area	Depth/area – volume Backwater development Lateral connectivity Floodplain development and flooding Refuge deeps

term supply of data. Similar approaches in rivers that are not fished, or are only fished by catch-and-return recreational fisheries, would have to rely on intensive sampling to obtain sufficient information.

The relationship between flood intensity and fish populations has been used to develop a model based on the density-dependent relationships that arise during the hydrological cycle. This was first developed by Welcomme and Hagborg (1977) and refined by Halls et al. (2001). The model predicts the population response to changes in the duration and depth of flooding of the floodplain and can be used as a valuable assessment tool (Fig. 7.2).

There appears, however, to be a fundamental difference between rivers that flood their riparian wetlands and those that remain confined within the river channel. Clearly, relationships between flooding and catch cannot be used as assessment tools if the floodplain is no longer inundated, so that alternative relationships between catch and processes within the river channel have to be derived. While the abundance of fish populations in rivers that flood appears to be primarily conditioned by the intensity of the high-water phase, in rivers that do not flood the abundance is probably more closely linked to minimum flows and low-flow spells.

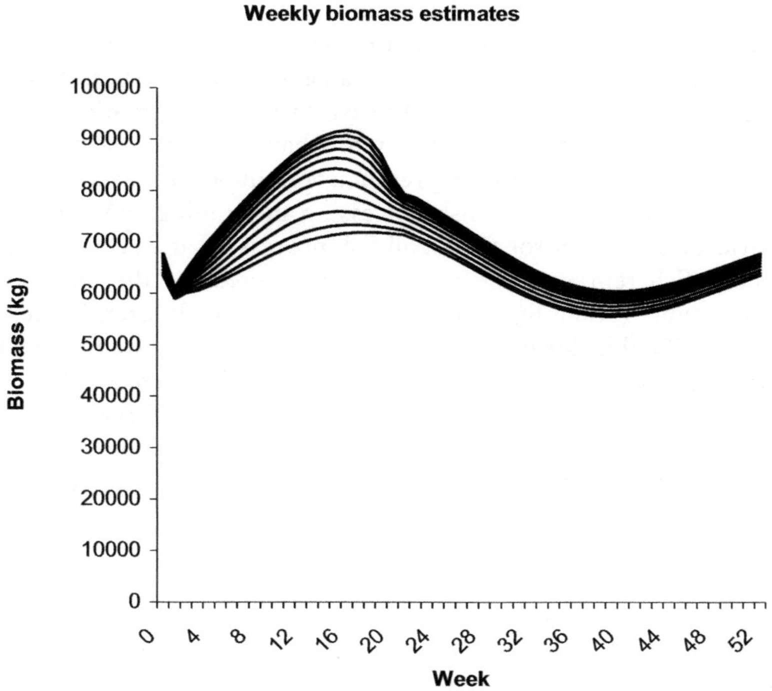

Fig. 7.2 Weekly variations in biomass related to flood height as sample outputs from the Halls and Welcomme model of fish responses to flow variations

7.3 The Social and Economic Role of River Fisheries

An important dimension in the assessment of river water requirements is the social and economic role of river fisheries in the livelihoods of riparian populations in developing countries. In the Lower Mekong basin, for instance, it is estimated than more than 60×10^6 people depend directly on this river and its tributaries as a source of food and income (Coates 2002).

River fisheries contribute to rural population livelihoods through food supply, income and employment. Even so, the exact contribution of river fisheries to the livelihood of the rural people is difficult to evaluate accurately (Béné 2006). Along these rivers, a high percentage of the population is engaged in full-time fishing. These are usually accounted in national statistics. It is thus recognised that, worldwide, more than 4.5×10^6 people depend directly on river fishing for their livelihood (FAO 2000). However, not included in these estimates are the many millions engaged in seasonal and occasional fishing activities. For communities living along African or Asian rivers, in particular those with major floodplains, fish are consistently the most important wild resource that is harvested, and in household economies, fish lie second overall only to agriculture. For example, in the Zambezi Basin, recent studies have shown that the value of inland fisheries (estimated through its share of the household's cash income) systematically generates more cash than cattle and sometimes more than crops (Turpie et al. 1999). For women in particular, fisheries represent a very important livelihood support through their involvement in fish processing and trading. In West and Southern Africa, women control almost all fish processing and retailing activities. For them, fish represent the primary, and sometimes the only, source of income to support their livelihoods and those of their children. In the Western Province of Zambia, for instance, 75 % of the women involved in fish trading activities are widows or divorced heads of households (Jere and Béné 2004).

River fisheries and associated activities (such as small-scale fish processing and trading) are important as a safety-net for the poorest rural households. Nevertheless, the importance of inland fisheries in poverty alleviation, rural, and national economies is still largely ignored by national decision-makers and planners (e.g. Ratner 2003). A recent Food and Agricultural Organization–United Kingdom Department for International Development review, based on 11 African countries, concluded that "*small-scale fisheries are rarely taken into account in PRSPs* [Poverty Reduction Strategy Paper] *formulation*" despite the importance of this activity for the rural poor (FAO/DFID 2003). This marginalization is especially acute for river fisheries for two main reasons: (a) the failure to properly account for the contributions that freshwater fisheries play in providing food and income to resource-poor households (Béné and Neiland 2003) due to the lack of information on their comprehensive socio-economic values (Coates 2002; Cowx et al. 2003; Neiland et al. 2004); and (b) the large majority of the rural communities living along rivers are remote from major urban centres. This physical isolation is further exacerbated by institutional and political marginalization. In effect, the majority of these rural communities are totally excluded from any involvement in policy-making processes or planning.

Competition for water is one of the most critical challenges that river communities in developing countries, in particular West Asia and sub-Sahara Africa, will face in the next two decades (Dugan 2006). To resolve the challenge will require the development of systems of governance that will allow the water needs of all users, including fishery dependent communities, hydropower, irrigated agriculture, and the urban sector, to be taken into account. Decentralization is one way in which governance can be improved. To achieve this, better methods for valuing the real contribution of river fisheries to rural economies need to be developed (Béné and Neiland 2003).

7.4 Methods for Estimation of Environmental Flow Requirements

Any requirement that flows be maintained in rivers for the conservation of fish and fisheries means that estimates have to be made as to how much water is needed in the river, its seasonal pattern, and the shape of the resulting flow curve, according to the needs of the fish species present. Environmental flow (EF) assessment methods were originally developed in the 1970s in the United States to predict the amount of water required to provide habitat for angling fish (Pusey 1998; Tharme 2003). Now more than 50 countries employ over 200 different approaches to assessing EFs in terms of the aspects of the ecosystem they consider, the amount of data they require, and the type and reliability of the answers they provide (Tharme 2003).

In general, procedures for the assessment of the water requirements of fish in rivers should: (a) take into account the complexity of ecological requirements of all life stages of fish in rivers; (b) be easy to understand and use; be cost effective; (c) be compatible with expertise available; (d) be legally robust; and (e) be generally accepted by all levels of fisheries and water user stakeholders.

Many procedures have been developed to attempt to establish the water requirements of fish and fisheries at site or reach level (Pusey 1998; Parsons et al. 2002; Dyson et al. 2003; Tharme 2003; Arthington et al. 2004). Environmental flow assessments for fish have been carried out at several levels of precision. Earlier methods used simple hydraulic rating methods, aimed mainly at establishing minimum flows that could be incorporated into legislation as statutory requirements, or used to grant licenses for abstraction. The deficiencies of these hydraulic methods as predictors of the responses of fish to flow soon became apparent and more detailed and effective methods were developed, incorporating improved data on the response of the aquatic environment to flow changes and on the habitat requirements of fish. These resulted in predictive models, such as PHABSIM (Bovee 1982), that model the performance of river habitats under a range of hydrological conditions, combined with a database of species-specific habitat suitability criteria. However, as habitat simulation models are expensive to implement, less costly, visual habitat scoring methods such as HABSCORE (Pusey 1998; Barbour et al. 1999) are often preferred.

Few if any of these habitat methods fulfil all the criteria listed above and most suffer problems of verification, which makes them vulnerable to legal challenge when allocating water rights and licenses to abstract. Increasing experience may make it possible to verify some of the methods (e.g., for comments on PHABSIM, see Pusey 1998; Dunbar 2003) and thus increase their legal robustness. More serious is the fact that most EF assessment methods were developed for use on small, headwater streams, often specifically for salmonids. Less attention was paid to larger lowland reaches and there is an urgent need for methods that will assess the flow needs of large, floodplain rivers. There, the large number of species present and the lack of knowledge of habitat suitability curves make any assessment of flows for individual species difficult. Established methods, such as wetted perimeter, habitat simulation, or visual habitat scoring methods, may be adapted for aggregated fish communities by incorporating environmental criteria specific to the slow-flowing lower reaches of a river (potamon; Table 7.5) but, in general, other approaches may have to be developed based on the linkages between hydrological regime, fish habitat availability, recruitment, and fish catch.

Large lowland rivers are complex and require more cost effective approaches than the labour-intensive habitat simulation and predictive models. Furthermore, the weaknesses of individual species assessment methods mean that several methods may have to be used together. This implies the need for framework approaches that are capable of synthesizing the best information available, be it drawn from precise site- and reach-based studies or from more general models and professional advice. Frameworks may be objective-oriented, such as the CAMS (Catchment Assessment Management Strategies) procedure in use in the United Kingdom (Environment Agency, UK) or scenario-based, such as the DRIFT (Downstream Response to Imposed Flow Transformations) procedure developed in South Africa (Arthington et al. 2003; King et al. 2003), and the Benchmarking Methodology used in Australia (Brizga et al. 2002) The latter type of method allows for evaluation of many alternative flow management options and estimation of their ecological, social, and economic consequences.

7.5 Guidelines for the Selection and/or Development of Tools for Determining Environmental Flows for Rivers and Wetlands

7.5.1 Legislation, Policy, and Practice Supporting Environmental Flows Should Focus on People

Several examples from Africa illustrate this principle. This is true where there is, as in South Africa, a well developed environmental flow legislation that allocates flows to ecosystems to safeguard their present and future use. In Tanzania, water resource managers drafting flow legislation suggested that the term "Environmental Flows" be changed to "Flows for People and the Environment" (Acreman et al. 2000). In Lesotho, environmental flow studies focussed on determining the costs of compensation and mitigation of the impacts felt by downstream rural people as a result of the flow and ecological changes caused by the Lesotho Highlands Water Project (Arthington et al. 2003; King et al. 2003; Metsi 2002). In Mozambique, there are lobbies on the Zambezi Delta for the introduction of similar legislation (Bento and Bielfuss, personal communication), see Box 7.1.

7.5.2 There is a Need to Understand the Ecosystem First, Before the Impacts on People can be Predicted

River-associated livelihoods are directly linked to the hydrology, structure, and functioning of the ecosystem. In some cases the links are obvious, such as a fish or plant that is eaten, water that is drunk, or a deep pool that is used for ceremonies (King et al. 2003). Riparian agriculture makes use of fertile soils and water supplied by the river. Some small industries, such as brick making,

Box 7.1 The Zambezi Delta

The Zambezi Delta (and Lower Zambezi Valley as a whole) is vital for the local, regional, and national economy of Mozambique. Here, livelihoods linked to the river include fisheries, floodplain agriculture, livestock grazing, and other land use practices. Lucrative commercial activities based in the delta include the prawn industry, sugar plantations, and wildlife hunting safaris, each of which contributes significantly to the gross domestic product (GDP) of Mozambique. The quality and quantity of the resources of the Zambezi Delta are, however, vulnerable to the cumulative impact of hydrological development throughout the Zambezi Basin. Upstream of the delta, Lake Kariba regulates about 40 % of the Zambezi catchment runoff and traps sediments, and the Cahora Bassa Dam regulates about 80 %. Flow changes as a result of these two hydropower schemes reduce the extent and duration of downstream flooding, and alter their seasonal distribution.

Since 1972, there have been widespread reports of ecological degradation in the delta due to these hydrological changes negatively affecting the viability of subsistence floodplain fisheries (Bielfuss et al. 2002), the carrying capacity for wildlife, and the productivity of economic prawn species (Bento and Beilfuss personal communication.).

Thus, activities in the delta depend to a large extent on the international and regional management of Zambezi waters, as well as on local management in the Zambezi Delta. Any environmental flow assessment methodology must allow for the inclusion of social data and for the transfer of information on the impact of flow changes on people.

reed weaving, and sand mining, are also dependent on the river for supplies of raw materials. Other links are less obvious. In the case of the Zambezi Delta, sediments supplied by the river supplement the prawn banks in the Indian Ocean (Beilfuss et al. 2002). Studies elsewhere have linked trace elements supplied by riparian vegetables to the health of riparian people (Metsi 2000).

In some cases, human pressure can alter the sensitivity of a resource to flow change. For instance, when employment opportunities are low in the formal sector, the number of fishers on the Zambian side of Lake Kariba rises markedly, with a consequent increase in pressure on the in-shore fishery. Long-term data on species composition and catch volumes indicate that the greater the pressure on aquatic resources, the less the carry-over of biomass from year to year, and thus the more the productivity of the fishery is dependent on breeding success in the preceding year (Kolding, personal communication). From this, two things emerge. The first is that when jobs are few, reliance on the natural fish resources increases. The second is that future changes to the hydrology (either natural or anthropogenic) are likely to directly affect the availability of these resources as pressure on the fishery reduces the buffer that would have been provided by longer-lived individuals, species, and natural recruitment processes.

7.5.3 There is No Such Thing as a Single Flow with a Single Flow Condition

The concept that there is a minimum amount of water that will 'maintain' a river ecosystem is false. The three driving variables determining the character of a river are climate, geology, and topography. These factors dictate the flow regime, the general geomorphological character of the river, the shape and size of the river channel, the size of the bed particles, and the basic water chemistry and temperature. These in turn determine habitat and resource availability and the fauna and flora that inhabit the river. Flow regulation affects this complex balance and, in general, the more the flow regime is altered, the greater the impact on the downstream ecosystems.

7.5.4 Tradeoffs are an Integral Part of Decision-Making and Scenario Generation is Vital

It is essential that EF methods are able to supply information on a range of possible outcomes, each of which is achieved with a different volume and temporal distribution of water, so that these can be offset against the anticipated benefits of using the water elsewhere.

Figure 7.3 illustrates the relationship between the anticipated yield of goods from Phase 1 of the Lesotho Highlands Water Project (LHWP) and the

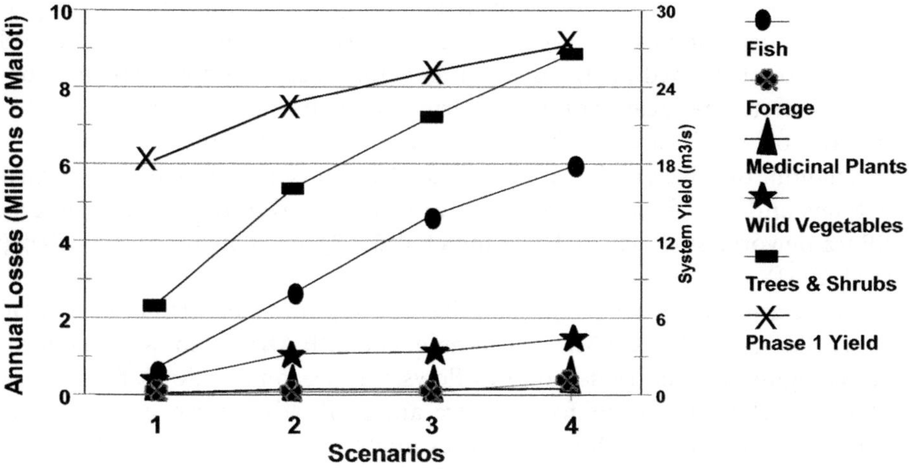

Fig. 7.3 The relationship between system yield from Phase 1 of the Lesotho Highlands Water Project (*Phase 1 Yield*) on the right-hand Y-axis and various ecosystem benefits expressed in monetary terms on the left-hand Y-axis for four EF scenarios (Metsi 2000)

cost[1] of compensation for the loss of various ecosystem benefits for four EF scenarios. Increased yield, and thus increased off-stream benefits resulting from water use, are off-set by decreased availability of fish, forage, medicinal plants, wild vegetables, trees, and shrubs, all of which are collected and used by the people living alongside the rivers in Lesotho downstream of Phase 1 of the LHWP (Metsi 2000). Initially, agreements between Lesotho and South Africa were based on Scenario 4 but, when the cost of compensating downstream communities for lost resources was also considered, Scenario 3 was preferred (Fig. 7.3).

Similarly, initial estimates of the economic benefits, excluding 'conservation benefits', of reinstating some of the flows to the Zambezi Delta in a country emerging from the ravages of civil war suggest that these would outweigh the lost hydropower revenues (Bento and Beilfuss, personal communication).

7.5.5 The River Ecosystem and Its Flow Regime Must be Compartmentalized to Provide the Required Scenario Information

Flow data can be analysed from an ecological perspective to better clarify the effects of water management on ecosystems. The different magnitudes of flow that form part of the natural flow regime affect riverine ecosystems in

[1] Excluding the cost of administering such a compensation system.

a variety of ways (see Fig. 7.1). For instance, the higher flows dictate the general geomorphological character, shape, and size of a river channel, as well as the size of the bed particles which generally decrease downstream. Upon this template, the local hydraulic, thermal, and chemical conditions create a mosaic of small areas of roughly uniform environmental conditions throughout the river, which are colonized by plants and animals. All of these habitats and their biotic components are considered in the more structured EF frameworks such as DRIFT (King et al. 2003) and Benchmarking (Brizga et al. 2002).

There are many ways to characterize flow regimes (see Richter et al. 1996, 1997; Olden and Poff 2003). For example, in South Africa, ten flow categories are recognised, viz.: dry-season low flows, wet-season low flows, four classes of inter-annual floods, and four of intra-annual floods (King et al. 2003). Elsewhere, a different set of flow categories might be relevant – for instance, the Lower Zambezi might not display much in the way of intra-annual floods but rather a gradual rise and fall of flow over the whole wet season. Ecologists who know a river well should be able to describe the important flow categories maintaining that river.

Although the evidence is sparse as yet, possible links between specific flow categories and ecosystem characteristics are emerging. For example, in many rivers in southern Africa, small pulses of high flow in the dry and wet season maintain the wetbank vegetation along the river margins (Fig. 7.4) and have been shown to trigger fish spawning (Cambray et al. 1997; King et al. 1998). Larger intra-annual floods sort sediments and maintain physical heterogeneity (Fig. 7.4). Conversely, the larger inter-annual floods may at times reduce the heterogeneity of the riverbed by dumping fine sediments within the channel, but may also increase heterogeneity by clearing vegetation and creating conditions for new growth.

Different parts of the flow regime have different consequences for the riverine ecosystem (King et al. 2003). It is possible to identify and isolate these different parts within a long-term hydrological data set of daily flows and to describe in isolation the probable biophysical consequences of partial or whole removal of any one of these parts. The parts of the flow regime can be re-combined in various ways to describe the river condition under any flow regime of interest (the biophysical part of the scenario). The social impacts of each river condition can then be described (the socio-economic part of the scenario).

In addition to the individual biophysical consequences of isolated parts of the flow regime, ecologists who know a river well can also identify how modification of several flow components in combination will alter the river ecosystem. For example, supplementation of the normal low flows in a river by release of dry-season irrigation flows will often enhance the growth of beds of aquatic plants, whereas storage of large floods in a reservoir will reduce wet-season scouring. These changes in combination increase the availability and

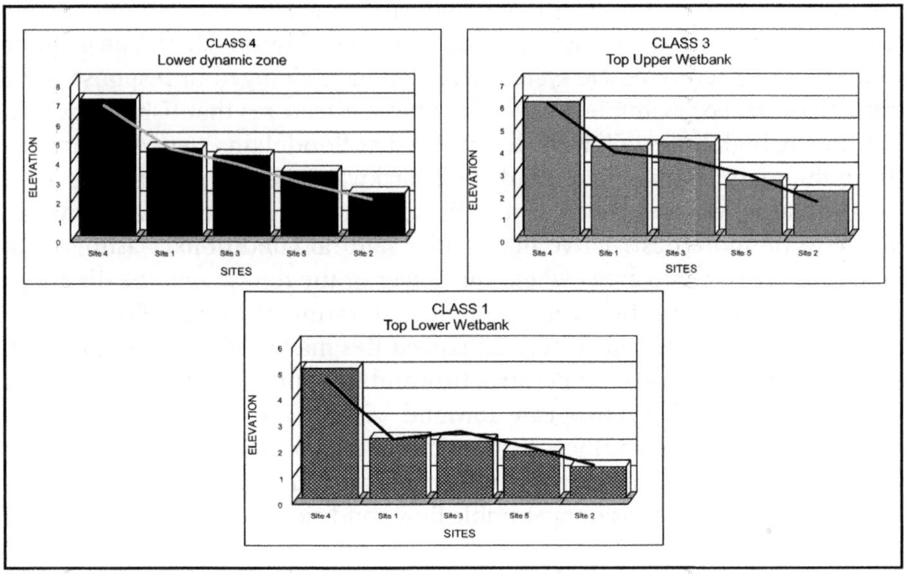

Fig. 7.4 Preliminary data supporting a relationship between different flood classes and the existence of riparian vegetation zones on the Breede River, Western Cape, South Africa

stability of the physical habitat and exacerbate the growth of plant beds (Mackay et al. 2003). The Benchmarking Methodology developed in Australia is a particularly effective way to assess such interactions.

7.5.6 Present-Day Conditions Offer the Best Starting Point

Most environmental flow requirements are established with respect to a baseline condition. Conservation and fisheries policies usually require the baseline to be as unmodified a river as possible, but social and economic considerations usually prefer it to be the current state of the river. Selection of the developmental objectives for any particular river is usually a compromise between the two, on the basis of the perceptions and requirements of society. However, in continents such as Europe, where there has been a long history of river regulation and flow modification, the unmodified (pristine) conditions of a river are virtually unknown and can be reconstructed solely on the basis of an incomplete historical record and experiences with similar rivers elsewhere in the world. This emphasises the importance of gathering as much data as possible now as to the nature and functioning of those unmodified and partly modified systems that still exist (Arthington et al. 2004).

The water requirements of fish in modified systems are likely to differ considerably from those of fish living in the pristine river. The fish community may have changed from its initial composition, usually to one dominated by eurytopic and introduced species (Bunn and Arthington 2002). The geomorphological structure of the system will also have changed in response to the flow changes. For example, many of the habitat features that fish use as refugia during high flows will have disappeared as floodplains were disconnected from the main channel and the diverse backwater and pool structure of the main channel was destroyed. Under these circumstances, excessive flows are likely to be more destructive than under natural conditions, washing adult and juvenile fish into areas where the power of the flow cannot be dissipated by overbanking. In other words, simple restoration of natural flow patterns alone cannot restore fish faunas. Enhanced EFs may need to be coupled with restoration of physical channel structure and floodplain connectivity. Experiences in the major European lowland rivers, for example, show that limnophilic and rheophilic components of the fish population were not restored despite improvements to water quality, and that they only responded when marginal wetlands or favourable flow conditions were restored (Aarts et al. 2004).

For these reasons, most river scientists take the nature and condition of the river at the time of their studies as the baseline. Predictions of flow-related change then describe how the river will change from present condition. In some instances, information on past characteristics can be used to inform environmental flow prescriptions and river restoration strategies.

Long-term changes in the climate will have major implications for both the use and the protection of water resources. For instance, much of southern Africa is predicted to become drier than the present, and so freshwater resources of the region are likely to be more stressed in the future. In many areas, relatively small increases in temperature will result in large additional water losses through evaporation (Chenje and Johnson 1996). Similarly, long-term climatological phenomena such as the El Nino (Smolders et al. 2000, 2002, 2004) and the Sahelain droughts (Lae 1992) can have recurring impacts on the general availability of water for human populations and for river fisheries. Predicting future water availability from past hydrology will become less certain in the face of a declining water resource and changes in the variability and extremes of river flows.

This does not mean that information on the natural condition of a system, and the changes that have taken place over time, are unimportant in EF assessments. They are vitally important in order to assess the response of the ecosystem to various impacts, to set objectives and strategies for river restoration, and to develop the capacity to predict how future changes are likely to affect the river ecosystem.

7.5.7 Methods Should be Usable in Both Data-Rich and Data-Poor Situations

River ecosystems are complex and research on them is under-funded. In Africa, at best, state funding has supported the collection and analysis of hydrological data, but long-term ecosystem monitoring has rarely occurred. In the Zambezi Basin, for instance, there are few national environmental data collection programmes, including hydrological data collection, that have not seriously faltered at some point over the past three or so decades.

In the face of such data scarcity, EF tools must make use of scientific and local expert knowledge. Indigenous knowledge systems need to be properly documented and incorporated in the EF methods, and community expertise accorded its proper place and value in the mix of required expertise (Brown and King 2002). Consideration and incorporation of local knowledge greatly enhances the community trust in the EF process.

The southern African Dialogue on Water and Climate (Dialogue on Water and Climate 2003) concluded that the region is vulnerable to climate and flow changes, and acknowledged that action is needed urgently. Here, local information is likely to prove vital, given the lack of scientific information available. In this way, EF methodologies should be able to deal with data poor situations. For example, the fish component of DRIFT sets out a template for the use of basic biological and ecological data on fish species and their flow dependencies, and a method for evaluating the consequences of alternative flow modifications, including the confidence that can be placed in assessments reliant on limited data and expert knowledge (Arthington et al. 2003).

7.5.8 Uncertainty is a Reality – Adaptive Management is Crucial

Difficult as it is to accurately predict hydrological, habitat, or chemical conditions in a changing river, it is immeasurably more difficult to predict the complex, inter-linked reactions of its hundreds or thousands of species, many of which are still unknown to science. Feedback loops are key attributes of EF methods intended for application in southern African, as these traits greatly enhance the ability to 'learn by doing' (Dialogue on Water and Climate 2003). Most of the structured holistic EF frameworks incorporate such feedback loops, or more explicit adaptive management procedures (see Arthington and Pusey 2003; Poff et al. 2003).

7.5.9 Implementation is Central to Promoting and Improving Environmental Flows

Success will depend especially on community trust in the management process and on complex ecosystem response to *actual* flow releases, none of which can happen if the agreed EFs are not delivered. All too often, the work done in determining EFs is undone by consequent day-to-day operation of the water management system. Flood flows not released during the wet seasons when they are required and requested can rarely be released later during the dry season without damage to the ecosystem.

Similarly, the feasibility of releasing different discharges from a proposed dam is a vital input to an EF assessment, since it is pointless describing floods within an EF requirement that simply cannot be released through a dam. Conversely, dam design should not be completed until the flood, and other requirements, of the EF regime are well understood, as it is possible that an EF regime may require outlets that are larger or more numerous than those required for routine dam operation. Release structures should not be designed based on construction costs alone, as the cheapest construction option may well be the most damaging one for the river and thus have the greatest long-term environmental costs. The challenge should be to design dams that can release water in ways that do least damage to rivers.

7.6 Discussion and Conclusion

Flow affects fish in rivers through several different mechanisms (Bunn and Arthington 2002). Flood discharges can sweep away juvenile and even adult fish. This problem is especially severe in species with semi-pelagic eggs, where critical flows are needed to convey eggs and larvae in the drift to the point where they can reach the juvenile nursery grounds on floodplains or in backwaters. If the flow is too fast, the larvae are accelerated past the appropriate area (Fuentes 1998) and, if too slow, they fall to the bottom. In either case they die. Flow is also required to aerate eggs laid in gravel beds and to maintain oxygen levels in the main channel during the dry season.

Increased flows are essential to the maturation of many stream fish species and act as triggers for spawning and migration (Lucas and Baras 2001). Critical factors here are not only the magnitude of the discharge but its timing relative to the season, temperature and, in some cases, day length or lunar phase. The ionic concentrations and sediment transport in the river can also affect breeding physiology (Smolders et al. 2002) and these abiotic processes are largely affected by discharge.

Changes in discharge impact on the form of the river system by changing erosion/deposition processes and, in many cases, either reduce the area inundated during seasonal floods, or suppress the floodplain completely. Changes in the form of the system, as a result of independent activities such as poldering, act reciprocally on flow patterns by isolating the floodplain and accelerating main channel flow. Changes in floodplain form of this type influence the composition of the fish fauna, its biomass, and productivity. There is much discussion as to the ways in which differences in the extent of flooding change species composition. Aarts et al. (2004) reviewed the fish populations in European large rivers and concluded that the presence or absence and the abundance of individual species is linked to the succession from main channel (eupotamon) to isolated water bodies of the former floodplain (paleopotamon) that is a feature of the stabilized and disconnected European floodplains. The way in which fish behave in the dynamic floodplains of the tropics and in arid-zone rivers is far more complicated and is still being actively discussed. Welcomme (1985) and, more recently, Arthington et al. (2005) and Pouilly and Rodríguez (2004) concluded that many factors influence species presence, absence, and relative abundance, including floodplain characteristics, depth, transparency, and the habitat complexity of floodplain lakes and lagoons, the nature of the bottom substrate, or even completely random processes (Chapman and Chapman 1993). Some authors, such as Pouilly and Rodriguez (2004) and Arthington et al. (2005), have found that the distance of separation and frequency of connection of channel and floodplain habitats does have some influence on fish communities. Novoa (1989), for example, found a strong correlation between *Prochilodus* catches and flood height in the Orinoco River, because the rare higher flood liberated fish that had been isolated in distant pools.

Not only is the velocity and volume of flow critical for fish and fisheries, but the timing of the flow events is equally important. Indeed, the form of the whole hydrograph must be considered when evaluating the impacts on fish of any changes that are made to it. The situation, however, is far from simple because of the intimate reciprocal linkages between flow and the geomorphology of the system. Evaluation of the impacts of flow must be carefully considered within the context of the developmental stage of the river and any independent interventions that are taking place. The situation is further complicated by the fact that fishing, through the fishing-down process (Welcomme 1999), produces effects on fish populations that may be easily confused with those resulting from changes in flow patterns and their consequences for recruitment.

Many of the simpler methods (hydraulic and hydraulic rating methods, according to Tharme 2003) for assessing the environmental flow requirements of fish do not adequately address these issues, in that they tend to focus on absolute amounts of water in the main channel of the system. Habitat simulation and the simpler holistic methods do not usually take these issues into account either but can be adapted to address some of the concerns about timing and velocity of flow at the site and reach level in river channels. These methods can also be extended to include floodplain criteria. However, the knowledge and trained personnel needed to allow for a fully scientific approach in formalized systems such as PHABSIM (Physical Habitat Simulation) is often lacking and there has to be considerable reliance on professional judgement. The more sophisticated ecosystem frameworks such as DRIFT and Benchmarking do attempt to include floodplain influences on fish (Arthington et al. 2003, 2004) and have the capacity to incorporate many types of quantitative data and predictive models. A valuable contribution here is the development and validation of population dynamics models which are capable of providing satisfactory solutions at the aggregated population level (Halls et al. 2001; Welcomme and Halls 2004).

The varied states of development of rivers, the range of social and economic demands that they are required to satisfy, and the rapidity of the accompanying developmental changes mean that careful choices have to be made in the allocation of the available water among user groups. Each of the sectors must carefully define its objectives and the overall allocation process should be arrived at by informed consideration of the costs and benefits of alternatives.

The complexity of the relationships between flow, habitat, fish, fisheries, the environment, and other user groups in large rivers is such that scenario-based framework methods, such as DRIFT and benchmarking, can best integrate knowledge gained from various sources and present alternatives for management to policy makers. Unfortunately, however, the valuation of floodplain river fisheries is largely lacking or at least unsophisticated, as is the understanding of the complex livelihood strategies of riparian peoples, including those who fish either full-time or part-time. This means that, whereas maintenance of fish diversity may be considered, fisheries interests are usually not represented adequately in decisions about water allocation.

As a general conclusion, the increasing amounts of water being abstracted for agriculture and the growing number of small and large dams all point to increasing control over water and the accompanying changes to the natural hydrographs of rivers. There is a growing consciousness of the need for understanding of EFs in all geographical regions, but in particular in the tropics where river fish are still a valuable food resource. Existing tools have been developed mainly for the needs of small temperate rivers and are generally inadequate for the assessment of impacts of flow changes in large, lowland, and floodplain rivers. Some methods are being developed and applied for the assessment of EFs in these larger systems (e.g., the Mekong) but agreement needs to be reached as to the harmonization and enhancement of approaches so that fisheries issues can be clearly represented within the general development and allocation of water in large river basins.

Acknowledgements We wish to thank two anonymous referees for their suggestions for improvements to the manuscript and our host institutions for support.

References

Aarts BG, Van Den Brink FWB, Nienhuis PH (2004) Habitat loss as the main cause of the slow recovery of fish faunas of regulated large rivers in Europe: the transversal floodplain gradient. River Res Appl 20:3–23

Acreman M, Farquharson FAK, McCartney MP, Sullivan C, Campbell K, Hodgson N, Morton J, Smith D, Birley M, Knout D, Lazenby J, Wingfield R, Barbier EB (2000) Managed flood releases from reservoirs: issues and guidance. (Report to DFID and the World Commission on Dams) Centre for Ecology and Hydrology, Wallingford

Albaret JJ, Lae R (2003) Impact of fishing on fish assemblages in tropical lagoons: the example of Ebrie Lagoon, West Africa. Aquat Living Resour 16:1–46

Arthington AH, Balcombe SR Wilson GA, Thoms MC, Marshall J (2005) Spatial and temporal variation in fish assemblage structure in isolated waterholes during the 2001 dry season of an arid-zone river, Cooper Creek, Australia. Mar Freshwater Res 56:1–11

Arthington A, Pusey BJ (2003) Flow restoration and protection in Australian rivers. River Res Appl 19:377–395

Arthington AH, Rall JL, Kennard MJ, Pusey BJ (2003) Environmental flow requirements of fish in Lesotho rivers using the DRIFT methodology. River Res Appl 19:641–666

Arthington AH, Tharme R, Brizga SO, Pusey BJ, Kennard MJ (2004) Environmental flow assessment with emphasis on holistic methodologies. In: Welcomme R, Petr T (eds) Proceedings of the second international symposium on the management of large rivers for fisheries, vol II. (RAP publication 2004/17) FAO Regional Office for Asia and the Pacific, Bangkok, pp 37–65

Bain MB, Finn JT, Booke HE (1988) Streamflow regulation and fish community structure. Ecology 69:382–392

Balon EK (1975) Reproductive guilds of fishes: a proposal and definition. J Fish Res Board Can 32:821–864

Baran E, Van Zalinge N, Bunn NP, Baird I, Coates D (2001) Fish resources and biological modelling approaches in the Mekong Basin. ICLARM, Penang and Mekong River Commission Secretariat, Phnom Penh, 60 pp

Barbour MT, Gerritsen J, Snyder BD, Stribling JB (1999) Rapid bioassessment protocols for use in streams and wadeable rivers: periphyton, benthic macroinvertebrates and fish, 2nd edn. (EPA 841-B-99-002) US Environmental Protection Agency, Washington, D.C.

Baron JS, Poff NL, Angermeier PL, Dahm CN, Gleick PH, Hairston NG Jr, Jackson RB, Johnston CA, Richter BG, Steinman AD (2002) Meeting ecological and societal needs for freshwater. Ecol Appl 12:1247–1260

Beilfuss R, Chilundo A, Isaacmen A, Mulwafu W (2002) The impact of hydrological changes on subsistence production systems and socio-cultural values in the lower Zambezi valley. (Working paper 6) Zambezi Wetlands Conservation Program, Zambezi

Béné C (2006) Contribution of inland fisheries to rural livelihoods and food security in Africa: an overview. In: Thieme ML, Abell RA, Olson DM (eds) Freshwater ecoregions of Africa: a conservation assessment. Island Press, Washington, D.C. (in press)

Béné C, Neiland A (2003) Valuing Africa's inland fisheries: overview of current methodologies with an emphasis on livelihood analysis. NAGA Q 26:18–21

Bovee KD (1982) A guide to stream habitat analysis using the instream flow incremental methodology. (Instream flow information paper 12, FWS/OBS-82/26) US Department of Fisheries and Wildlife Service, Fort Collins, 248 pp

Brenner T, Buijse AD, Lauff M, Luquet JF, Staub E (2004) The present status of the river Rhine with special emphasis on fisheries development. In: Welcomme R, Petr T (eds)

Proceedings of the second international symposium on the management of large rivers for fisheries, vol I. (RAP Publication 2004/16) FAO Regional Office for Asia and the Pacific, Bangkok, pp 121–148

Brizga SO, Arthington AH, Pusey BJ, Kennard MJ, Mackay SJ, Werren GL, Craigie NM, Choy SJ (2002) Benchmarking, a 'top-down' methodology for assessing environmental flows in Australian rivers. In: Proceedings of international conference on environmental flows for rivers (published on CD). University of Cape Town, Cape Town

Brown CA, King JM (2002) Environmental flows: requirements and assessment. In: Hirji R, Johnson P, Maro P, Chiuta TM (eds) Defining and mainstreaming environmental sustainability in water resources management in southern Africa. (SADC, IUCN, SARDC and IBRD technical report) IUCN, Gland

Bunn SE, Arthington AA (2002) Basic principles and ecological consequences of altered flow regimes for aquatic biodiversity. Environ Manage 30:492–507

Cambray JA, King JM, Bruwer C (1997) Spawning behaviour and early development of the Clanwilliam yellowfish (*Barbus capensis*; Cyprinidae), linked to experimental dam releases in the Olifants River, South Africa. Regul Rivers: Res Manage 13:579–602

Chapman LJ, Chapman CA (1993) Fish populations in tropical floodplain pools: a re-evaluation of Holden's data on the river Sokoto. Ecol Freshwater Fish 2:23–30

Chenje M, Johnson P (eds) (1996) Water in southern Africa. (SADC, IUCN and SARDC technical report) IUCN, Gland

Christensen MS (1993) The artisanal fishery of the Mahakam River floodplain in East Kalimantan, Indonesia. III Actual and estimated yields, their relationship to water levels and management options. J Appl Ichthyol 9:202–209

Coates D (2002) Inland capture fishery statistics of Southeast Asia: current status and information needs. FAO Regional Office for Asia and the Pacific, Bangkok

Cowx I, Almeida O, Béné C, Brummett R, Bush S, Darwall W, Pittock J, Brakel M van (2003) Value of river fisheries. In: Welcomme R, Petr T (eds) Proceedings of the second international symposium on the management of large rivers for fisheries vol I. (RAP publication 2004/16) FAO Regional Office for Asia and the Pacific, Bangkok, pp 1–20

De Graaf G, (2003) Dynamics of floodplain fisheries in Bangladesh, results of 8 years fisheries monitoring in the compartmentalization Pilot Project. Fish Manage Ecol 10:191–199

Dialogue on Water and Climate (2003) Climate changes the water rules: how water managers can cope with today's climate variability and tomorrow's climate change. Printfine, Liverpool

Dugan P (2006) Fish out of water: competing for water for Africa's freshwater ecosystems. In: Thieme ML, Abell RA, Olson DM (eds) Freshwater ecoregions of Africa: a conservation assessment. Island Press, Washington, D.C.

Dunbar MJ (2003) Testing PHABSIM predictions of physical habitat. In: Environmental flows for river systems; incorporating the fourth international ecohydraulics symposium. Cape Town

Dyson M, Bergkamp G, Scanlon J (eds) (2003) Flow. The essentials of environmental flows. IUCN, Gland, 118 pp

FAO (2000) The state of world fisheries and aquaculture. Food and Agriculture Organization, Rome

FAO/DFID (2003) Report of the consultation on integrating small-scale fisheries in poverty reduction planning in West Africa. (GCP/INT/735/UK) Food and Agriculture Organization, Rome

Fremling CR, Rasmussen JL, Sparks RE, Cobb SP, Bryan CF, Claflin TO (1989) Mississippi river fisheries: a case history. In: Dodge D (ed) Proceedings of the international large river symposium (LARS). Can Spec Pub Fish Aquat Sci 106:309–351

Fuentes CM (1998) Deriva de larvas de sábalo, *Prochilodus lineatus*, y otras especies de peces de interés comercial en el río Paraná Inferior. PhD thesis, Universidad de Buenos Aires, Buenos Aires 136 pp

Gregory R, Guttman H (2002) The ricefield catch and rural food security. In: Edwards P, Little DC, Demaine H (eds) Rural aquaculture. CAB, Oxford, pp 1–13

Halls AS (1998) An assessment of the impact of hydraulic engineering on floodplain fisheries and species assemblages in Bangladesh. PhD thesis, Imperial College, London, 532 pp

Halls AS, Kirkwood GP, Payne AI (2001) A dynamic pool model for floodplain river fisheries. Ecohydrol Hydrobiol 1:323–339

Hesse L, Schmulbach JC, Carr JM, Keenlyne KD, Unkenholz DG, Robinson JW, Mestl GE (1989) Missouri River fishery resources in relation to past, present and future stresses. In: Dodge D (ed) Proceedings of the international large river symposium (LARS). Can Spec Pub Fish Aquat Sci 106:352–371

Hohensinner S, Habersack H, Jungwirth M, Zauner G (2004) Reconstruction of the characteristics on a natural alluvial river–floodplain system and hydromorphological changes following human modifications: the Danube River (1812–1991). River Res Appl 20:25–41

Holcik J, Kmet T (1986) Simple models of the population dynamics of some fish species from the lower reaches of the Danube. Folia Zool 35:183–191

Jere J, Béné C (2004) Assessing the potential for a fishery component in the Zambia Initiative. [Report of the evaluation mission on the Zambia Initiative (Government of Zambia/UNHCR) in Western Province] WorldFish Center, Cairo, 23 pp

Junk WJ, Wantzen KM (2003) The flood pulse concept: new aspects, approaches and applications – an update. In: Welcomme R, Petr T (eds) Proceedings of the second international symposium on the management of large rivers for fisheries, vol II. (RAP publication 2004/17) FAO Regional Office for Asia and the Pacific, Bangkok, pp 117–140

Junk WJ, Bayley PB, Sparks RE (1989) The flood pulse concept in river–floodplain systems. In: Dodge D (ed) Proceedings of the international large river symposium (LARS). Can Spec Pub Fish Aquat Sci 106:110–127

Karenge L, Kolding J (1994) Inshore fish population changes in Lake Kariba, Zimbabwe. In: Pitcher TJ, Hart PJB (eds) Impact of species changes in African lakes. Chapman and Hall. London

King J, Brown C, Sabet H (2003) A scenario based holistic approach to environmental flow assessments in rivers. River Res Appl 19:619–639

King JM, Cambray JA, Impson ND (1998) Linked effects of dam-released floods and water temperature on spawning of the Clanwilliam yellowfish *Barbus capensis*. Hydrobiologia 384:245–265

Kolding J (1993) Population dynamics and life-history styles of Nile tilapia, *Oreochromis niloticus*, in Ferguson's Gulf, Lake Turkana, Kenya. Environ Biol Fish 37:25–46

Kolding J (1994) Plus ca change, plus c'est la meme chose; on the ecology and exploitation of fish in fluctuating tropical freshwater systems. PhD thesis, University of Bergen, Bergen

Krykhtin KL (1975) Causes of periodic fluctuations in the abundance of the non-anadromous fishes of the Amur River. J Ichthyol 15:826–829

Lae R (1992) Influence de l'hydrobiologie sur l'evolution des pecheries du delta Centrale du Niger, de 1966 a 1989. Aquat Living Resour 5:115–126

Lambert de Brito Ribeiro MC, Petrere M (1990) Fisheries ecology and management of the jaraqui (*Semaprochilodus taeniurus*, *S. insignis*) in Central Amazonia. Regul Rivers 5:195–215

Loneragan NR, Bunn SE (1999) River flows and estuarine ecosystems: implications for coastal fisheries from a review and case study of the Logan River, southeast Queensland. Aust J Ecol 24:431–440

Lucas M, Baras E (2001) Migration of freshwater fishes. Blackwell Science, Oxford, 420 pp

Mackay SJ, Arthington AH, Kennard MJ, Pusey BJ (2003) Spatial variation in the distribution and abundance of submersed aquatic macrophytes in an Australian subtropical river. Aquat Bot 77:169–186

Metsi (2000) Final report: summary of main findings. (Report LHDA 648-002) Metsi Consultants/Lesotho Highlands Development Authority, Lesotho

Metsi (2002) Development of a mitigation and compensation framework for communities downstream of LHWP phase 1 structures. (Report LHDA 678-004) Metsi Consultants/Lesotho Highlands Development Authority, Lesotho

Moses BS (1987) The influence of flood regime on fish catch and fish communities of the Cross River floodplain ecosystem, Nigeria. Environ Biol Fish 18:51–65

Neiland AE, Béné C, Bennett E, Turpie J, Chong CK, Thorpe A, Ahmed M, Valmonte-Santos RA, Balasubramanian H (2004) River fisheries valuation: a global synthesis and critical review. WorldFish Center, Penang

Nilsson C, Reidy CA, Dynesius M, Revenga C (2005) Fragmentation and flow regulation of the world's large river systems. Science 308:405–408

Novoa DF (1989) The multispecies fisheries of the Orinoco river: development, present status, and management strategies. In: Dodge DP (ed) Proceedings of the international large rivers symposium. Can Spec Pub Fish Aquat Sci 106:422–428

Oberdorff T, Guegan J-F, Hugueny B (1995) Global patterns of fish species richness in rivers. Ecography 18:345–352

Olden JD, Poff NL (2003) Redundancy and the choice of hydrological indices for characterising streamflow regimes. River Res Appl 19:101–121

Parsons M, Thoms M, Norris R (2002) Australian river assessment system: review of physical river assessment methods – a biological perspective. (Monitoring river health initiative, technical report 21) Environment Australia, Cooperative Research Centre for Freshwater Ecology, Brisbane

Payne AI, Harvey MJ (1989) An assessment of the *Prochilodus platensis* Holmberg population in the Pilcomayo river fishery, Bolivia using scale-based and computer-assisted methods. Aquacult Fish Manage 20:233–248

Piskozub A (1982) Effect of regulation on the Vistula River channel. Wyd Komuniakcji i Lacznosci, Warsaw

Poff NL, Allan JD, Bain MB, Karr JR, Prestegaard Kl, Richter BD, Sparks RE, Stromberg JC (1997) The natural flow regime a paradigm for river conservation and restoration. BioScience 47:769–784

Poff N, LeRoy J, Allan D, Palmer MA, Hart DD, Richter BD, Arthington AH, Meyer JL, Stanford JA, Rogers KH (2003) River flows and water wars: emerging science for environmental decision-making. Front Ecol Environ 1:298–306

Pouilly M, Rodríguez MA (2004) Determinism of fish assemblage structure in neotropical floodplain lakes: influence of internal and landscape lake conditions. In: Welcomme R, Petr T (eds) Proceedings of the second international symposium on the management of large rivers for fisheries, vol II. (RAP publication 2004/17) FAO Regional Office for Asia and the Pacific, Bangkok, pp 243–266

Pusey BJ (1998) Methods addressing the flow requirements of fish. In: Arthington AH, Zalucki JM (eds) Comparative evaluation of environmental flow assessment techniques: review of methods. (Occasional paper 27/98) Land and Water Resources Research and Development Corporation, Canberra, pp 64–103

Pusey BJ, Arthington AH, Read MG (1998) Freshwater fishes of the Burdekin River, Australia: biogeography, history and spatial variation in assemblage structure. Environ Biol Fish 53:303–318

Quiros R, Cutch S (1989) The fisheries and limnology of the lower Plata basin. In: Dodge DP (ed) Proceedings of the international large rivers symposium. Can Spec Pub Fish Aquat Sci 106:429–443

Ratner B (2003) The politics of regional governance in the Mekong river basin. Global Change 15:59–76

Regier HA, Welcomme RL, Steedman RJ, Henderson HF (1989) Rehabilitation of degraded river systems In: Dodge DP (ed) Proceedings of the international large river symposium. Can Spec Pub Fish Aquat Sci 106:86–89

Revenga C, Kura Y (2003) Status and trends of biodiversity of inland water ecosystems. (Technical series 11) Secretariat of the Convention on Biological Diversity, Montreal, 120 pp

Richter BD, Baumgartner JV, Powell J, Braun DP (1996) A method for assessing hydrologic alteration within ecosystems. Conserv Biol 10:1–12

Richter BD, Baumgartner JV, Wigington R, Braun DP (1997) How much water does a river need? Freshwater Biol 37:231–249

Smolders AJP, Velde G van der, Reolofs JGM (2000) El Nino caused collapse of the sabalo fishery (*Prochilodus lineatus* Pisces: Prochilodontidae) in a South American river. Naturwissenschaften 87:30–32

Smolders AJP, Guerrero Hiza MA, Velde G van der, Roelofs JGM (2002) Dynamics of discharge, sediment transport, heavy metal pollution and Sabalo (*Prochilodus lineatus*) catches in the lower Pilcomayo River (Bolivia). River Res Appl 18:415–427

Smolders AJP, Hudson-Edwards KA, Velde G van der, Roelofs JGM (2004) Controls on water chemistry of the Pilcomayo river (Bolivia, South America). Appl Geochem 19:1745–1758

Stankovic S von, Jankovic D (1971) Mechainismus der fischproduktion im gebeit des mittleren Donaulaufes. Arch Hydrobiol Suppl 36:299–305.

Tharme RE (2003) A global perspective on environmental flow assessment: emerging trends in the development and application of environmental flow methodologies for rivers. River Res Appl 19:397–396

Turpie JK, Smith B, Emerton L, Barnes JI (1999) Economic value of the Zambezi Basin wetlands. Unpublished report for IUCN

University of Michigan (1971) The fisheries of the Kafue River flats, Zambia, in relation to the Kafue Gorge dam. (Report prepared for UN/FAO: FI:SF/ZAM 11, technical report 1) University of Michigan, Ann Arbor, 161 pp

Vannote RL, Minshall G.W, Cummins KW, Sedell JR, Cushing CE (1980) The river continuum concept. Can J Fish Aquat Sci 37:130–137

Welcomme RL, Hagborg D (1977) Towards a model of a floodplain fish population and its fishery. Environ Biol Fish 2:7–24

Welcomme RL, Halls A (2004) Dependence of tropical river fisheries on flow. In: Welcomme R, Petr T (eds) Proceedings of the second international symposium on the management of large rivers for fisheries, vol II. (RAP publication 2004/17) FAO Regional Office for Asia and the Pacific, Bangkok, pp 267–284

Welcomme RL, Halls A (2001) Some considerations of the effects of differences in flood patterns on fish populations. Ecohydrol Hydrobiol 1:313–321

Welcomme RL (1979) Fisheries ecology of floodplain rivers, Longman, London, 317 pp

Welcomme RL (1999) A review of a model for qualitative evaluation of exploitation levels in multi-species fisheries. J Fish Ecol Manage 6:1–20

Welcomme RL (1985) River fisheries. (Food and Agriculture Organization fisheries technical paper 262) FAO, Rome, 330 pp

Welcomme RL (2001) Inland fisheries: ecology and management. Fishing News Books, Oxford. 358 pp

Welcomme RL, Winemiller KO, Cowx IG (2006) Fish environmental guilds as a tool for assessment of ecological condition of rivers. River Res Appl 22:377–396

World Commission on Dams. (2000) Dams and development: a new framework for decision-making. (Report of the Word commission on dams) Earthscan/Thanet, London

8 Water Management and Wise Use of Wetlands: Enhancing Productivity

R.L. Welcomme, R.E. Brummett, P. Denny, M.R. Hasan, R.C. Kaggwa, J. Kipkemboi, N.S. Mattson, V.V. Sugunan, K.K. Vass

8.1 Introduction

After 50 years of steady growth, the inland fisheries of the world are now showing signs of having peaked. Since 1998, catches recorded by FAO (FAO 2005) have shown signs of decline. In some cases, such as those of North America and Europe, catches have fallen largely for social and economic reasons. In other regions there are signs that over-fishing is driving the decline, while in many other cases the decline is primarily for environmental reasons. The form and function of many of the world's rivers are changing rapidly in response to the demands of a range of users. The trends noted by Arthington and Welcomme (1994) at the First World Fisheries Congress in 1992 have continued and intensified. Over 70 % of North Temperate rivers have already been heavily modified (Dynesius and Nilsson 1994) and more recent studies by Nilsson et al. (2005) show that most of the river systems in the tropics have at least one dam controlling flow. Further modifications are being made to rivers for navigation, flood control and the reclamation of lateral floodplains. In addition, quantities of water are being abstracted from rivers, lakes and reservoirs to satisfy the needs of irrigated agriculture, domestic use and industry, and natural hydrographs in rivers are increasingly being altered (Bunn and Arthington 2002; Dyson et al. 2003; Welcomme and Halls 2004). By way of illustration, these changes are classified into four different stages in Table 8.1.

Lakes and reservoirs are affected by human interventions mainly by eutrophication. Changes in morphology are less common, although siltation and local alterations to the riparian zone can result from urbanisation, marina construction and deforestation. In lakes and particularly reservoirs used for water supply and power generation, rapid drawdown can occur that damages riparian vegetation and fish populations.

Ecological Studies, Vol. 190
J.T.A. Verhoeven, B. Beltman, R. Bobbink, and D.F. Whigham (Eds.)
Wetlands and Natural Resource Management

Table 8.1 Characteristics of lowland rivers at various stages of development

Development stage	State of river channel	State of floodplain	Floodplain use	Fisheries
Unmodified	Freely meandering or anastomosing, often with islands. Habitat structure diverse	Natural flooding. Usually forested, interspersed with floodplain water bodies	Wild life. Hunting/gathering seeds and fruits	Fish assemblage intact. Capture fisheries on wild fish stocks
Slightly modified	Freely meandering or anastomosing, often with islands. Habitat structure diverse	Natural floodplain. Some forests, usually savannah with floodplain grasses	Seasonal cattle rearing, draw-down agriculture, floating rice culture	Larger fish species may diminish in abundance. Capture fisheries on wild stocks, Some management of floodplain water bodies, brush park and drain-in fisheries
Modified	Locally regulated with some damming and leveeing but with some reaches still relatively unregulated. Some backwaters persist. Habitat structural diversity locally reduced	Floodplain partially modified, deforested: Seasonal floods reduced, floodplain water bodies sometimes isolated. Local poldering and flood control structures	Seasonal flood rice culture, vegetable gardening, dry season crops, cattle rearing	Some rheophilic and limnophilic species locally threatened, larger fishes disappeared. Capture fisheries in main channel, and in rice fields. Some rice–fish culture, brush park fisheries, control of water level in some floodplain water bodies
Highly modified	Often heavily dammed, sometimes in cascades. Fully regulated and channelised, often with revetted banks and dredged navigation channels. Backwaters eliminated. Diversity low	Floodplain dry or flooding completely controlled with extensive drainage and irrigation canals. Off-channel water bodies largely eliminated or isolated. May be heavily poldered	Intensive, irrigated; rice culture and other dry season crops	Limnophilic and rheophilic species absent: assemblage dominated by eurytopic species. Capture fisheries in river channel only. Cage culture in river and irrigation channels. Small-scale pond farming and rice–fish farming on floodplain, stocked fisheries in surviving water bodies

It is unlikely that any of these changes will be reversed for many centuries and it may be anticipated that the pace of river and lake modification will intensify in the coming years in response both to the economic imperatives of development and to global warming. This means that fisheries that are already affected negatively by heavy fishing will be placed under increasing pressure by environmental degradation. They can respond to this in two ways. Inland fish can disappear as a resource for human diets and food fisheries will be replaced by recreational or conservationist goals, as has already happened in much of Europe. Alternatively fisheries will continue but in a very different manner from the present day, with much reduced catches from a relatively restricted spectrum of species. If fisheries are to continue to provide appreciable quantities of valuable protein food some form of intensification of management will have to occur, similar to that which has already taken place in many of the world's lakes and reservoirs (Welcomme and Bartley 1998). This paper reviews some of the techniques that are already current in some rivers as candidates for wider use in mitigating the impacts of human development on rivers.

8.2 Trends in Capture Fisheries

8.2.1 Increasing Pressure – Decreasing Catch

Inland fisheries are responding to changes in the form and function of rivers, the increasing eutrophication and silting of lakes and to the high fishing pressures that are current in Asia, Africa and parts of Latin America by progressing through the 'fishing down' process (Welcomme 1999). This means that there is ongoing reduction in the mean size of the fish caught and a disappearance of large and migratory species, sometimes accompanied by falling catches. In many cases the evidence for this is anecdotal because the data are lacking for a full statistical analysis of the declines in catch, but nearly all 96 rivers from the 61 river basins (from all continents and climatic zones) presented at The International Symposium on the Management of Large Rivers for Fisheries held in Phnom Penh in February 2003 (Welcomme and Petr 2004) registered some degree of damage to their fish stocks. In many lakes and rivers, this process is likely to culminate in fish populations that consist mainly of small, eurytopic, native and introduced species.

8.2.2 Fisheries Management

Currently, fisheries management has proved inadequate to the task of maintaining fish stocks and fisheries. It is felt that, if some alternative management system can be installed, the health of the fishery can be restored. Proposals have been made on many inland fisheries for co-management to substitute for the previously centralised approaches (Hartmann et al. 2004). Such systems should involve local people to a greater degree in decision-making, resulting in more flexible management systems. They can also formulate and enforce regulations that correspond best to the needs of the fishery at local level. In addition to the general management of the fishery, proposals are also made for setting up conservation and harvest reserves. Hoggarth et al. (2003) cite several successful examples from Indonesia, together with guidelines for setting up such reserves on rivers. Fish conservation zones (FCZs) are also a common feature of fisheries management in the Lower Mekong Basin. For example in Khong District, southern Lao PDR, 63 villages institutionalised co-management for aquatic resources and included the establishment of FCZs as an important component of management regulations. The FCZs are often deep areas of the Mekong mainstream that vary in size between 0.25 ha and 18.0 ha, where fishing is not allowed (Baird 1999). Although fishermen generally support the FCZs, it has not been possible to demonstrate increases in the catch (Chomchanta et al. 2000).

The central problem of fisheries management, that of excess effort, is difficult to resolve equitably. Fisheries can only become sustainable when the numbers of fishermen and their effort are controlled. Such control calls for considerable social and economic reorientation, for at present most fisheries operate under open-access regimes and fishing is an essential part of the livelihood strategies of riparian people. This situation is aggravated by the growing demand on inland fish by increasing numbers of fishermen. As a result, fishing is targeting a resource that is declining because of environmental change. Therefore, the way forward should include better regulation of conventional capture fisheries in inland waters as well as seeking methods for the intensification of yield that are better adapted to the changing environment.

8.3 Methods for the Enhancement of Inland Fisheries

As conventional capture fisheries appear to have reached their limits in many rivers and lakes, alternatives are needed that will enhance present levels of yield and will mitigate for losses to the resource base caused by overfishing and environmental change. Many such methods have been proposed and some that have already been developed by traditional societies in various rivers and lakes may be transferred elsewhere.

8.3.1 Species Introductions

Numerous species have been transferred around the world to establish new fisheries or to fill a vacant niche (Welcomme 1988). Reservoirs particularly have benefited from introductions, as many native river faunas have not been able to fully utilise the trophic and spatial resources and adequately colonise the new lentic environments (Fernando and Holcik 1982). Rivers and downstream lakes have also been colonised by escapes from reservoirs, aquaculture installations and the aquarium fish trade, and these are often adapted to the more lacustrine conditions of regulated rivers. Thus, species such as the tilapias and common carp, that are already common in lakes and reservoirs, have colonised lowland rivers, sometimes achieving pest proportions. Even though not prominent in many existing faunas, exotic elements are present in most river and lake basins and these could well become more dominant as conditions change.

8.3.2 Stocking

Stocking is generally carried out to:
- maintain stocks of species that are now unable to reach their breeding grounds because of dams;
- support recreational fisheries by increasing stocks of target species for the sport fisherman;
- increase or sustain yield of threatened species by compensating for lack of spawners resulting from overfishing or environmental damage;
- enhance yields in lakes, controlled river reaches, floodplains and floodplain pools (for summaries, see Cowx 1998).

Stocking in Rivers and Riverine Wetlands

Stocking programmes to compensate for breeding failures are mainly used to maintain stocks of salmonids and sturgeons in temperate river systems. However, with increasing numbers of commercially valuable migratory species being threatened by dams in the tropics, stocking programmes may be initiated here too. Stocking programmes to raise productivity of rivers have been tried in some cases. Stocking drain-in ponds has been done for many years in some traditional fisheries and recently both open floodplains and enclosed oxbow lakes have been stocked in Bangladesh (Middendorp et al. 1998; Petr 1998).

Stocking open floodplains might seem risky, as the chances of recapturing the stocked fish are low. However, many floodplains are now poldered and access to the impounded plain is limited. In such cases, stocking on the rising flood carries the fish onto the floodplain where they are contained. In Bangladesh, 59 900 ha of floodplain were stocked with some 50 million fingerlings of carp species over a 5-year period at levels equivalent to 17.3 kg ha^{-1}. The stocked fish increased about 8-fold by weight between stocking and harvesting and gave yields of 300–800 kg ha^{-1}, of which 25 % were stocked fish and 75 % wild fish (Payne and Cowan 1998). Stocking was estimated to provide an increment of 20 000 t (Islam 1998). It can be concluded that stocking small, closed floodplains can be cost-effective (Ali 1998) and might be considered for similar poldered floodplains elsewhere in the world.

Stocking floodplain lakes, especially those that are isolated from the river or only flood infrequently, can provide more control over recapture. Pinter (1983) recorded early attempts in Hungary where relic floodplain water bodies were stocked to give yields that compared favourably with aquaculture in ponds. In Bangladesh, Ahmed (1998) described the stocking of small floodplain lakes ('beels') so that the fry could spread over local, contained floodplains. Here, 905 ha of nursery beels yielded a cost/benefit ratio of about 6:1 and the practice may provide an alternative to transporting fry into the floodplain from distant hatcheries. Oxbows, fully managed as 'put, grow and take' fisheries have also been tried in Bangladesh. Here, yield in a group of small beels increased from 120 kg ha^{-1} to 982 kg ha^{-1} giving returns of 167–182 % on total investment and 23–264 % on operating costs. Carp yields in 17 such ponds were positively correlated with stocking density by the equation (Hasan et al. 1998):

$$\text{Yield} = 15.88 + 0.184 \text{ stocking density}$$

In India, the potential fish yield in Assamese floodplain lakes was estimated as a function of the primary productivity (Sugunan and Sinha 2001). Even at only 1 % of primary production, the potential in the Barak valley is 981–1313 kg ha^{-1}, 817–1889 kg ha^{-1} in Lower Assam, 791–1399 kg ha^{-1} in Central Assam and 476–2324 kg ha^{-1} in Upper Assam. The average natural fish yield in the beels of Assam is about 173 kg ha^{-1} $year^{-1}$.

Scientific management of West Bengali beels raises fish yield from a normal level of 100–150 kg ha^{-1} $year^{-1}$ to 1000–1500 kg ha^{-1} $year^{-1}$. Open beels in West Bengal were shown to be less productive (42–150 kg ha^{-1} $year^{-1}$), whereas yields were higher in closed beels (66–4333 kg ha^{-1} $year^{-1}$). Many beels are actively managed by co-operative societies and are regularly stocked with carp and exotic species.

The natural average fish yield in the floodplain wetlands of Bihar is reported to be 40–200 kg ha^{-1} $year^{-1}$, whilst their potential calculated from energy assimilation and available energy reserves is 1500 kg ha^{-1} $year^{-1}$. The potential fish yield of beels in Uttar Pradesh varies between 637.5 kg ha^{-1} $year^{-1}$ and 1326.5 kg ha^{-1} $year^{-1}$, while the actual yield is 43.0–357.0 kg ha^{-1} $year^{-1}$.

It can be concluded that systematic management of closed floodplain pool systems is a highly practical method to increase yields of marketable fish. However, this approach generally implies social changes, in that control over the resource needs to be assigned to a limited group of operators rather than the previously open-access fishery, because it is unlikely that any individual or group would stock fish into a water body without a guarantee of a return on the investment. An alternative is for such programmes to be publicly funded, much as are the stocking of reservoirs in Thailand, in which case the expenses accrued can be regarded as a subsidy to subsistence fishing to compensate for losses in natural yield caused by environmental changes.

Stocking in Lakes and Reservoirs

Many lakes, small dams and reservoirs are now stocked as part of regular management. There are two basic strategies for stocking for recreational and production fisheries in lakes: (a) the establishment of reproducing, balanced populations and (b) unbalanced put-and-take fisheries.

Reproducing populations can either be internally or externally balanced. In externally balanced systems, fishers remove a certain percentage of the fish, often targeting larger individuals. This system is most commonly observed in developing countries where stocking programs are undertaken to address problems of food insecurity or declining capture fisheries. Properly managed, these systems can be highly productive. Inappropriate management, including overfishing, taking fish that are too small or fishing in the breeding season, can pose a major threat to their sustainability. In the tropics, stocking small water bodies in Burkina Faso with 20 kg (800 fingerlings) of *Oreochromis niloticus* per hectare increased yield from 23 kg ha^{-1} to 269 kg ha^{-1} (de Graaf and Waltermath 2003). India reports up to 10-fold increases in yield from stocking programs (Sugunan 1995). Five-fold increases have been reported from Thailand, Indonesia, the Philippines and Malaysia (Fernando 1997).

Internally balanced systems are composed of reproducing predators and prey species and are most popular for the enhancement of sport fisheries. The fishery is self-sustaining and does not require restocking if exploitation is properly regulated. However, the ratio of predator to prey species is critical to success and is often difficult to achieve, particularly in areas where food insecurity and poverty create incentives to poach or violate bag limits. Examples include the widely successful stocking programs for mixed centrarchid sport fisheries in the United States and Europe (Davies 1973). Such programs stimulate economic growth and generate millions of dollars in fishing fees, tackle and equipment sales, hotel and restaurant revenues and guide services annually and create important recreation opportunities (Pitcher and Hollingsworth 2002). Introduction of the largemouth bass, the red swamp crayfish (*Procambarus clarkii*) and two tilapia species into Lake Naivasha, Kenya, have increased the yield of food and sport fish from virtually zero prior to the introduction up to 300 t, despite very poor management.

Unbalanced, or put-and-take fisheries, are based on species that cannot spawn in the lake or dam environment, for example grass, silver and bighead Chinese carps. Trout stocked in dams that have no access to streams are unbalanced. Weed control programs involving grass carp are another example. The advantage of these systems is that they do not rely on any particular level of fishing pressure to maintain balance. However, they must be restocked regularly.

Stocking rate is critical to the success of stocking programmes and is generally based on estimates of productivity generated through ecological studies. In general, yield rises with increasing stocking density, up to a limit determined by the abundance of food organisms, i.e. the carrying capacity (Welcomme and Bartley 1998). The number and sizes of fish to stock is based on estimates of carrying capacity, expressed in terms of weight/unit area. Within a given carrying capacity, the total weight of fish can be in the form of either many small fish or fewer larger fish. Typically, the number of fish to stock is calculated on the basis of desired fish size and the productivity of the water body, simply put:

Number to stock (ha^{-1}) = [Carrying capacity (kg ha^{-1})]/[Minimum average weight desired (kg)]

If mortality and growth rates are known, an inversion of the standard mortality formula can be used to give a more accurate estimate (Welcomme 1976):

$$N_o = N_c^{[z\,(c-o)]}$$

where N_o is the number to stock, N_c is the number of fish desired at age of capture (c), o is the age at stocking and z is the total mortality.

Naturally occurring species of fish and other aquatic fauna or flora can have strong impacts on the success of small water body management. Typically, these impacts are seen by managers as negative. Indigenous species compete for food, disrupt nesting/breeding behaviour and can predate stocked fish. By muddying water and uprooting plants, common carp can disrupt food webs and render ineffective the stocking of planktivorous species. Stocking of tilapias in Lake Nasser, Egypt, to increase yield was foiled by the Nile Perch (*Lates niloticus*), which ate the bulk of the stocked fish, often at the point of stocking. Stocking with forage species in cases where natural reproduction occurs, is often superfluous (Welcomme and Bartley1998).

Predators or competitors can be removed through the use of poisons or selective fishing, but the effectiveness of this is directly proportional to the size of the water body in question and the degree to which it can be physically manipulated through draining, weed control, etc. When control of unwanted species is not possible, or when multiple users demand different species for different purposes, fish population structure becomes a critical management issue. Balancing species with different ecological niches can be complicated and subjective and requires careful control of exploitation intensity and strategy.

Support to culture-based fisheries of small water bodies not only brings direct benefits in terms of increased recruitment of valuable species, but also tends to catalyse institutional development and encourages investment into common pool resources. However, the level of uncertainty in this type of intervention is high, both in terms of the direct benefits of increased fish yields and in terms of institutional development. To address this uncertainty, an adaptive approach to management, or learning-by-doing, has proved effective (Garaway et al. 2001).

The Thai Village Fish Pond Development Programme, a community-based fisheries management project, seeks to strengthen community cohesiveness by training and supporting village committees in managing village fish ponds. Economic proceeds from the ponds are generally used for community activities. However, possibly the most important benefit is the development of local resource management institutions. There are some 4400 village fish ponds across the country, of which more than three-quarters are in the northeast and north Thailand. The ponds are on average 8 ha and the average yield is 435 kg ha^{-1} $year^{-1}$ (Lorenzen et al.1998b; Virapat et al.2000).

The Small-Scale Wetlands Indigenous Fisheries Management project (SWIM) was carried out between 1999 and 2001 in Champassak province, Lao PDR (Anon 2001). SWIM activities included stocking the back-swamps as well as the Sedone River. The project was able to demonstrate benefits of involving local resource users in the research. Villagers participated in the breeding and nursing of *Barbonymus gonionotus* for stocking and also in tagging and catch per unit effort (CPUE) data collection. The project highlighted the importance of small wetlands to the livelihoods of villagers. Apart from contributing to household fish consumption and cash, funds for community use were also enhanced.

In the Central Highlands of Vietnam, stocking of reservoirs is usually carried out as an economic activity. Local authorities award an individual or group a contract to manage the fishery. Typically, these reservoirs are small (generally less than 200 ha) and depauperate in fish. The contractor pays for the fingerlings and imposes whatever regulations are deemed necessary. The species stocked are usually alien and include silver carp (*Hypophthalmichthys molitrix*), bighead carp (*Aristichthys nobilis*), grass carp (*Ctenopharyngodon idella*), rohu (*Labeo rohita*), mrigal (*Cirrhinus cirrhosus*), common carp (*Cyprinus carpio*) and tilapia (*Oreochromis* sp.). Of these, silver and bighead carp are the most important for reservoir stocking. However, some water bodies also support considerable self-recruiting capture fisheries (Phan and De

Silva 2000; Thai Ngoc Chien et al. 2001). Similar mixes of species are used in stocking Sri Lankan tanks, where yields of between 53–1800 kg ha^{-1} year^{-1} (mean 450 kg ha^{-1} year^{-1}) have been recorded (Wijenayake et al. 2005). These include a considerable proportion of naturally occurring predatory snakeheads.

It is often difficult to evaluate stock enhancement programmes, in particular when stocking species that already are naturally recruited in the water body. The cases where benefits have been evaluated usually involved the stocking of alien species, or indigenous species that were not present in the water body before stocking, since all recaptures of these species can be assigned to the stocking. An example is the stocking of *Macrobrachium rosenbergii* into Thai reservoirs (Sripatrprasite and Lin 2003). Stocking of small water bodies that are isolated from the floodplain and where there is little or no natural recruitment are amenable to detailed analysis of benefits (Lorenzen et al. 1998a, b).

8.3.3 Extensive Culture Methods

Drain-In Ponds

Fish holes, drain-in ponds or fingerponds are ponds dug into floodplains of rivers and lakes that exploit two natural phenomena: (a) the high natural productivity of the land–water interface zone and (b) the trapping of fish in depressions following flood retreat after seasonal rains. Such artificial structures are known from the pre-Columbian Amazonian (Erikson 2000) and have been recorded from the Mekong system in Cambodia (Chevey and Le Poulain 1940; Daep et al. 2003), the Chone River in Ecuador, (Bonifaz et al. 1985) the Sudd, Lake Chad (Sarch and Birkett 2000), and the Oueme River in the Republic of Benin (Welcomme 1971). They are akin to the 'hortillonnage' systems of Mexico (Micha et al. 1992) and Rwanda (Barbier et al. 1985), although these are normally stocked from aquacultural sources and water levels are controlled to maximise yields for trade. The Oueme fish holes, known locally as 'whedos', are long channels 4–5 m in width and up to 1.5 m deep cut into the surface of the floodplain (Fig. 8.1). They are usually covered with floating vegetation. They cover approximately 3 % of the total 1000 km^2 of the Oueme and So river delta; and in some zones as much as 13 % of the area may be in the form of drain-in ponds. Yields were high from 1955 to 1958 at 2.12 t ha^{-1} year^{-1}, but had reduced somewhat by 1968–1970, when there were 1.57 t ha^{-1} year^{-1} (Welcomme 1971) and had further reduced to 0.8 t ha^{-1} year^{-1} by 2003. This fall in yield possibly corresponds to a reduction in the populations of fish in the Oueme due to very heavy fishing. Daep et al. (2003) list similar structures from the Mekong, usually measuring 3–15 m in length, 2–5 m wide and 2 m deep. These are filled with floating vegetation or brush and woody debris. The pre-Columbian ponds in Ecuador were

Fig. 8.1 Fish holes (Whedos) on the Oueme floodplain, Republic of Benin. **A** Aerial view of a group of fish holes. **B** Vegetated fish hole before fishing. **C** Fish hole after fishing, with vegetation removed.

extensive zigzag earthworks that served as fish weirs to guide fish into large circular ponds (Erikson 2000). These structures still survive and capture a range of species, including *Hoplosternum*, *Cichla*, *Sarrasalmus*, *Prochilodus* and *Erythrina* spp. This tradition continues in the Chone River, where similar structures attract and rear the migratory *Dormitator latifrons* to form the basis of the 'chamera' fisheries, some of which have no connection to the river, even during the flood, and must be filled and emptied by pumping water. In Bangladesh, approximately 30 000 ponds were created on the floodplain when soil was excavated to form islands for human residence; additional ponds of this type have been created from residual floodplain water bodies, which have been cleaned and rehabilitated throughout the Ganges/Brahmaputra system.

These installations all operate primarily on a drain-in principle, whereby fish present on the floodplain during the flood are concentrated into the ponds at falling water and remain until harvested later in the dry season. They, therefore, only function properly in floodplains that are still regularly flooded. In some systems where natural flooding is restricted, ponds are filled with water pumped from the river. Dragnets that concentrate the fish at one end of the pond normally harvest them, but in some systems the water is pumped out at the end of the dry season. Drain-in ponds favour air-breathing species such as *D. latifrons*, clariid catfishes, snakeheads and anabantids, because of the anoxic conditions that develop in them. Yields from drain-in systems may also be enhanced by stocking with juveniles from the river, from other floodplain depression lakes, as in the 'chameras' and some of the derelict water fisheries of India, or from hatcheries.

Drain-in ponds represent an important technology for increasing fish yield on floodplains and the technology deserves to be transferred to floodplains where their use is not current. In their natural form, they add to the diversity of the floodplain and form a reserve of fish that can be exploited towards the end of the dry season when catches by many other gears are falling. In their more managed form, they represent a transition to regular aquaculture in ponds on the reclaimed floodplain. Attempts to apply this technology to new areas of lacustrine floodplain illustrate some of the problems with extension of the technology.

Attempts are being made to extend the drain-in pond principle to lakes and lacustrine wetlands under the integrated agro-piscicultural fingerpond projects in Lake Victoria. Here, the three main characteristics of the ponds are that: (a) water levels are un-regulated, (b) natural waste products from the compound are used to fertilise the ponds and (c) the ponds are stocked with fish naturally from the surrounding wetlands as poor societies cannot afford feeds or the handling and transport costs for fingerlings. The ponds are installed at the landward edge of fringing wetlands which are flooded during the rains by rising lake and river levels. The floodwater flushes out deoxygenated water from the swamps and carries with it the diversity of life from the lake/swamp interface. This interface zone is a place of rich diversity and high productivity where organisms, including fish, thrive and multiply. As the floodwaters recede, fish become trapped in the fingerponds where they are cultured and available for harvest until the ponds dry out at the end of the dry season. Natural fertiliser additions and management of the fingerponds ensures continuous fish growth and supply in the dry season.

The swamp vegetation between the ponds is cleared and the soil from the excavation of the ponds is spread over the area for agricultural plots. These plots are covered in the floods but are planted with cash crops as soon as they are exposed by the receding water. In the experiments carried out so far, natural stocking of the ponds has been successful, with the trapping of large numbers of small fish in most ponds. Stocks are dominated by haplochromines and tilapias. Tilapias constitute 80–90 % of the total biomass and are dominated by three species: *Oreochromis niloticus*, *O. variabilis* and *O. leucostictus*. Initial numbers of fish in each pond vary enormously from an average of around 500 individuals per pond up to more than 2000 individuals, whilst total biomass values are usually 2–3 kg, but can be up to 10 kg per pond (100–150 kg is equivalent to 500 kg ha^{-1}). The variability in stock numbers and the large number of small fish can create a problem for management; and premature breeding of under-sized fish in overstocked conditions can aggravate the situation.

When the fish are trapped in the ponds, fertiliser application is essential. This uses natural waste, including green manure, kitchen waste (such as banana skins and sorghum bran) and chicken and cow dung. At the end of the season when the ponds have nearly dried out, the accumulated sediment is removed and applied to the horticultural plots. Weeds and crop wastes are composted.

Artificial Control of Floodplain Lagoons

Blocking drainage channels with dams, weirs and sluices can control the rate at which floodplain water bodies are emptied. In some cases, such as the fisheries of the Logone and Chari floodplain depressions (Drijver et al.1995), fish in stored waters are fished with weirs and bag-nets on outflowing channels as they return to the river. The lagoons themselves are fished later, usually by communal fisheries. In others, the extension of the period of flooding is used to increase the survival and growth of the fish. Reed et al. (1967) and Reizer (1974) suggested improvements to indigenous systems of the Niger and Senegal Rivers, where yields of impounded lagoons of around 185 kg ha^{-1} $year^{-1}$ (similar to those of unimpounded permanent water bodies) could be raised to 500 kg ha^{-1} $year^{-1}$ when the lagoons were stocked with additional juvenile fish and were fed. This type of management is well adapted to floodplains that are poldered as the extensive levees enclosing the floodplain cell, together with sluice gates and canals (Fig. 8. 2), give a high degree of controllability to the water regime. Control of the water regime is most commonly used in the interests of agriculture, usually rice culture, but can also be used to advantage by the fishery. Recent work in Bangladesh has shown that regulation of sluice gates giving access to poldered areas can control the quantity of fish entering and leaving the enclosed system (Halls 2005).

Brush Parks

Brush parks consist of masses of branches and/or vegetation placed in compact masses in rivers, shallow lakes and coastal lagoons (Fig. 8.3). The parks

Fig. 8.2 Entrance to floodplain lagoons and poldered areas. **A** Sluice on inflow canal to an agricultural polder, Bangladesh. **B** Entrance to a floodplain lagoon complex (fishing lot) with nets to capture fish leaving the lot, Cambodia

vary in size between a few metres in diameter to up to a hectare and may be installed free-standing, recessed into the banks or in drain-in ponds. They are common in Sri Lanka, Madagascar, Benin, Nigeria, Cameroon, Chad, Ivory Coast, India, Bangladesh, Cambodia, China, Mexico and Ecuador. Welcomme (2002) reviewed the available literature on the construction, functioning and the environmental and social risks of this fishing method.

Brush parks are fished once to several times a year, depending on the size and type of installation. The parks are commonly thought of as simple fish-attracting devices and, as such, are classed in some areas as exploitative gears that are banned by law. However, they can operate as a true aquaculture system that rears fish, as well as acting as harvest reserves by contributing fish to the open water in which they are installed. Whether or not they are exploitative depends on the frequency with which they are fished and the type of fish species colonising them. If the fish, typically tilapias and other cichlids, do breed in them, large quantities of fry are generated at low fishing frequencies. These leave the park and benefit other fisheries. Yields from brush parks can

Fig. 8.3 Brush parks in different rivers and lagoon. **A** Mas Athu, Negombo Lagoon, Sri Lanka. **B** Katha, Bangladesh. **C** Samrah being installed, Tonle Sap, Cambodia. **D** Acadja Oueme River, Benin Republic

be very high. For example, the acadja systems of Benin yield between 1.2 t ha^{-1} $year^{-1}$ and 15.5 t ha^{-1} $year^{-1}$ (average 5.7 t ha^{-1} $year^{-1}$), depending on the fishing regime (Laleye, personal communication); and parks in Bangladesh and Sri Lanka give similar yields. In addition to the perception that brush parks are a damaging fishing method, it is thought that they may accelerate silting. This may be true in some lakes, but in rivers the seasonal flushing flows of the high floods probably counteract this. Indeed, parks have been installed in many areas for considerable periods of time without recorded damage. Brush parks have been transferred to areas outside those in which they are traditionally used, mainly in West Africa, but have caused a number of social problems. Nevertheless, they continue to attract interest as part of epiphyte-based culture systems.

Rice–Fish Culture

The floodplains of many tropical rivers have been converted to rice culture and the practice is increasing. The increased control of water that is needed for modern rice crops impacts heavily on the natural reproductive cycle of many fishes. Rice fields that retained connection to the river through canals are colonised by many species of fish and form the basis for rich fisheries that contribute to the total yield. Rice fields have a natural yield of 50–150 kg ha^{-1} $year^{-1}$ (Gregory and Guttman 2002). Natural yield levels can be improved upon by modifying the rice fields to favour fish, principally by incorporating fish refuge or dry-season ponds in the rice fields (Sinha 1991). Halwart and Gupta (2004) quote figures of 233 kg ha^{-1} and 184 kg ha^{-1} for rainfed and irrigated seasons in Bangladesh.

Rice culture covers an area approaching 12×10^6 ha in the Lower Mekong Basin (MRC 2003). Of this area only about 88,000 ha is estimated to be used for rice-fish culture (Phillips 2002). There is great potential to increase fish yield from this type of system. Using a conservative estimate of 100 kg ha^{-1} $year^{-1}$, the potential yield of fish from rice–fish culture may exceed 1×10^6 t. In intensive, irrigated rice culture, fish may be held in the rice fields for two or three consecutive rice crop cycles; and the fish yield can reach 370 kg ha^{-1} $year^{-1}$ (Phillips 2002).

There are several reasons to why rice–fish culture is not more popular in the Lower Mekong Basin. One is that rice fields in the floodplain are naturally 'stocked' with various aquatic organisms, so that even without investment there is a substantial yield, which may rival the rice crop in value. This wild resource is traditionally available for anyone that collects or catches it; and this practice tends also to apply to stocked fish. Therefore, what could be called 'theft' may be seen as a traditional right. Flooding, apart from bringing in wild fish, can also allow stocked fish to escape. Further, there is pressure on rice farmers to intensify the culture systems by adopting more high-yielding varieties and increasing the use of pesticides. Even if a farmer elects to practice rice–fish culture, the fish may be affected by pesticides that drift into stocked paddies with the wind or enter when nearby paddies are drained. Studies have shown that rice–fish culture generally generates more profit than rice alone (FAO/ICLARM/IIRR 2001). Rice–fish culture has attracted a considerable literature, with several manuals giving details of methodology, yields, etc. (e.g. FAO/ICLARM/IIRR 2001). The practice is fully compatible with the major use of regulated floodplains and should be more widely disseminated.

Pond Culture

Small ponds may be constructed on floodplains where there is sufficient control of flooding to avoid the ponds being submerged. This approach has been developed mainly in China but is also common in the Mekong countries of Southeast Asia and in Bangladesh (Fig. 8.4). Ponds are generally small, about 0.12 ha in Lao PDR, with yields equivalent to about 850 kg ha^{-1} (Funge-Smith 1999; Lorenzen et al. 2000), although much higher yields are recorded in China and in Vietnam, where about 5 t ha^{-1} $year^{-1}$ have been recorded by Luu et al. (2003). Here, however, the ponds are intensively managed by integration of crops, fish and livestock.

Pen and Cage Culture

Pen culture is used in many inland waters. The most notable development is in Laguna de Bay, Philippines, where large amounts of milkfish, *Chanos chanos*, are produced. Adaptations of drain-in ponds or fish canals in lacustrine and riverine wetlands, where the channel is barred at its egress by fencing, serve the same purpose in many areas, including the Mekong in South-East Asia and lakes Aheme and Nokoue in Africa. Investigations conducted by the CIFRI, Barrackpore, demonstrated that the culture of fishes and prawns in pens erected in the marginal areas of beels was an effective management tool for increasing fish yield and income for poor fishers. The pen culture technology developed

Fig. 8.4 Fish ponds incorporated into rice fields on a Bangladeshi floodplain

by the Institute has the additional advantage of being economical and giving a reasonable return on investment when low-cost materials like bamboo, cane, wooden logs, etc. are used in their construction. Pen culture of freshwater fin fishes and prawns has been successfully demonstrated by the Institute in the wetlands of Bihar, West Bengal and Assam, where there are vast areas of wetlands. In Assam, the rate of adoption of pen culture technology is particularly high. A combination of indigenous and exotic carps along with giant freshwater prawn is highly successful. Species for mixed culture include *Catla catla, Hypophthalmichthys molitrix, Labeo rohita* and *Cirrhina mrigala.* Mixed culture of carps and prawns (*Macrobrachium rosenbergii*) also gives good yields. Species ratios of *Catla catla* (20 %), *Hypophthalmichthys molitrix* (15 %), *Labeo rohita* (20 %) and *Cirrhina mrigala* (45 %) are common. *Cirrhina mrigala* can be replaced with *Macrobrachium rosenbergii* in mixed culture. Mixed culture of carps gives yields as high as 4000–5000 kg ha^{-1} $year^{-1}$. In mixed culture of carps and prawns, the yield of carps is 2000–2500 kg ha^{-1} $year^{-1}$ and prawns 500–800 kg ha^{-1} $year^{-1}$. In monoculture, *M. rosenbergii* is the most profitable, giving a yield of 1308 kg ha^{-1} and a return on investment of 29.5 % within four months in some pens.

An increasingly popular approach to fish production in both small and large lakes is the introduction of aquaculture cages. Provided there is sufficient water movement (e.g. currents, wind), the carrying capacity of cages within a reservoir can approximate that of the entire water body. Very high yields are thus possible from these cages. In lake Kariba, Zimbabwe, for example, tilapia is cultured in 500 m^3 cages at a stocking density of 50 kg m^{-3}. Costa-Pierce (1997) reports standing stocks of 200 kg m^{-3} of carp and tilapia in cages in Saguling Reservoir, West Java, Indonesia.

Cage culture is not highly developed in rivers, although it is common in some areas, particularly in the Lower Mekong (Fig. 8.5). It is best developed in China, Cambodia and Viet Nam, where operators mainly fatten wild-caught catfish with small, trash fish obtained from the local fishery. It has also been introduced into Bangladesh on an experimental basis, with some technical success. In China, the area under cages reached 281 500 ha in 1978; and by 1993 this area had increased to 359 926 ha. Fish yields increased from 308.5 kg ha^{-1} in 1978 to 980 .0 kg ha^{-1} in 1993 (Hu and Liu 1998).

In India, the fry of minor carps such as *Labeo gonius* are reared in polythene net cages up to the releasable size (10 cm, 10 g). Species of this group possess better taste, faster growth potential and better market demand than other minor carps. Since natural stocks in beels are not sufficient to sustain a stable fishery, they need to be enhanced through stocking with fish at 10 cm length or at 10 g weight, although supplies of seed of this size in sufficient quantity are difficult and expensive for fishers.

Another trial conducted at Assam to rear major carp seed to advanced fingerling stage in Nylon net cages gave encouraging results. Fry grew to fingerling size (10 cm, 12–28 g) in just two months, with a survival of 68–77 % at a

Fig. 8.5 Cages in rivers. **A** Bangladesh. **B** Tonle Sap, Thailand

stocking density of 30 fry m^{-2}. Up to three crops per year of these could be produced. After the monsoon, the same cages were utilised to produce table fish, especially *Labeo gonius*, which attained 50–100 g in 112 days.

The large amounts of feed and medicines required for intensive culture can be responsible for local degradation of the environment in lakes and shallow marine areas. In some lakes such as Kariba, where the natural productivity is low through lack of nutrients, the enrichment of the areas around the cages may be an advantage. Contamination of the area around the cages is less likely to be a problem in flowing rivers, where the current may keep accumulations under control but, in the dry season when the flow may cease, local deoxygenation may result. Furthermore, much of the cage culture in rivers is relatively extensive, although there is scope to intensify operations in future. Cage culture has actual and potential value for transfer to small dams, rivers, floodplain lakes and irrigation canals to increase the overall production of valuable table fish and to mitigate the effects of environmental changes on overall production.

8.4 Social and Economic Implications

The changes that are occurring in rivers inevitably lead to fundamental alterations in livelihood strategies of riparian peoples and the way in which their societies are organised. The decline in fish stocks will eventually lead to declines in the contribution of fishing to income and diets; and the intensification of agriculture will lead to changes in the ways in which land and water is allocated. Many of the enhancement strategies discussed in this paper will have similar effects. Better control of the fishery to make it more sustainable depends on a decrease in overall effort, which will lead to inequities as fisheries move from open-access resources to ones in which access is strictly regulated. Most of the enhancement techniques involve a similar move from open access to privately controlled systems. Brush parks, cage culture, farm ponds and stocked water bodies all lead to some form of control or ownership over the riverine environment that excludes other fishers. This does not appear to be a problem in old established systems where custom has adjusted to the cohabitation of culture and capture fisheries, but conflict often arises where a new culture system is introduced. Enhancement of the fishery therefore has to proceed as a negotiated part of the general reorganisation of floodplain river society that accompanies regulation of flow and the environment. Where this has occurred, usually through co-management schemes, benefits have resulted to the communities. A ten-year effort in Bangladesh to introduce community-based fisheries management of waterbodies (40–500 ha) produced improvements in productivity of between 25 % and 70 %; and new institutional arrangements were used to implement seasonal closures, access restrictions, installation of fish aggregating devices and the establishment of protected areas for the benefit of local communities (Thompson et al. 1999). Exclusivity of ownership, transferred from government (public) to local management entities that are comprised of, and operate on behalf of, fishing communities, was found to be a critical factor in success. In a review of traditional inland fisheries management systems in Africa, COFAD (2002) identified exclusivity of access and "a locally evolved and collectively owned cognitive base and established, accepted and functioning local institutions" as key elements of success.

Workers in Burkina Faso found that simple management strategies for seasonal dams could increase fish standing stock from 60 kg ha^{-1} to over 600 kg ha^{-1} (Baijot et al. 1994). In Malawi, local management entities proved themselves capable of managing seasonal water bodies for fish production (Chikafumbwa et al. 1998). What started as a micro-project with two villages has, over the past five years, expanded to over 12 villages with no additional external input and has been so successful that demand for fingerlings to stock the ever-increasing number of community lakes in the area has become a serious constraint.

Table 8.2 Management issues for a range of small waterbodies in Burkina Faso (de Graaf and Waltermath 2003)

Large (20–700 ha)	Medium (5–20 ha)	Small (1–5 ha)
Management and ownership by more than one village	Management and ownership by only one village	Management and ownership by one (extended) family group
People view management as the business of the government	Local management is possible in cases where legal framework exists	Distribution of benefits regulated locally through traditional mechanisms
Difficult to obtain exclusive access for local management entities	Exclusive access can be organised	Access rights are normally exclusive to the family group
Impact of investments in management are difficult to monitor	Impacts of investments are relatively easy to monitor	Impacts of investments obvious to owners
Inequitable distribution of benefits among stakeholders	Benefits more equitably distributed among stakeholders	Benefits distributed according to family norms
Management generally not sustainable outside of a project context	Can be sustainable if developed under a proper legal framework	Highly sustainable

In Burkina Faso, training and group formation for fisheries management committees helped to develop flexible management plans for a range of small waterbodies (de Graaf and Waltermath 2003). Some of the key issues in regard to the development of sustainable local management strategies are shown in Table 8.2. As in Bangladesh, the main finding is that exclusive access is a critical component of improved management.

8.5 Discussion

Capture fisheries have reached or passed their sustainable yield in many lakes, reservoirs and rivers. Even with improved management, catches are likely to fall rather than rise, although their overall sustainability and the quality of the fish they produce would be improved. Any increases in yield must come from enhancement methods. The most successful example of this approach is China, which sustained increases in inland fisheries yield over the past few decades. This was achieved despite serious environmental problems through the systematic intensification and control over the fish populations. It

Table 8.3 Enhancement methods appropriate to different developmental stages of rivers

Development stage	Productivity method
Unmodified	Conventional fisheries management; brush parks in main channel and floodplain lakes; cage culture in main channel
Slightly modified	Conventional fisheries management; drain-in ponds, brush parks, controlled discharge form floodplain water bodies; cage culture in main channel and in floodplain lakes
Modified	Conventional fisheries management; stocking polders and oxbows, drain-in ponds (pumped and naturally flooded), rice–fish culture, brush park fisheries, cage culture in main channel
Highly modified	Stocking main channel and residual water bodies; rice–fish culture, traditional pond culture; cage culture in main channel and irrigation canals

now seems inevitable that, if freshwater fish is to remain a major item in the diets of tropical countries, some parallel process of intensification of yield must occur. A number of techniques have been developed in various parts of the world that may be suitable for more general introduction into new areas. These are not all adapted to the same developmental stage of the river and Table 8.3 proposes appropriate methods for each level.

The various methods all result in yields that far exceed the 50–100 kg ha^{-1} $year^{-1}$ characteristic of unmodified floodplains and can thus go some way to mitigating the losses that occur in fish yield as flooding is increasingly controlled.

Introduction of most of these methods will lead to considerable social change with a movement away from open access and public ownership towards controlled access and private control of the resource. In this, fisheries reflect the overall destabilisation and reorganisation of riparian communities following river regulation.

References

Ahmed MN (1998) Fingerling stocking in open waters. In: Middendorp HAJ, Thompson PM, Pomroy RS (eds) Sustainable inland fisheries management in Bangladesh. ICLARM Conf Proc 58:201–208

Ali MI (1998) Society and community based management under the new fisheries management policy. In: Middendorp HAJ, Thompson PM, Pomroy RS (eds) Sustainable inland fisheries management in Bangladesh. ICLARM Conf Proc 58

Anon (2001) Small-scale wetlands indigenous fisheries management project (SWIM): reports 1–7. Living Aquatic Resources Research Centre (LARReC), Vientiane

Arthington A, Welcomme RL (1994) The condition of large river systems of the world. In: Armantrout NB (ed) Condition of the world's aquatic habitats. (Proceedings of the world fisheries congress theme 1) Oxford/IBH, New Delhi, pp 44–75

Baijot EJ, Moreau J, Bouda S (1994) Aspects hydrobiologiques et piscicoles des retenues d'eau en zone soudano-sahélienne. Centre Technique de Coopération Agricole et Rurale, Wageningen

Baird IG (1999) Towards sustainable co-management of Mekong River inland aquatic resources, including fisheries, in southern Lao PDR. In: ICLARM (ed) International workshop on fisheries co-management. ICLARM, Penang

Barbier P, Kalimanzira C, Micha JC (1985) L'aménagement de zones marécageuses en écosystèmes agro-piscicoles. Le project de Kirarambogo au Rwanda (1980–1985). FUCID, Namur

Bonifaz N, Campos M, Castelo R (1985) El chame. Fundacion Ciencia, Quito

Bunn SE, Arthington AA (2002) Basic principles and ecological consequences of altered flow regimes for aquatic biodiversity. Environ Manage 30:492–507

Chevey P, LePoulain F (1940) La peche dans lex eaux douces du Cambodge. Mem lnst Oceanogr Indochin 5

Chikafumbwa FJK, Katambalika K, Brummett RE (1998) Community managed „thamandas" for aquaculture in Malawi. World Aquacult 29:54–59

Chomchanta P, Vongphasouk P, Chanrya S, Soulignavong C, Saadsy B, Warren TJ (2000) A preliminary assessment of Mekong fishery conservation zones in Southern Lao PDR, and recommendations for further evaluation and monitoring. (LARReC technical report 1) The Living Aquatic Resources Research Center (LARReC), Vientiane

COFAD (2002) Back to basics; traditional inland fisheries management and enhancement systems in sub-Saharan Africa and their potential for development. Deutsche Gesellschaft für Technische Zusammenarbeit, Eschborn

Costa-Pierce B (1997) From farmers to fishers; developing reservoir aquaculture for people displaced by dams. (Technical paper 369) World Bank, Washington, D.C.

Cowx IG (1998) Stocking and introduction of fish. Fishing News/Blackwell, Oxford

Daep L, Degen P, Zalinge N van (2003) Fishing gears of the Cambodian Mekong. (Cambodian fisheries technical paper series, vol IV) Inland Fisheries Research and Development Institute of Cambodia (IFReDI), Phnom Penh

Davies WD (1973) Managing small impoundments and community lakes. Meet Southeast Assoc Game Fish Commiss 27:14–17

Drijver CA, Van Wettten JCJ, Groot WT de (1995) Working with nature: local fishery management on the Logone floodplain in Chad and Cameroon. In: Van den Breemer JPM, Drijver C, Venema B (eds) Local resource management in Africa. Wiley, New York, pp 29–45

Dynesius M, Nilsson C (1994) Fragmentation and flow regulation of river systems in the northern third of the world. Science 266:753–762

Dyson M, Bergkamp G, Scanlon J (2003) Flow – the essentials of environmental flows. IUCN, Gland

Erikson CL (2000) An artificial landscape-scale fishery in the Bolivian Amazon. Nature 408:190–193

FAO (2005) Fishstat plus. Available at: www.fao.org/fi/statist/fisoft/fishplus.asp

FAO/ICLARM/IIRR (2001) Integrated agriculture–aquaculture: a primer. (Fisheries technical paper T407) FAO, Rome

Fernando CH (1997) Reservoir fisheries of Southeast Asia: past, present and future. Proc IPFC 17:475–489

Fernando CH, Holcik J (1982) The nature of fish communities: a factor influencing the fishery potential and yields of troical lakes and reservoirs. Hydrobiologia 97:127–140

Funge-Smith S (1999) Small scale rural aquaculture in Lao PDR. FAO, Vientiane

Garaway CJ, Lorenzen K, Chamsingh B (2001) Developing fisheries enhancement in small water bodies: lessons from Lao PDR and Northeast Thailand. In: De Silva SS (ed) Reservoir and culture-based fisheries: biology and management. Australian Centre for International Agricultural Research, Canberra, pp 227–234

Graaf G de, Waltermath M (2003) Fisheries enhancement and participatory fisheries management in small water bodies in Burkina Faso. Lessons learned 1, Deutsche Gesellschaft für Technische Zusammenarbeit, Eschborn

Gregory R, Guttman H (2002) The rice field catch and rural food security. In: Edwards P, Little DC, Demaine H (eds) Rural aquaculture. CAB, Oxford, pp 1–13

Halls AS (2005) The use of sluice gates for stock enhancement and diversification of livelihoods. (Fisheries assessment report R8210). MRAG, London

Halwart M, Gupta MV (eds) (2004) Culture of fish in rice fields. FAO/WorldFish Center, Rome

Hartmann W, Dugan P, Funge-Smith S, Hortle KG, Kuemlangan B, Lorenzen K, Marmulla G, Mattson N, Welcomme RL (2004) People and fisheries management In: Welcomme RL, Petr T (eds) Proceedings of the second international symposium on the management of large rivers for fisheries vol I. (RAP publication 2004/16) FAO Regional Office for Asia and the Pacific, Bangkok, pp 61 – 92

Hasan MR, Bala N, Middendorp HAJ (1998) Secchi disc as a tool to determine stocking density and predict fish yield in culture based fisheries. In: Middendorp, HAJ, Thompson PM, Pomroy RS (eds) Sustainable inland fisheries management in Bangladesh. ICLARM Conf Proc 58

Hoggarth DD, Koeshendrajana S, Aeron-Thomas M, Garaway C, Halls AS, Nasution Z, Samuel S, Sarnita A (2003) An integrated assessment on Indonesian river fishery reserves, introduction and study design. Indones Fish Res J 9:1–6

Hu B, Lui Y (1998) The development of cage culture and its role in fishery enhancement in China. In: Petr T (ed) Inland fishery enhancements. (FAO fisheries technical paper) FAO, Rome, pp 374–463

Islam MZ (1998) Enhancement of floodplain fisheries: experience of the third fisheries project. In: Middendorp HAJ, Thompson PM, Pomroy RS (eds) Sustainable inland fisheries management in Bangladesh. ICLARM Conf Proc 58:209–218

Lorenzen K, Juntana J, Bundit J, Tourongruang D (1998a) Assessing culture fisheries practices in small waterbodies: a study of village fisheries in north-east Thailand. Aquacult Res 299:211–224

Lorenzen K, Garaway CJ, Chamsingh B, Warren TJ (1998b) Effects of access restrictions and stocking on small water body fisheries in Laos. J Fish Biol 53[Suppl A]:345–357

Lorenzen K, Nguyen Khoa S, Garaway C, Arthur R, Kirkwood G, Chamsingh B, Litdamlong D, Innes-Taylor N, Siebert D (2000) Impacts of irrigation and aquaculture development on small scale aquatic resources. (R 7235) Department for International Development, London

Luu LT, Trang PV, Cuong NX, Demaine H, Edwards P (2003) Promotion of small-scale aquaculture in the Red River Delta, Vietnam. In: Edwards P, Little DC, Demaine H (eds) Rural aquaculture. CAB, Oxford, pp 55–75

Micha JC, Halen H, Rosado Couoh J-L (1992) Changing tropical marshlands into agro-pisciculture ecosystems. In: Maltby E, Dugan P, Lefeuvre JC (eds) Conservation and development: the sustainable use of wetland resources. IUCN, Gland, pp 179–186

Middendorp HAJ, Thompson PM, Pomroy RS (1998) Sustainable inland fisheries management in Bangladesh. ICLARM Conf Proc 58

MRC (2003) State of the basin report: 2003. Mekong River Commission, Phnom Penh

Nilsson C, Reidy C, Dynesius M, Revenga C (2005) Fragmentation and flow regulation of the world's large river systems. Science 308:405–408

Payne AI, Cowan V (1998) Review of stock enhancement in the floodplains of Bangladesh. In: Petr T (ed) (1998) Inland fishery enhancements. FAO Fish Tech Pap 374:153–158

Petr T (1998) Inland fishery enhancements. FAO Fish Tech Pap 374

Phan PD, De Silva SS (2000) The fishery of the Ea Kao reservoir, southern Vietnam: a fishery based on a combination of stock and recapture, and self-recruiting populations. Fish Manage Ecol 7:251–264

Phillips MJ (2002) Freshwater aquaculture in lower Mekong basin. (MRC technical paper 7) Mekong River Commission, Phnom Penh

Pinter K (1983) Hungary. In: Brown EE (ed) World fish farming, cultivation and economics. AVI, Westport, pp 297–304

Pitcher TJ, Hollingsworth CE (2002) Fishing for fun: where's the catch? In: Pitcher TJ, Hollingsworth CE (eds) Recreational fisheries: ecological, economic and social evaluation. Blackwell Science, Oxford, pp 1–16

Reed WJ, Burchard AJ, Hopson AJ, Jenness J. Yaro I (1967) Fish and fisheries in Northern Nigeria. Ministry of Agriculture, Zaria

Reizer C (1974) Definition d'une politique d'amenagement des ressources halieutiques d'un ecosystems aquatique complex par i'etude de son environemenl abiotique, biotique et anthropique. Le fleuve Senegal moyen et inferieur. PhD thesis, Fondation Universitaire de Luxembourg Arlon, Arlon

Sarch M-T, Birkett C (2000) Fishing and farming at Lake Chad: response to lake-level fluctuations. Geogr J 166:156–172

Sinha VRP (1991) Paddy-fish production systems. In: Sinha VRP, Srivastava HC (eds) Aquaculture productivity. Oxford/IBH, New Delhi, pp 373–382

Sripatrprasite P, Lin C K (2003) Stocking and recapture of freshwater prawn (*Macrobrachium rosenbergii* De Man) in a run-of-river type dam (Pak Mun Dam) in Thailand. Int Symp Manage Large Rivers Fish2:11–14

Sugunan VV (1995) Reservoir fisheries of India. (FAO fisheries technical paper 345) FAO, Rome

Sugunan VV, Sinha VRP (2001) Sustainable capture and culture-based fisheries in freshwaters of India. In: Pandian TJ (ed) Sustainable Indian fisheries. National Academy of Agricultural Sciences, New Delhi, pp 43–70

Thai Ngoc Chien, Sollows JD, Nguyen Quoc An, Phan Dinh Phuc, Nguyen Quoc Ngai, Troung Ha Phoung (2001) Is Lak lake overfished? In: De Silva SS (ed) Reservoir and culture-based fisheries: biology and management. Australian Centre for International Agricultural Research, Canberra, pp 71–80

Thompson PM, Sultana P, Islam MN, Kadir MM, Hossain MM, Kabir MS (1999) Overview of the community-based fisheries management project: achievements and implications. In: Thompson PM, Grover J (eds) Community based fisheries management and future strategies for inland fisheries in Bangladesh. Department of Fisheries, Dhaka

Virapat C, Phimonbutra U, Chantarawaratid C (2000) Fishery and fisheries management in Thai reservoirs: review and assessment. (Internal report for the MRC project 'Management of reservoir fisheries in the Mekong basin') MRC, Vientiane

Welcomme RL (1971) Rapport au Gouvernement du Dahomey sur l'evolution de la pêche intérieure, son état actuel et ses possibilités. [FAO/UNDP (TA) 2938] FAO, Rome

Welcomme RL (1976) Approaches to resource evaluation and management in tropical inland waters. (IPFC/76/SYM/48) FAO, Rome

Welcomme RL (1988) International introductions of inland aquatic species. FAO Fish Tech Pap 294

Welcomme RL (1999) A review of a model for qualitative evaluation of exploitation levels in multi-species fisheries. J Fish Ecol Manage 6:1–20

Welcomme RL (2002) An evaluation of tropical brush and vegetation park fisheries. J Fish Manage Ecol 9:175–188

Welcomme RL, Bartley DM (1998) An evaluation of present techniques for the enhancement of fisheries. J Fish Ecol Manage 5: 351–382

Welcomme RL, Halls A (2004) Dependence of tropical river fisheries on flow. In: Welcomme RL, Petr T (eds) Proceedings of the second international symposium on the management of large rivers for fisheries, vol II. (RAP publication 2004/17) FAO Regional Office for Asia and the Pacific, Bangkok, pp 267–284

Welcomme RL, Petr T (eds) (2004) Proceedings of the second international symposium on the management of large rivers for fisheries, vol I. (RAP publication 2004/16) FAO Regional Office for Asia and the Pacific, Bangkok

Wijenayake WM, Jayasinghe EA, Amarasinghe US, Athula JA, Pushpalatha KB, DeSilav SS (2005) Culture based fisheries in non-perennial reservoirs in Sri Lanka: production and relative performance of stocked species. J Fish Manage Ecol 12:249–258

Section III

Wetland Biogeochemistry

9 Hydrological Processes, Nutrient Flows and Patterns of Fens and Bogs

WLADIMIR BLEUTEN, WIEBE BORREN, PAUL H. GLASER, TAKEO TSUCHIHARA, ELENA D. LAPSHINA, MARKKU MÄKILÄ, DON SIEGEL, HANS JOOSTEN, MARTIN J. WASSEN

9.1 Introduction

The total area of northern peatland has been estimated to be 5.8×10^6 km^2 (Table 9.1) A great deal, particularly in Europe, has been drained for agriculture, forestry, housing and other human use. The ability of wet, pristine peatland (mires) to accumulate atmospheric carbon, to clean water and to buffer river floods and their values of characteristic biodiversity urges for conservation action and the restoration (rewetting) of drained peatland. Hydro-ecological field data from pristine mire ecosystems will enhance modeling approaches for the prediction of effective water and land use management aimed at mire conservation and restoration.

The scarce information on pristine mires leads to unjustified negative opinions and sometimes even biased scientific results. For an estimation of the world surface area of peatland, most publications refer to the data compiled by Gorham (1991). This leads to the effect that the importance of Russian peatland is strongly underestimated in climate change modeling, i.e. the depletion of atmospheric carbon by terrestrial sinks is much larger because the surface area of Russian peatland is about double the size of 1.5×10^6 km^2 given by Gorham (1991). Merely in Western Siberia, the surface area of peatlands is more then 1×10^6 km^2 (Yefremov and Yefremova 2001). Further, more then 90 % of the Russian peatland is still unaffected by draining activities and is sequestering atmospheric carbon.

Most of the bogs and fens in other regions have been drained by human activities aimed at lowering the groundwater table, extracting the groundwater for water supply and irrigation, regulating the river channel and discharge, etc. and can be designated today as former mires or peatland. In Europe and

Ecological Studies, Vol. 190
J.T.A. Verhoeven, B. Beltman, R. Bobbink, and D.F. Whigham (Eds.)
Wetlands and Natural Resource Management

Table 9.1 Surface area of peatlands in Northern America and Eurasia

Location	Surface area of peatlands (km^2)	Reference
Asian Russia	3 121 000	Vomperskiy et al. (1999)
Europe	916 262	Bragg (2002)
United States of America	550 000	Gorham (1991)
Canada	1 270 000	Waddington et al. (1998)
Total northern peatland	5 857 262	

beyond, the wish to restore fens and bogs is growing because of the high biodiversity values; but because of the slow adaptation of vegetation to changed hydrological conditions, the level of degradation is difficult to establish. Therefore, predicting the effectiveness of rewetting aimed at nature restoration is hampered by a lack of knowledge of the pristine hydrological and hydrochemical (nutrient) conditions of fen ecosystems.

This chapter reports on empirical and modeling studies performed in relatively undisturbed mires in the northern hemisphere of America, Europe and Asia. First, the major mire vegetation types of Western Siberia are presented, next we go deeper into regional and local hydrological studies performed in bogs and fens and finally we summarize the results of an ecohydrological analysis performed in the Ob valley.

9.2 Appearance of Pristine Fens and Bogs

9.2.1 General

Mires reveal close relations between topography, hydrology, hydrochemistry and vegetation (Kulczynski 1949; Moore and Bellamy 1973; Ratcliffe 1977; Gore 1983). In general, differences in the topographical and hydrological position of mires give rise to different types of mire water, especially concerning nutrient and calcium contents and acidity, which in turn determine the floristic composition and diversity of the mire vegetation (Sjörs 1950; Succow and Joosten 2001).

Peat formation is effective where plant litter decay is reduced by low redox potential, as happens in water-saturated conditions. Such conditions can develop in climates suitable for plant growth (trees, herbs, mosses, lichens) at places where precipitation exceeds evapotranspiration and which have a low soil infiltration capacity. Peat bog mires occur in particular at places with a flat topography and large distances between rivers and drainage channels, where water-table mounds can develop. Such mounds favor the growth of raised bogs by: (1) driving local recharge (downward) flow systems that isolate the surface waters from the influx of inorganic solutes derived from mineral soil and (2) allowing the water-table to mound upward and saturate new layers of peat.

Mires fully dependent on nutrient supply by dry and wet (snow included) atmospheric deposition are typified as 'ombrotrophic'. Mires which in addition to atmospheric deposition receive dissolved nutrients and minerals by groundwater flow are mostly named 'minerotrophic' or 'rich' fens. Poor (oligotrophic) fens receive water and nutrients from the atmosphere and from adjacent ombrotrophic bogs. Starting from groundwater-fed mires, raised (domed) bogs can develop while losing contact with groundwater.

Big mire complexes present in Canada, United States and Siberia can include various mire types. Raised bogs, developed mostly at or near the water, divide the discharge surplus water to downstream stretches of large poor fen 'lawns', occasionally with teardrop-shaped bog islands. Further down and between the teardrop islands, patterned ridge-hollow (string-flark) bogs develop, with elongated hollows perpendicular to the water flow direction. Also, water tracks may be present in these poor fens. At locations where water tracks cut into the mineral subsoil, where the peat layer is thin or at the margins of mire complexes groundwater seeps up, feeding the minerotrophic fens.

9.2.2 Climate and Mire Vegetation of the Western Siberian Taiga

By its characteristics of vegetation and climate, the taiga zone of Western Siberia between the Ural mountains and the Yenisey river has been split into three main zones: Southern Taiga (56–58° N), Middle Taiga (58–61° N) and Northern Taiga (61–64° N). From south to north, precipitation increases (0.51–0.57 m year^{-1}) and there is a decrease in both radiation (0.32–0.25 kCal m^{-2}) and summer temperature (average July: 18–14 °C).

As a result of this climate gradient, higher evapotranspiration rates facilitate slightly higher average ion concentrations in the Southern Taiga compared to the Northern Taiga. The electroconductivity of peat water diminishes from 5–12 mS m^{-1} in the south to 3–7 mS m^{-1} in the north, mainly caused by a change in hydrocarbonate content. Nitrate and orthophosphate contents are generally lower in the north (0.4–0.1 g NO_3^- m^{-3}, 1.1–0.2 g PO_4^{3-} m^{-3}); but only for the bog type 'ryam' (Table 9.2) nitrate decreases exponentially with latitude (r^2=0.77). Nutrient concentrations were explained better by different mire types then by climatic zones (Bleuten and Lapshina 2001).

Table 9.2 gives the vegetation, species and species diversity of the major mire types of the Southern Taiga, ordered along increasing nutrient availabil-

Table 9.2 Total number of species by layer (tree/herbs/mosses and lichens) and dominant plant species of main mire types of Western Siberian taiga zones, based on 1278 Braun-Blanquet vegetation relevées

Mire type	Description	Species data	Species
1	Pine dwarf shrubs bogs ('ryam')	76 species (25/32/19)	*Ledum palustre, Chamaedaphne calyculata, Oxycoccus microcarpa, Rubus chamaemorus, Drosera rotundifolia, Eriophorum vaginatum, Sphagnum fuscum, S. magellanicum, S. angustifolium*
2	Hollows and *Sphagnum* lawns	33 species (1/18/14)	*Carex limosa, Scheuchzeria palustris, Eryophorum vaginatum, E. russeolum, Drosera rotundifolia, D. anglica, Rhynchospora alba, Sphagnum balticum, S. papillosum, S. lindbergii, S. jensenii, S. majus, Cladopodiella fluitans*
3	Poor through-flow fens and water tracks		
3.1a	Poor sedge *Sphagnum* moss fens	43 species (–/16/27)	*Carex rostrata, Chamaedaphne calyculata, Oxycoccus palustris, Menyantes trifoliata, Drosera anglica, Sphagnum fallax*
3.1b	Poor sedge–herb *Sphagnum* moss fens	106 species (–/49/57)	*Carex lasiocarpa, Comarum palustre, Lysimachia thyrsiflora, Galium trifidum, Calliergon stramineum, Sphagnum obtusum, S. centrale, S. teres, S. subsecundum, S. platyphyllum, S. contortum, Schistichylopsis laxa*
3.2	Poor sedge brown moss fens	89 species (–/36/53)	*Carex lasiocarpa, C. elata* ssp. *omskiana, C. chordorrhiza, C. heleonastens, Utricularia intermedia, Juncus stygius, Scorpidium scorpioides, Warnstorfia exannulata, W. fluitans, Meesia triquetra, Campylium polygamum, Pseudocalliergon trifarium*
4	Rich sedge brown moss fens	97 species (–/30/67)	*Carex diandra, Thelypteris palustris, Saxifraga hirculus, Stellaria crassifolia, Epipactis palustris, Dactylorhiza incarnata, Parnassia palustris, Drepanocladus sendtneri, D. aduncus, Hamatocaulis vernicosus, Bryum pseudotriquetrum, Tomentypnum nitens*
5	Wooded swamps ('sogra')	373 species (14/153/206	*Carex elongata, C. loliacea, C. disperma, Dryopteris cristata, Corallorhiza trifida, Listera cordata, L. ovata, Malaxis monophyllus,* et al., many liverworts, as well as *Mnium stellare, Thuidium recognitum, T. philibertii, Timmia megapolitana, Rhizomnium punctatum, R. pseudopunctatum*
6	Tussock–sedge flood plain mires	153 species (–/68/85)	*Carex cespitosa, C. juncella, Salix cinerea, S. rosmarinifolia, Spiraea salicifolia, Calamagrostis purpurea, C. phragmitoides, Lathyrus palustris, Veronica longifolia, Lysimachia vulgaris, Scutellaria galariculata*

ity (from ombrotrophic to minerotrophic). Bog mires of *Sphagnum*–dwarf shrub communities with stunted (0.5–4.0 m) pines, in Russian language called 'ryam' (type 1), cover large areas and are present as ridges in patterned bog complexes, together with hollows and lakes. Ryam-type present in the Southern Taiga resembles the permafrost types like flat palsas and ridges of patterned bogs of the Northern Taiga. Actual vegetation differs by the low coverage of lichens in the Southern Taiga (ca. 6 %) compared to the Middle Taiga (14 %) and the Northern Taiga zone (69–85 %). Ryam types with higher trees are missing in the North. Hollows and *Sphagnum* lawns (type 2) are limited to ombrotrophic conditions of watershed plains and some parts of old river terraces. The species diversity of hollow vegetation types, which sometimes cover large waterlogged areas on the flat top of watershed mires, appear to be very low: only 33 species including mosses and lichens have been recorded.

Through-flow fens (type 3) receive water from oligotrophic bog complexes and from 'teardrop' ryam bog islands. Water flows very slowly over the upper peat layer and through the living plant and moss layer of these types. Locally, this water converges into water tracks. Two sub-types and two variants have been discerned.

The first variant (variant sub-type 3.1a) consists of oligotrophic sedge–*Sphagnum* moss of through flow fens within large ombrotrophic peatland. The second variant (variant sub-type 3.1b) comprises slightly minerotrophic sedge–herb–*Sphagnum* moss fens, also called transitional mires, developed on the periphery of large watershed mires and along the uplands of river terraces.

Poor sedge–brown moss fens (sub-type 3.2) are widespread in the relatively drier climate parts of the Southern Taiga zone watershed plain between the Ob and Irtysh rivers areas.

Very wet rich brown moss fens (type 4) fed by carbonate-rich groundwater have developed in non-flooded parts of river valleys and on lower river terraces. Rich brown moss fens, as well as poor brown moss through-flow fens are only present in the Southern Taiga and sub-Taiga zone (transition between Southern Taiga zone and Steppe zone), caused by a subsoil of carbonate-rich loam sediments. Probably the very low net precipitation may cause more exchange with deeper groundwater, as also was reported from the Glacial Lake Agassiz Peatland study (Glaser et al. 2004a, b).

A wide variety of woodland fens on shallow to thick (up to 6 m) peat, common in Western Siberia is represented in this classification by only one type (type 5). These fens are mostly fed by groundwater, are characterized by a high species richness and include forests of tall trees. In the nutrient-rich flooded parts of the river plain, extensive mires (type 6) with dense shrubs and sometimes high (up to 1 m) tussocks of sedges are present.

9.3 Hydrology of Bogs: Examples from Canada, United States and Western Siberia

9.3.1 Aspects of Large-Scale Hydrology

For peat accumulation, the soil surface should be water-saturated for a great part of the growing season. Such conditions can be found in shallow freshwater wetlands as lakes, marshes, river valleys and also on broad flat land of water divides. The Red Lake Peatland (1200 km^2, 48° N, 95° W) in the United States (Glaser et al. 1981) and the Great Vasyugan Bog Complex in Western Siberia (57 000 km^2, 56–58° N, 76–83° E) which both developed on a flat water divide area (Bleuten and Lapshina 2001) are examples of the latter type of areas. There, the classic sequence of mire types, as described above, from ombrotrophic bogs to minerotrophic fens, is present in a downhill sequence.

Deviations from these generally accepted relations between groundwater flow and raised bog formation have been found for the Hudson Bay Lowlands (HBL: 51° N, 83° W) along the Albany River (Canada) and the Glacial Lake Agassiz Peatland (GLAP: 48° N, 94° W) region of Minnesota (USA). The large bog mire complex of the Hudson Bay Lowlands (51° N, 83° W) along the Albany River (Canada) have been built up in an area the drainage of which was disturbed by post-glacial isostatic upheaval (Glaser et al. 2004a, b). There, the rate of uplift increased spatially toward an uplift dome centered over the present mouth of James Bay. The emerging coastline of the lowlands has therefore been rising at a faster rate than regions of the interior throughout the post-glacial period, thereby: (1) decreasing the regional gradient, (2) raising the regional water-table and (3) driving the formation and development of peatland. This hypothesis is supported by the synoptic analysis of vegetation, water chemistry, hydrogeology and peat stratigraphy of representative peat landform types across the study area, which straddles a peat-covered till plain and subtly mounded moraine (Glaser et al. 2004a, b). Broad fen water tracks, in contrast to a normal position, developed down-slope from the moraine complex and in the lower reaches of the local catchments on the till plain, where mineral solutes are transported upward through the peat cover by advective dispersion along lateral flow paths or groundwater discharge.

In the Glacial Lake Agassiz Peatland (48° N, 94° W) region of Minnesota (USA), however, peatland development was largely driven by climate-driven flow systems. The actual regional climate, however, is unexpectedly dry with only a slight moisture surplus and periodic droughts that perturb groundwater flow systems. Initial local and synoptic hydrogeologic surveys in this region showed that spring-fens and some fen water tracks were consistently located over discharge zones for groundwater, whereas most of the larger fen water tracks developed in areas of lateral flow. One unexpected result of this work was that the raised bogs also developed over discharge zones for groundwater (Siegel 1983; Romanowicz et al. 1995; Siegel et al. 1995; Glaser et al. 1997). During moist periods, water-table mounds within these bogs drove surface waters downward and deflected the upwelling groundwater to the bog margins, but during droughts, the water-table mounds dissipated, permitting the groundwater to rise closer to the peat surface.

9.3.2 Local Scale Hydrology of Bogs

Water flow systems of bog mires include three main processes: (1) overland flow through the green part of the bog ('acrotelm') above the mean lowest groundwater level, (2) sub-surface shallow groundwater flow through the upper peat layers and (3) deeper groundwater flow. Of these three, overland flow is considered the most important. In order to test this hypothesis, an intensive measurement program monitoring pH and electrical conductivity (EC) has been completed in the Haukkasuo concentric bog (60°49' N, 26°57' E) in southeastern Finland (Mäkilä 1997). The monitor network included 710 measurements of temperature and EC in two perpendicular and vertical cross-sections (length: 1900 m and 1150 m). The distance between the probing sites was 100 m and the measurements were made from the bog surface to the bottom, at 0.25 m spacing. The longer transect was measured six times at different seasons. In four sites, the peat column was sampled down to the bottom by taking 89 volumetric samples with a special piston sampler.

Acidity

The acidity of the Haukkasuo bog decreases quite smoothly from pH 3.2–3.5 at the surface towards pH 5.4–6.0 at 6 m below the peat surface at the central parts of the bog. Near the bog margins, the acidity decrease is faster with depth: pH 6.0 is reached at 2.0–2.5 m below the peat surface. The source of acidity is located near the bog surface, where free hydrogen ions are produced by the cation exchange ability of *Sphagnum* moss and peat, and where organic acids are released by the aerobic decomposition of *Sphagnum* (Shotyk 1988). In addition, the bog surface receives acid precipitation water (pH=4.6). The highest pH values (pH>6.0) are reached near the mineral soil substratum of the Haukkasuo bog, where the supply of bases is sufficient to neutralize the acidity.

Conductivity

On most profiles, the conductivities slowly decrease, in parallel with acidity, from the bog surface to the boundary between *Carex* and *Sphagnum* peats (Figs. 9.1, 9.2). Below this boundary, the conductivities increase with mineral content down to the mineral subsoil. The increasing trend of conductivities continues from the *Carex* layer into the clay substratum, indicating the trans-

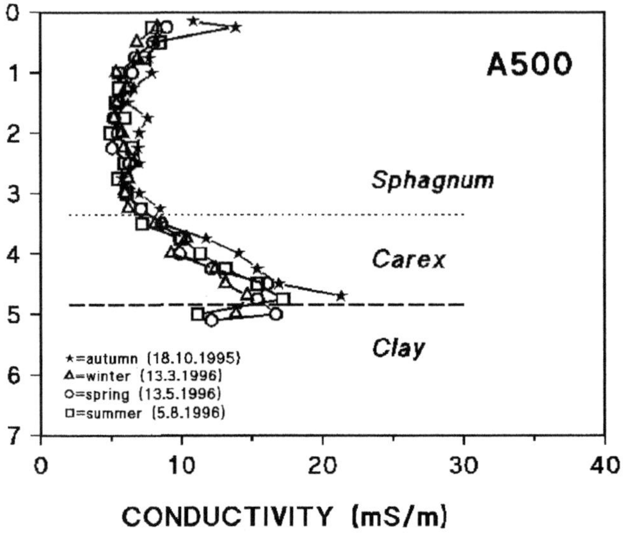

Fig. 9.1 Vertical profile of measured electrical conductivity (mS m^{-1}) by season. *Y-axis* Depth below peat surface (m)

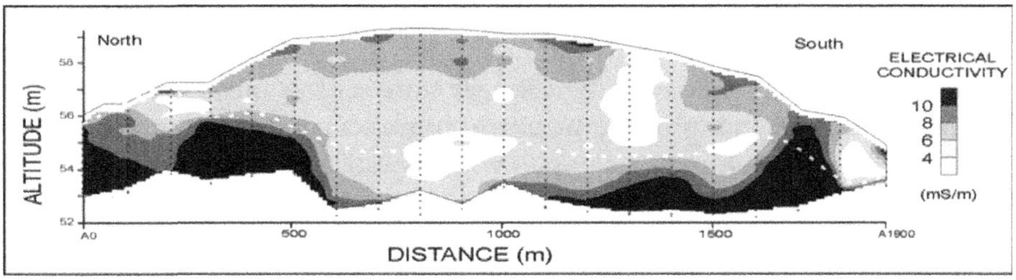

Fig. 9.2 Cross-section through the Haukkasuo bog mire (Finland) with electrical conductivity gradients. Above the *white hatched line*, peat is formed by *Sphagnum* sp., below this line peat is formed by *Carex* sp

fer of ions by diffusion and advection across the peat–clay interface. The conductivity of the basal layers is lower in the central parts of the bog in comparison with the margin, which is clearly seen in the conductivity cross sections of Haukkasuo (Fig. 9.2). The lowest conductivities (about 4 mS m^{-1}), being somewhat higher than the mean conductivity of precipitation in the area (2.3 mS m^{-1}), have been measured in the interior parts of Haukkasuo, extending to the base of the bog in the center. The basal layers of the marginal areas of the Haukkasuo bog are characterized by higher conductivities, rising to the bog surface near its northern and southern margins (Fig. 9.2). In the basal *Carex* peat layers, the conductivity of peat pore water and the pH values are clearly lower near the bog center as compared to the intervening and marginal areas.

The increase of conductivity towards the base is weak near the bog center, where infiltrated surface water flows downwards, but is strong near the margins, where solute-rich groundwater is probably upwelling from the sediments below the peat.

Water flowing in the lower parts of the bog seems to dissolve mineral material and becomes more and more mineralized on the way to the discharge sites at the bog margins. The conductivity pattern proves that a consistent water flow system exists within the bog mire. The measured peat densities of Haukkasuo (50–100 kg m^{-3}) correspond approximately to a hydraulic conductivity range of 5×10^{-6} m s^{-1} to 5×10^{-5} m s^{-1}, according to the relations presented by Päivänen (1973) for Finnish peat. This means that water flux through the lower peat layers in the bog is small and that most of the precipitation and snow melt water discharges through the acrotelm and the loose top peat layers. This confirms the generally accepted concept of slow water movement through bogs.

9.3.3 Modeling a Western Siberian Bog

A three-dimensional (3D) dynamic model for forecasting Holocene and twenty-first century peat and carbon accumulation has been developed for the Bakchar Bog complex, situated in the Southern Taiga zone of Western Siberia. This bog complex consists of various mire types and covers an area of about 800 km^2, located at the watershed area between the rivers Bakchar and Iksa (56°49' N, 82°48' E), which are tributaries of the river Ob. The model is based on the (1D) concepts of Ivanov (1981) that: (1) water level doming depends on conductivity and distance to drainage systems (e.g. rivers), (2) peat accumulation follows this doming and (3) development of new peat layers changes the hydrological transmissivity, promoting higher water levels and new peat accumulation, ultimately until water level doming enters equilibrium with discharge capacity. For the 3D approach, a lower conductivity of deeper and older peat layers was supposed. As a result of these concepts, the computed water flow is dominated by surface and sub-surface water flow, according to the findings described above (Mäkilä 1997).

The model was built using the environmental modeling language PCRaster (Wesseling et al. 1996) and the ground-water modeling code MODFLOW-2000 (Harbaugh et al. 2000). The used time step was 100 years and the horizontal resolution 100 m. Vertically, the model allowed for several layers, i.e. the mineral subsoil and several peat layers. To estimate superficial flow conditions, a vegetation layer (acrotelm) with high conductivity was added. The peat accumulation process started around 6500 years B.P. (Borren et al. 2004).

Besides mineral soil, the bog complex includes four mire types within the study area: marshy peat, through-flow fen, ridge-hollow complex and ombrotrophic raised bog ('ryam': see Section 9.2.2). All these five have been implemented in the model.

Note that the peat body is located on a watershed area between rivers, which means that the influence of groundwater and river water is negligible. At places where the water table is above the mineral surface level, marshy peat is formed. As long as the water table is higher than the surface level, the accumulation process continues. As soon as the initial peat deposits are thicker than 0.5 m, a change to other peat types takes place. The transition from one type to another depends on nutrient supply, which in the model was simulated by superficial water-flow conditions. It was assumed that the nutrient supply to vegetation was higher if more water flowed through the vegetation. The main model inputs were hydraulic conductivity, peat growth rate, carbon content, precipitation, evapotranspiration, rivers and digital elevation models (DEM) of the mineral subsoil and the present peat surface. The peat surface DEM is the top of the catotelm.

Table 9.3 shows the used peat growth rates, organic matter densities and carbon contents of the mire types (Bleuten and Lapshina 2001; Borren et al. 2004). Holocene precipitation was derived from the record of Klimanov (personal communication; Fig. 9.3). The net recharge was calculated by multiplying the precipitation with an evapotranspiration factor, which differed

Table 9.3 Properties of various mire and soil types. Peat growth rate, dry organic matter density and carbon content (% of dry organic matter), according to Bleuten and Lapshina (2001) and Borren et al. (2004). Horizontal hydraulic conductivity and evapotranspiration factor (% of precipitation) estimated in this study

Mire or soil type	Peat accumulation rate ($\times 10^{-3}$ m year^{-1})	Dry organic matter density ($\times 10^{3}$ g m^{-3})	Carbon content (% of dry organic matter)	Horizontal hydraulic conductivity (m day^{-1})	Evapo-transpiration factor (%)
Marshy peat	0.5	63.75	54.0	0.0871[a]	85[a]
Through-flow fen	0.35	76.48	53.7	0.0871[a]	85[a]
Ridge–hollow complex	0.39	46.67	53.3	0.0871[a]	85[a]
Ryam	0.42	61.53	47.7	0.0871[a]	85[a]
Mineral soil				3.17	61
Vegetation				337	

[a] One value estimated for all peatland types

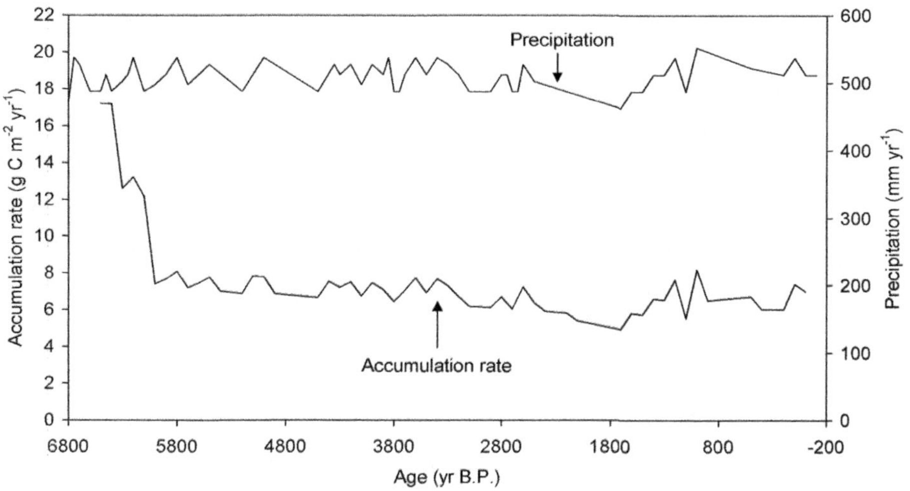

Fig. 9.3 Holocene precipitation record and modeled carbon accumulation; precipitation according to Klimanov (personal communication)

between peat-covered area and mineral soil. The evapotranspiration factors and the hydraulic conductivities of the various layers were estimated by calibrating the model to present hydrological conditions with the PEST algorithm. During a field expedition in August 2002 across the Bakchar Bog, the groundwater level was at the surface (hollows) or 0.1–0.2 m below the peat surface level (ridges, ryam).

Figure 9.4 shows a comparison of the peat distribution over the study area as derived from the DEM, which is the real present situation, and as simulated by the model. At 87 % of the cells, the model prediction, whether there is peat or not, is correct.

The modeled carbon accumulation rate in the study area shows a decrease from 6500 years B.P. to the present (Fig. 9.3). Over this period, a gradual decrease from 17.2 g C m^{-2} $year^{-1}$ to 7.0 g C m^{-2} $year^{-1}$ took place. Obviously, the domed water table between the rivers and streams cannot rise to any extent and will eventually reach a certain maximum (Ingram 1982; Winston 1994). The peat body can retain the incoming water to a limited extent and the lateral expansion will decrease as the peat cover approaches the rivers and streams. The expansion rate is highest at the start of the peat formation, which is shown in Fig. 9.3 by the high carbon accumulation rates.

Fluctuations in the carbon accumulation rate correspond to precipitation as shown by comparison with the precipitation record in Fig. 9.4. The total amount of accumulated carbon in the study area was estimated at 47.4×10^3 g C m^{-2}. This is lower than the mean estimate by Yefremov and Yefremova (2001) of 71.0×10^3 g C m^{-2} for peatlands in the Southern Taiga zone of West-

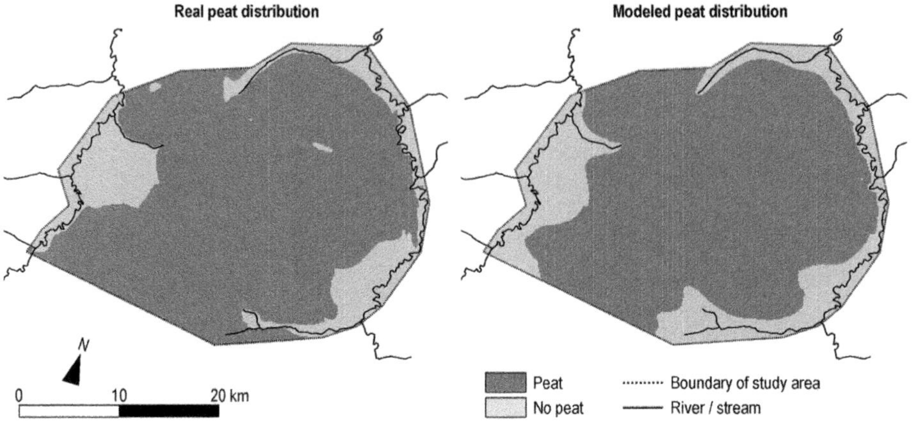

Fig. 9.4 Real and modeled peat distribution within the Bakchar Bog study area. The real situation is based on digital elevation models of mineral subsoil and peat surface

ern Siberia. This difference can be due to the late start of peat formation in our study area. Extrapolation of the model over the twenty-first century gives a near-future carbon accumulation rate of 7.0 g C m^{-2} $year^{-1}$. For the whole study area, this means a carbon accumulation of 4.4×10^9 g C $year^{-1}$.

By this forward modeling, starting from bar mineral soil, the peat thickness and peat presence are predicted quite well, which proves the model concept.

9.4 Fens: Analysis of a Large Pristine Fen in the River Ob Valley

9.4.1 General

Within the vast peatland of Western Siberia, chains (hundreds of kilometers long) of untouched minerotrophic fens are present in the Ob river valley on top of wide Pleistocene terraces which limit the Ob river floodplain. The elevation difference between terraces is 5–20 m. The fens at higher terraces are partly drained for agricultural land use. A section (3–5 km wide, surface area 13 km^2; 56° N, 84° E) belonging to a fen-terrace of ca. 50 km^2 developed on the lower Ob river terraces west from Tomsk was selected to study the hydrology and nutrient conditions (Fig. 9.5).

Peat thickness varies from 5.5 m, in a former river channel eroded in the terrace, near the terrace scarp, to 0.5 m near a secondary Ob channel. By peat accumulation, most of the surface elevation rises above highest Ob river flood

Fig. 9.5 Aerial photograph of the Ob River study site. *Upper left*, upland with forest patches (*dark gray*) and arable land (*light gray*). *Right*, the Ob River. Between the upland and the river is the studied fen on the lower Ob River terrace. The dark patch is a cloud. Numbers indicate the locations of piezometers. Note the darker gray peat ridges stretched half-way up the terrace

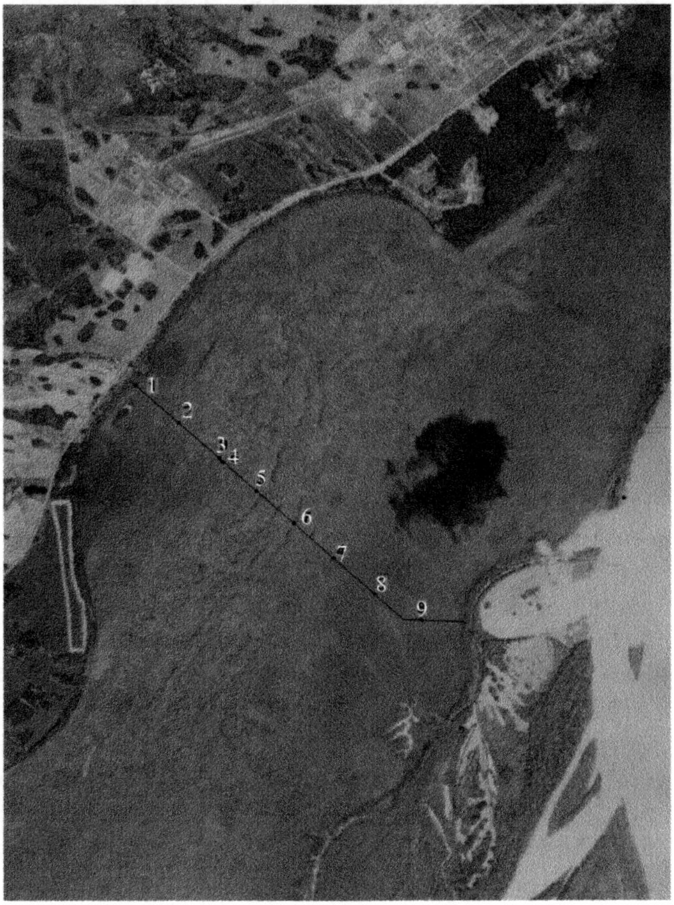

level. Climate is severe continental, with 5.5 months of growing season (average summer temperature, June–August: 18 °C) and precipitation in excess of 0.21 m year^{-1}, of which 60 % as snow (Bleuten and Lapshina 2001).

Because of the extremely hostile conditions of dense bushes, high water table and weak peat soil, the field study had to be limited to a 3-km transect across the fen from the mineral upland (higher terrace edge) to the nearest river Ob tributary (Fig. 9.5). Along the transect, vegetation relevées have been made and samples taken for estimating the biomass productivity for nutrient content. In piezometer tubes, after coring for peat layer measurements placed in sets of three at different depths, the water level (heads) have been measured weekly from May to September 2002. Transect surface elevation was measured with an optical leveler. Water samples were taken at 14-day intervals and analyzed for main ion components, metals, acidity, alkalinity and conductivity.

9.4.2 Vegetation, Nutrients and Productivity

A close relation between topography, hydrology, hydrochemistry and vegetation has been observed in valley mires where distinct vegetation zones are present parallel to the valley axis (Pałzcyński 1984; Heathwaite 1993; Ratcliffe 1977). The width and characteristics of such vegetation zones depend on site conditions, such as the ratio between the supplies of groundwater, surface water and precipitation (Wassen and Joosten 1996; Wassen et al. 2002). In the studied valley fen, a clear vegetation zonation parallel to the valley axis was present. Eight vegetation communities were distinguished, according to their floristic composition (Table 9.4). Based on vegetation characteristics, the studied transect (Fig. 9.6) can be divided into three main zones. Close to the valley edge, a minerotrophic reed zone is located, showing both high-productive, species-poor vegetation dominated by *Phragmites australis* (community 1) and low-productive, species-rich vegetation dominated by *Betula pubescens* (community 2). The open sedge fen located further from the scarp comprises sampling locations 2, 3, 4 and 5 and contains herbaceous vegetation, mainly consisting of sedges (e.g. *Carex lasiocarpa*, *Carex rostrata*) and *Menyanthes trifoliata*. The mixed sedge–shrub and woodland fen (sampling locations 6, 7, 8, 9) is characterized by the dominance of sedge tussocks (*Carex cespitosa*), trees (*Pinus sylvestris*, *Betula pubescens*) and shrubs (*Salix rosmarinifolia*) and shows a relatively high total plant species richness. Community 2 combines a high number of species with a low productivity. Communities 3, 4 and 5 show predominantly phosphorus limitation. The peat chemical data show that soil pH is slightly higher near the valley edge (communities 1, 2) and in the open sedge communities (communities 3, 4, 5) than in the forested communities (communities 6, 7, 8).

9.4.3 Hydrology and Modeling

Water flow in fens is of the 'through-flow' type('Durchströmungsmoore' in Succow and Joosten 2001), which means that most of the horizontal waterflux is at or close to the surface. The groundwater discharges usually into the fens by vertical flow. This concept has been tested by analyses of hydrological field data and modeling.

Using transect data of the cross-section peat profile together with climate data from a nearby meteorological station and river stage data, a 2D steady-state groundwater model was developed based on the MODFLOW code, later enhanced into a 3D model for the entire 13 km^2 fen area. The 2D model was calibrated with measured groundwater heads and the 3D version with a digital elevation model based on available map data. The measured heads were very close to surface elevation and the dynamics of head variation were very

Table 9.4 Species composition and characteristics of vegetation types including peat chemical data and hydrochemical data per vegetation type along the Ob River terrace fen transect. For *vegetation data*, mean values with standard deviation are given (numbers within brackets indicate number of samples). Mean total numbers of species and mean number of vascular plants are based on relevées made in July 2002 in homogeneous areas of 25 m^2 (Dierschke 1994). Covers of vascular plants and mosses were estimated using a modified Braun–Blanquet scale (Wilmanns 1998). Biomass production was measured by calculating the dry weight of the total standing above-ground living biomass of vascular plants (harvested at 4 July and 9–12 July) after drying at 70 °C for 48 h. After extraction of N, P and K content of vegetation using a Kjeldahl digestion, nutrient limitation was determined by calculating the N:P, N:K and K:P ratios in vascular plants using critical ratios (Olde Venterink et al. 2003). The pH of peat (5–15 cm depth) was determined after Rowell (1997) in rewetted samples in 0.01 M $CaCl_2$ using WTW-pH96. The PO_4 of surface water was determined using ICP

	Vegetation type									
	1	2	3		4	5	6	7	8	
Sampling location	0	1	2	3	4	5	6	7	8	9
Distance from terrace (m)	0	80	480	860	960	1180	1630	2030	2500	2850
Vegetation data										
Mean total number of species	18	30	22		16	16	32	40	23	
Mean number of vascular plants	16	21	12		12	11	20	21	12	
Mean tree cover (%)	0	10	1		0.5	0	30	34	13	
Number of relevées	5	14	18		16	10	10	18	14	
Biomass production (g m^{-2})	718±180 (3)	322±79 (6)	361±146 (4)		420±71 (8)	530±211 (3)	no data	no data	no data	
Nutrient limitation (% P-limited)	33 (3)	0 (6)	75 (4)		50 (8)	67 (3)	no data	no data	no data	

Table 9.4 (*Continued*)

	Vegetation type							
	1	2	3	4	5	6	7	8
Species								
Phragmites australis	V/4	III/2a	–	–	–	–	I/1	–
Angelica archangelica	V/2b	–	–	–	–	–	–	–
Carex lasiocarpa	–	V/2a	V/2a	V/2a	III/2a	I/+	III/1	II/1
Carex rostrata	II/+	IV/1	IV/2a	V/2b	V/2a	V/2b	II/+	–
Menyanthes trifoliata	–	V/3	V/2a	V/3	IV/2b	V/2a	V/2b	–
Betula pubescens	–	V/2a	III/1	II/+	I/+	IV/2a	V/2a	V/2a
Petula humilis	–	III/+	V/2b	III/1	III/2a	V/2a	IV/1	–
Drepanocladus aduncus	I/+	II/1	IV/2a	III/1	V/2a	II/1	II/1	II/+
Parnassia palustris	–	IV/+	IV/+	III/+	II/+	III/+	II/+	–
Typha latifolia	–	–	–	–	V/3	–	–	–
Pinus sylvestris	–	–	–	I/+	–	V/2b	V/2b	III/+
Carex cespitosa	V/2b	I/+	–	–	–	IV/2a	IV/2a	III/2a
Salix rosmarinifolia	–	–	IV/1	–	II/+	IV/2a	IV/2a	V/2b
pH	7.2±0.13	7.2±0.21	7.0±0.15	6.8±0.27	7.0±0.14	6.5±0.23	6.8±0.23	6.7±0.15
Number of samples	5	14	18	18	10	14	18	14
PO_4 of surface water	3.78±4.10 (3)	1.31±0.89 (15)	2.44± 1.36 (17)	2.31±1.34 (15)	2.94±3.02 (10)	1.28±0.80 (8)	0.93±0.52 (17)	0.67±0.27 (6)

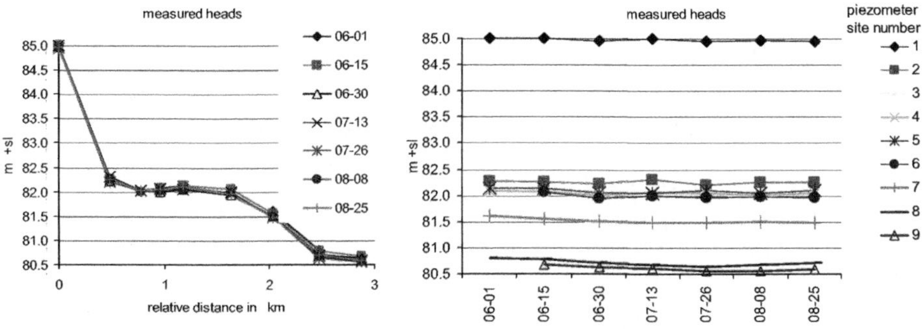

Fig. 9.6 Measured hydraulic heads (groundwater levels) in the Ob River terrace fen from June (*06*) through August (*08*). *Left* Along the transect (see Fig. 9.5), *right* time-courses by piezometer

small (0–0.05 m) during the measured period (Fig. 9.6). This means that the 'ground' water level is at or above the surface because 2-week-averaged precipitation was 0.003–0.0047 m day^{-1} and no drains are present. By using the (vertical 2D model) computed groundwater heads as upper boundary conditions, the groundwater flow lines have been calculated (MODPATH). Close to the terrace scarp, a strong groundwater discharge directed to the surface (seepage) takes place near piezometers 1–2 (Fig. 9.7). The seepage water then flows over 'land' in the direction of the Ob river towards piezometer 3, where the water is dammed up by peat ridges, which result in infiltration and downward percolation of surface (over land flow) water. Closer to the Ob, caused by the relative low river stage, infiltration conditions are also present. The measured EC is clearly affected by the seepage water coming from deep and distant upland (Fig. 9.8) The raised EC values could be explained by correlated salt (NaCl) and hydrocarbonate (HCO_3^-) concentrations. The nitrogen (NH_4^--N + NO_3^--N) concentration is generally low (0.1–1.6 g N m^{-3}) in seepage parts of the transect but is raised (2.5–4.0 g N m^{-3}) where infiltration dominates. The low redox conditions in the groundwater of the seepage parts of the cross-section promote denitrification. In the infiltration parts, the relatively oxygenized water nitrates can stay longer in solution.

With the 3D model, water fluxes by over land flow and seepage were calculated and, with PCRaster (Wesseling et al. 1996), were related to the recharge (net precipitation) flux. The outcome of this exercise was the generally low contribution of rainwater to the water balance (0.01–0.3 %) for most of the model area, except on the peat ridges (10–25 %). The contribution of seepage water to the cell water balance was more than 10 % near the terrace scarp and downstream of the peat ridges and, in the rest of the modeled area, lower than 1 %. This means that most water and dissolved nutrients are transported over land between the plants and sedge hummocks.

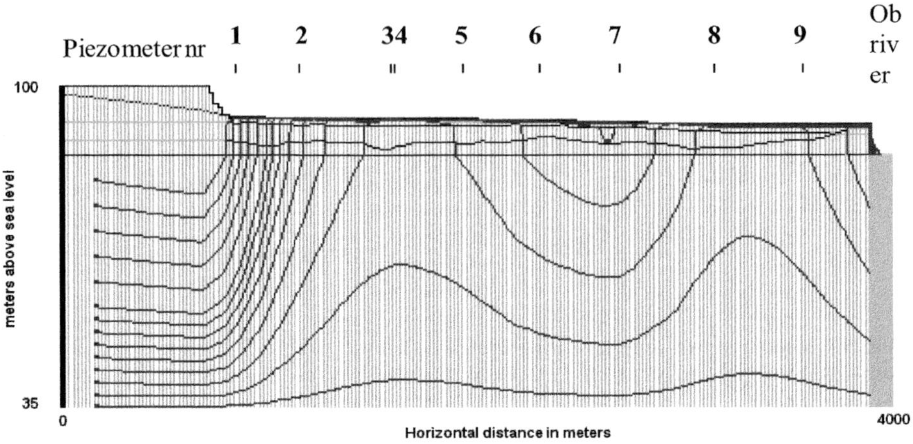

Fig. 9.7 Groundwater flow paths in the vertical cross-section along the measured transect on the Ob River terrace fen

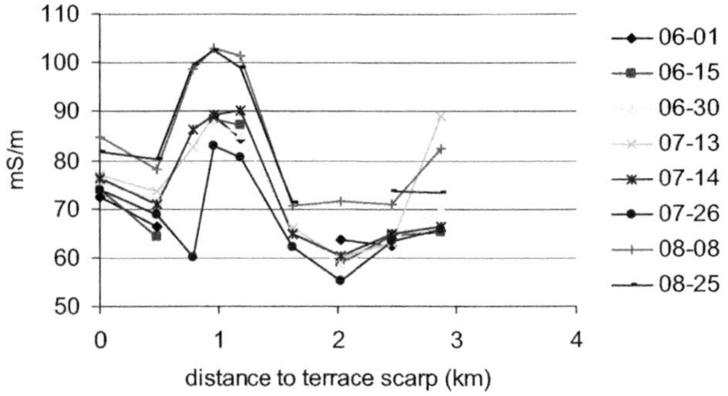

Fig. 9.8 Dynamic and spatial variation in measured peat water electrical conductivity at 1–2 m below the 'surface' along the Ob River terrace fen transect

9.4.4 Hydro-Ecological Integration

Vegetation zonation and hydrochemical characteristics can be explained by hydrological processes. Although there is a strong upward flux of groundwater close to the valley edge, the steep scarp here still is the cause of a significant unsaturated zone, causing relatively dry conditions (Figs. 9.6, 9.7). This may lead to high mineralization rates and more eutrophic conditions. The relatively eutrophic conditions give rise to high-productive, species-poor vegetation dominated by *Phragmites australis* at sampling location 0 and to a high abundance of *Betula pubescens* at sampling location 1. The open sedge fen is also characterized by a strong seepage flux, which in combination with the lower altitude results in relatively high water levels at the mire surface (Fig. 9.6). This leads to the presence of e.g. sedges and *Menyanthes trifoliata* and to the absence of trees, as they (at least periodically) require water levels below the mire surface (Moore and Bellamy 1973; Wassen et al. 1992). The groundwater seepage flux decreases with increasing distance from the valley edge (Fig. 9.7) and the absence of discharge of regional groundwater in the woodland fen (Fig. 9.7) prevents constant inundation of the mire surface. This results in water levels near or below the mire surface at sampling locations 6, 7 and 8 (Fig. 9.6) and enables trees to establish in this zone. This analysis shows a number of close relationships between hydrology, peat chemistry and vegetation zoning. Such relationships are blurred in peatlands dominated by human intervention and thus can only be studied in relatively undisturbed mires.

9.5 Discussion and Conclusion

In Europe and beyond, the wish to restore bogs and fens is growing because of the high biodiversity values; but because of the slow adaptation of vegetation to changed hydrological conditions (e.g. draining), the level of degradation is difficult to establish.

Therefore, predicting the effectiveness of rewetting aimed at nature restoration is hampered by a lack of knowledge of the pristine hydrological and hydrochemical (nutrient) conditions of fen ecosystems. Based on vegetation, a generalized mire type classification, including two bog types and seven fen types, has been developed. Mainly because of permafrost in the Northern Taiga and Forest Tundra zones, not all mire types are present in the entire south–north gradient across the Western Siberian plain. Related to latitudinal climate change, nutrient richness slightly decreases to the North. Nutrient differences between mire types are greater than by latitude.

Bog mires develop in Siberia and North America where groundwater stagnates. This occurs on flat water divide areas and where the drainage system has been changed by geotectonic uplift. In the latter conditions, the 'normal' sequence in a downhill direction from ombrotrophic bog mires via transitional mires, including poor fens and rich fens, to minerotrophic fens can be upset. Water flow on bog mires is predominantly of the 'through-flow' type, which means that most of the surplus precipitation water flows over the peat surface between the mosses and herb roots and tussocks. The dominance of surface runoff over deeper water flow paths has been proved, based on intensive measurements of EC, both vertically and across a Finnish bog mire. This principle has been applied successfully in a 3D MODFLOW-based peat accumulation model of an 800 km^2 large mire complex in the Southern Taiga of Western Siberia.

Also in fen mires, the dominant water flow is of the through-flow type. As hypothesized, close relations between hydrology, hydrochemistry and vegetation result in a zonation of vegetation in the Ob valley mires. Due to the dominance of massif influx of groundwater at the upland margin, differences in water chemistry are small throughout the study area and the differences in vegetation characteristics between the zones are correspondingly small. The dominance of groundwater prevents a marked influence of precipitation on vegetation characteristics in the Ob valley mires, resulting in vast areas with quite low-productive (300–500 g dry weight m^{-2}) species-rich (20–30 species per relevée) fen vegetation dominated by sedges such as *Carex lasiocarpa* and *C. rostrata*, forbs like *Menyanthes trifoliata* and *Parnassia palustris* and brownmosses in the major part of the study area.

In the studied Ob river valley fen, apparently a layer consisting of water was present between the top peat and the (floating) roots and green parts of the fen plants and mosses. Most of the nutrients are transported through this top-layer. The nutrient balance showed that discharge of groundwater is much more important than precipitation (by orders of magnitude). The slow adaptation of fen vegetation after cutting canals for drainage can be explained by this typical layer, which can take many years to disappear as horizontal water flow capability gradually decreases during the desiccation process. Vegetation change can be expected when water balance fluxes change, in particular when groundwater influx (seeping-up of groundwater) disappears (Wassen et al. 2002).

Acknowledgement Parts of this study has been supported by EU-INTAS projects (nrs: 99-1718, 03-51-6294)

References

Bleuten W, Lapshina ED (eds) (2001) Carbon storage and atmospheric exchange by West Siberian peatlands. (FGUU scientific report 2001-1) Utrecht University, Utrecht, 169 pp

Borren W, Bleuten W, Lapshina ED (2004) Holocene peat and carbon accumulation rates in the southern taiga of western Siberia. Quat Res 61:42–51

Boyer MLH, Wheeler BD (1989) Vegetation patterns in spring-fed calcareous fens: calcite precipitation and constraints on fertility. J Ecol 77:597–609

Bragg OM (2002) Hydrology of peat-forming wetlands in Scotland. Sci Total Environ 294:111–129

Dierschke H (1994) Grundlagen und Methoden der Pflanzensoziologie. Ulmer, Stuttgart

Glaser PH, Wheeler GA, Gorham E, Wright HE Jr (1981) The patterned mires of the Red Lake peatland, northern Minnesota: vegetation, water chemistry and landforms. J Ecol 69:575–599

Glaser PH, Siegel DI, Romanowicz EA, Shen YP (1997) Regional linkages between raised bogs and the climate, groundwater, and landscape of north-western Minnesota. J Ecol 85:3–16

Glaser PH, Siegel DI, Reeve AS, Janssens JA, Janecky DR (2004a) Tectonic drivers for vegetation patterning and landscape evolution in the Albany River region of the Hudson Bay Lowlands. J Ecol 92:1054–1070
Glaser PH, Hansen BCS, Siegel DI, Reeve AS, Morin PJ (2004b) Rates, pathways and drivers for peatland development in the Hudson Bay Lowlands, northern Ontario. J Ecol 92:1031–1053
Gore AJP (1983) Mires: swamp, bog, fen and moor. (General studies. Ecosystems of the world, 4a) Elsevier, Amsterdam
Gorham E (1991) Northern peatlands: role in the carbon cycle and probable responses to climatic warming. Ecol Appl 1:182–195
Harbaugh AW, Banta ER, Hill MC, McDonald MG (2000) MODFLOW-2000, the U.S. Geological Survey modular ground-water model user guide to modularization concepts and the ground-water flow process. (Open-File Rep 00-92) US Geological Survey, Washington, D.C., 121 pp
Heathwaite AL (ed) (1993) Mires: process, conservation and exploitation. Wiley, Chichester, 506 pp
Ingram HAP (1982) Size and shape in raised mire ecosystems: a geophysical model. Nature 297:300–303
Ivanov KE (1981) Water movement in mirelands. Academic, London, 276 pp
Kulczynski S (1949) Torfowiska Polesia. Peat bogs of Polesie. Mem Acad Pol Sci Lett Sci Mater Nat Ser B Sci Nat 15:1–359
Mäkilä M (1997) Holocene lateral expansion, peat growth and carbon accumulation on Haukkasuo, a raised bog in southeastern Finland. Boreas 26:1–14
Moore PD, Bellamy DJ (1973) Peatlands. Elek Science, Londen
Olde Venterink H, Wassen MJ, Verkroost AWM, Ruiter PC de (2003) Species richness-productivity patterns differ between N, P and K limited wetlands. Ecology 84:2191–2199
Päivänen J (1973) Hydraulic conductivity and water retention in peat soils. Acta For Fenn 129:1–70
Pałczyński A (1984) Natural differentiation of plant communities in relation to hydrological conditions of the Biebrza valley. Pol Ecol Stud 10:347–385
Ratcliffe D (1977) A nature conservation review, vol 1. Cambridge University Press, Cambridge, 401 pp
Romanowicz EA, Siegel DI, Chanton JP, Glaser PH (1995) Temporal variations in dissolved methane deep in the Lake Agassiz peatlands, Minnesota (USA). Global Biogeochem Cycles 9:197–212
Rowell DL (1997) Bodenkunde – Untersuchungsmethoden und ihre Anwendungen. Springer, Berlin Heidelberg New York
Shotyk W (1988) Review of the inorganic geochemistry of peats and peatland waters. Earth Sci Rev 25:95–176
Siegel DI (1983) Ground water and the evolution of patterned mires, glacial lake Agassiz peatlands, northern Minnesota. J Ecol 71:913–921
Siegel DI, Reeve AS, Glaser PH, Romanowicz E (1995) Climate-driven flushing of pore water in humified peat. Nature 374:531–533
Sjörs H (1950) On the relation between vegetation and electrolytes in North Swedish mire waters. Oikos 2:241–258
Succow M, Joosten JHJ (2001) Landschaftsökologische Moorkunde. Schweizerbart'sche Verlagsbuchhandlung, Stuttgart, 622 pp
Vomperskiy SE, Ciganova OP, Kovalev AG, Glukhova TV, Valyaeva NA (1999) Waterlogged areas of Russia as a factor of consolidation of atmospheric carbon. In: Scientific Council of the Federal Research Program of Russia (ed) Global changes in envi-

ronment and the climate. (Selected scientific papers. Special issue) Ministry of Education of Russia, Moscow, pp 124–145

Waddington JM, Griffis TJ, Rouse WR (1998) Northern Canadian wetlands: net ecosystem CO2 exchange and climatic change. Clim Change 40:267–275

Wassen MJ, Joosten JHJ (1996). In search of a hydrological explanation for vegetation changes along a fen gradient in the Biebrza Upper Basin (Poland). Vegetatio 124:191–209

Wassen MJ, Barendregt A, Palczynski A, De Smid JT, De Mars J (1992) Hydro-ecological analysis of the Biebrza mire (Poland). Wetlands Ecol Manage 2:119–134

Wassen MJ, Bleuten W, Bootsma MC (2002) Biebrza as a geographical reference. Ann Warsaw Agric Univ Land Reclam 33:27–47

Wesseling CG, Karssenberg D, Van Deursen WPA, Burrough PA (1996) Integrating dynamic environmental models in GIS: the development of a dynamic modelling language. Trans GIS 1:40–48

Wilmanns O (1998) Ökologische Pflanzensoziologie, 6th edn. Quelle and Meyer, Wiesbaden.

Winston RB (1994) Models of the geomorphology, hydrology, and development of domed peat bodies. Geol Soc Am Bull 106:1594–1604

Yefremov SP, Yefremova TT (2001) Present stocks of peat and organic carbon in bog ecosystems of West Siberia. In: Bleuten W, Lapshina ED (eds) Carbon storage and atmospheric exchange by West Siberian peatlands. Utrecht University, Utrecht/Tomsk, pp 73–78

10 Ecological Aspects of Microbes and Microbial Communities Inhabiting the Rhizosphere of Wetland Plants

Paul L.E. Bodelier, Peter Frenzel, Harold L. Drake,
Thomas Hurek, Kirsten Küsel, Charles Lovell,
Patrick Megonigal, Barbara Reinhold-Hurek, Brian Sorrell

10.1 Introduction

Wetlands are among the most important ecosystems on earth. Many fossil fuels on which we now depend were formed in ancient wetlands. At the present time, the high productivity and biodiversity of these ecosystems is of great value for life on our planet. The typical biogeochemical features of wetland soils are at the very centre of the productivity of these systems.

The permanent or periodic flooding of soils or sediments has a number of physical, chemical and biological consequences for the soil/sediment environment (e.g. Ponnamperuma 1984). The most important ramification of the water layer on top of flooded soils and sediments is the restricted entry of atmospheric oxygen. The diffusion of oxygen in water is 10 000 times slower than in air. Oxygen is rapidly depleted in the upper layers of inundated soil due to chemical and biological oxidation processes, resulting in a soil profile where the presence of oxygen is limited to the upper millimetres (e.g. Frenzel et al. 1992). Wetland plants obviously developed mechanisms to equip themselves for colonisation and growth in anoxic flooded soils and sediments, as reviewed elsewhere (Armstrong et al. 1994; Blom and Voesenek 1996). Changes in anatomy, morphology and metabolism are of paramount importance for surviving in anoxic root environments. So-called aerenchymatous tissue (see Section 10.2) in shoots and roots facilitates the transport of oxygen from the atmosphere to the roots, thereby supplying the roots with oxygen for respiration. However, roots also lose substantial amounts of oxygen to the surrounding anoxic soil, as demonstrated in numerous studies (cited by Armstrong et al. 2000). By means of this process of radial oxy-

Ecological Studies, Vol. 190
J.T.A. Verhoeven, B. Beltman, R. Bobbink, and D.F. Whigham (Eds.)
Wetlands and Natural Resource Management

gen loss (ROL), wetland plants create oxic–anoxic interfaces and thereby provide habitats for both aerobic and anaerobic microbes, facilitating nutrient recycling, as schematically depicted in Fig. 10.1. Next to oxygen, plants excrete organic carbon compounds which provide many of the microbial processes in the rhizosphere with the necessary energy to perform their various biogeochemical reactions.

This chapter will focus on the ecological aspects (physiology, diversity, community stability, biogeochemistry) of microbes and microbial communities occupying the niches provided by wetland plants with a focus on selected groups of microbes involved in biogeochemical reactions important for plant growth and trace gas emission from wetland systems. The first section will focus on the environmental conditions microbes face in this fluctuating oxic–anoxic habitat. The dynamics and dimensions of oxygen release will be highlighted from the angle of plant anatomical aspects.

The presence of oxic–anoxic interfaces in wetland soils implies that anaerobic as well aerobic bacteria have to adapt to this environment. Anaerobic acetate-producing (acetogenic) bacteria, which have also been reported to be closely associated with wetland plant roots, will be presented in Section 10.3

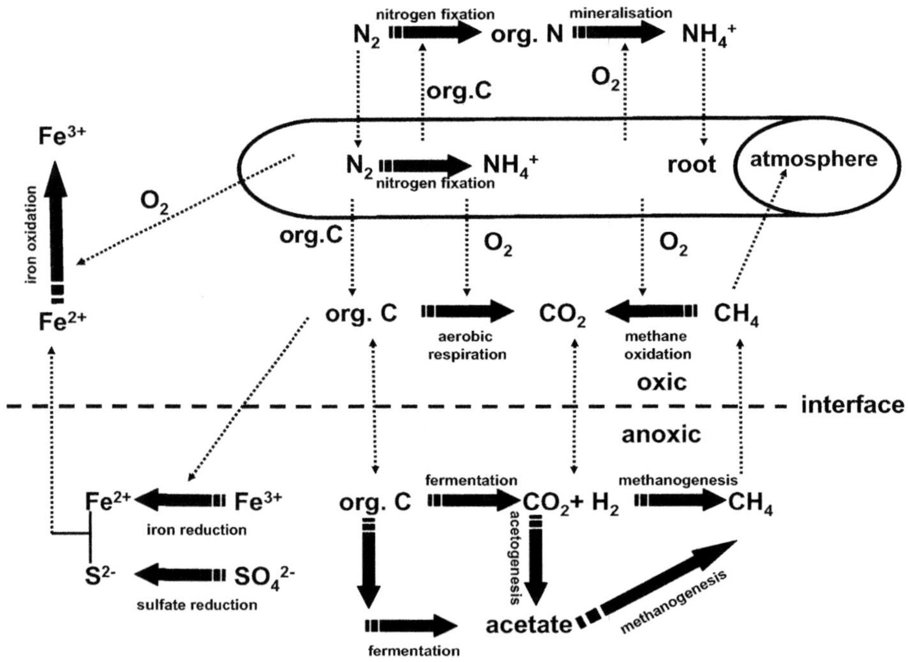

Fig. 10.1 Schematic presentation of the interactions between oxygen-releasing roots and the aerobic and anaerobic microbial processes involved in elemental cycling, as well as methane emission from flooded soils and sediments. *Block arrows* indicate microbial processes, *dotted arrows* depict diffusion processes

as an example of the adaptive mechanisms microbes acquired to function and survive in this highly dynamic habitat.

The requirement to adapt combined with the spatial as well as temporal, very dynamic and heterogeneous wetland rhizosphere environment suggest that there is ample opportunity for microbial diversification. Microbial diversity, especially in soil environments, is often accompanied by a high degree of functional redundancy. The latter is often believed to be of importance for maintaining biogeochemical functions under environmental change. The rhizosphere of graminoid wetlands plants is inhabited by a surprisingly diverse and functionally redundant group of free-living N_2-fixing microbes. Sections 10.4 and 10.5 will show how this high diversity and community maintains stability when subjected to environmental disturbances like fertilisation, clipping, shading and differences in plant-growth forms.

Specialised microbial groups of often low diversity are involved in important biogeochemical as well as environmentally important reactions. In the wetland rhizosphere system with its characteristic oxic–anoxic interfaces, obligatory aerobic as well as anaerobic microbes operate in close proximity, thereby facilitating elemental cycling. This is exemplified in Section 10.6 by presenting evidence on the importance of plant-mediated biological iron reduction and oxidation by microbes in relation to the chemical conversion of this important element.

Wetlands are also considered to be sources of the greenhouse gas methane. However, the oxic rhizosphere enables aerobic methane-consuming microbes to mitigate a substantial amount of the produced methane. Section 10.7 deals with aspects of methane-processing microbes as affected by wetland plant roots, from the level of the organisms up to the ecosystem-scale flux.

10.2 The Microbial Habitat in the Wetland Rhizosphere

It is hardly surprising that contemporary microbial ecology has identified a wide variety of microbial taxa, ranging from obligate aerobes to obligate anaerobes, intimately associated with higher plant roots in flooded soils. The oxygen concentration within and around roots varies from nearly atmospheric values to strict anoxia, with steep gradients in oxygen concentration in tissues and rhizospheres (Armstrong and Beckett 1987). Concentrations vary not only spatially but also temporally, with a constantly shifting mosaic of oxic and anoxic microsites as roots grow, explore soil, age and die. Plant growth in wetlands significantly oxidises the sediment where new, young tissue is colonising previously unvegetated soil, but also makes it more reducing when older biomass is incorporated into the soil organic matter. A mature root may have both oxic and anoxic habitats compartmentalised internally, while its external surface can form both oxidised rhizospheres and anoxic surfaces, depending on age and branching patterns.

10.2.1 Root Structure and Function

Oxygen transport within roots occurs solely by diffusion; early suggestions that significantly greater fluxes could be achieved by some form of mass transport proved not to be viable (Armstrong et al. 1991). Roots are radially symmetrical, with tissues arranged in concentric shells (Fig. 10.2A). Intercellular airspaces begin forming within a few microns of the dividing meristem, initially as small interstices amongst the cells of the cortex and subsequently proliferating to form the large aerenchyma that are such a striking and conspicuous feature of most wetland plant roots. As airspaces are only found in the cortex in most species, longitudinal transport is restricted to the cortical ring, because liquid-phase diffusion is too slow to allow it to occur to any significant extent in the other tissues. The non-porous stele, external 'wall' layers and rhizosphere therefore depend on radial diffusion in the liquid phase for their oxygen. Most plants that grow in flooded soils have a 'herringbone' root system, i.e. there is a dominant main root axis with smaller lateral branches (Bouma et al. 2001). The laterals are generally miniature versions of the main axis, but lack large aerenchyma and are less porous. Laterals provide additional demand on the axial oxygen supply and their formation restricts the oxygen flux to the main apex and shortens the main axis (Sorrell et al. 2000).

Rates of liquid-phase diffusion within the stele and wall layers are also affected by axial changes in tissue permeability as the root matures. In most roots, there is impregnation of the endodermis (Fig. 10.2A), initially with suberin and subsequently lignin, to provide a metabolic barrier to phytotoxins entering the root in the transpiration stream. This also decreases the oxygen diffusion coefficient and restricts stelar aeration. Many wetland plants differ from non-wetland species in suberising and lignifying one or more of the epidermal and hypodermal layers of the root wall, conserving oxygen internally to maximise axial fluxes and confining radial oxygen loss to the tip.

10.2.2 Oxygen Distribution within Roots

Meristematic tissue in roots is highly sensitive to anoxia. Only a few hours, or days at most of oxygen deprivation are usually lethal (Armstrong et al. 1991). Root structure is a solution for maximising fluxes to the meristem, with stelar aeration and rhizosphere oxidation being of secondary importance from the point of view of plant survival. Indeed, roots are somewhat expendable in the

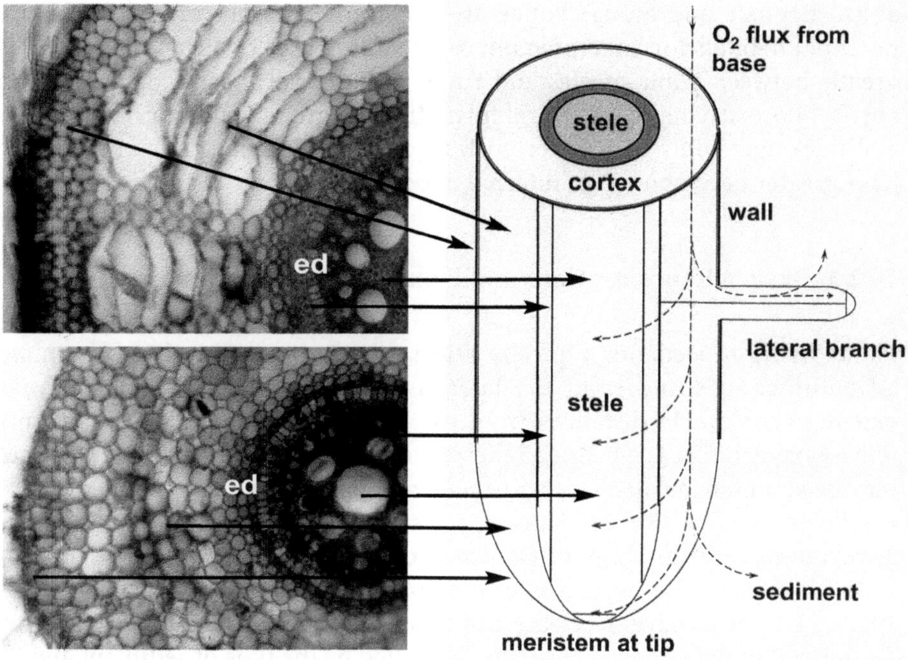

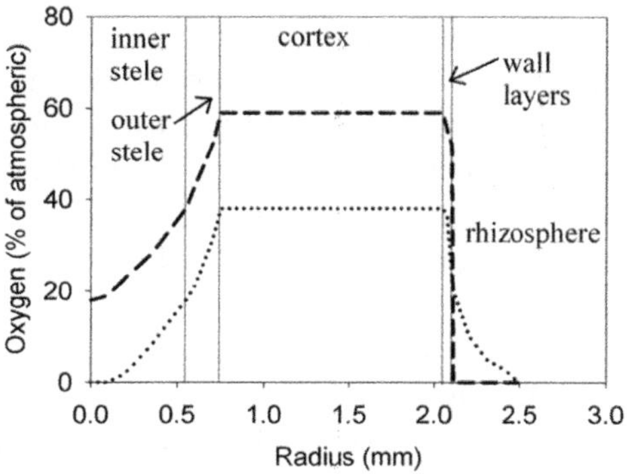

Fig. 10.2 **A** Schematic diagram of a monocotyledonous wetland root, showing oxygen flux in relation to the concentric shells of stele, cortex, and the 'wall layers' and sites of root oxygen loss. Photographs of transverse sections of mature (*upper*) and young (*lower*) roots of *Cyperus involucratus* show the location of these tissues and development of cortical aerenchyma. *ed* Endodermis. **B** Radial profiles of oxygen concentration from the centre of a 0.3-m root of *C. involucratus* through the root (radius = 2.1 mm) and into the rhizosphere, calculated from anatomy, porosity, permeability and respiration data, using the model of Armstrong and Beckett (1987), programmed as described by Sorrell et al. (2000). *Dashed* and *dotted lines* are profiles at 0.15 m and 0.29 m from the root base, respectively

long-term survival of wetland plants, which is mainly achieved by rhizomes and other underground shoots, which show much greater anoxia tolerance and ability to recover from anoxia than roots (Summers et al. 2000).

Profiles of oxygen concentration in roots in anoxic environments are usually similar to those shown in Fig. 10.2B, with relatively high concentrations maintained in the cortex and steep declines in the stele and wall layers. Subapical hypoxic and anoxic zones are a common feature of steles and are potential habitats for anaerobic microbes. The degree of stelar anoxia varies greatly between plant species and roots of the same species, depending on cortical porosity, degree of lateral formation, length and endodermal permeability. Longer roots with lower cortical oxygen concentrations are likely to have greater development of internal hypoxic and anoxic zones.

10.2.3 Oxygen Concentrations and Fluxes in the Rhizosphere

Plants living in wetlands typically release oxygen near the tip of the main adventitious axis and from the laterals (Fig. 10.2A). Oxidised zones may extend considerable distances from the tip in some roots. Roots not only release oxygen that is surplus to their own requirements. Near the tip, a root may be simultaneously releasing oxygen and have an anoxic stele (e.g. 0.29 m profile in Fig. 10.2B). In many species, laterals, with their high surface area:volume ratio and high resistance to diffusion (Sorrell 1994), are responsible for most of the oxygen release into the rhizosphere (Armstrong et al. 1992). The rate of oxygen release can therefore vary considerably, depending on the root morphology, and is also controlled by the type of sediment and its effect on diameter of the oxidised rhizosphere.

Aeration and rhizosphere oxidation are not only a function of root properties, but also depend on shoot oxygen transport to maintain high oxygen concentrations at the root base, especially when there is standing water. In many species, oxygen concentrations in shoot–root junctions decrease markedly as water depth around shoots increases. The decrease in belowground allocation that also usually occurs (Lenssen et al. 1999) reduces the amount of root biomass available for microbial colonisation. The pressurised convective flows that occur in the shoots and rhizomes of the deepest-growing species are essential for root aeration, as they maintain high basal concentrations despite not flowing directly through roots. Inhibition of pressurised flow in these species decreases plant aeration and radial oxygen loss and prevents rhizosphere oxidation (Armstrong et al. 1992).

10.3 Survival Strategies of Anaerobes in the Oxic Rhizosphere: Acetogens as an Example

From the previous section, it is obvious that the wetland rhizosphere as a micro-habitat for microbes is a highly dynamic environment where fluctuation of oxic and anoxic conditions occur in both space and time. Both obligatory aerobic as well as anaerobic microbes are exposed to unfavourable conditions. In order to survive these periods, adaptation or stress-tolerance is required.

Acetogens and sulfate reducers, so-called obligate anaerobes, can be detected in flooded rhizospheres (Hines et al. 1999; Küsel et al. 1999). Sea grass rhizosphere has higher numbers of acetogens than unvegetated anoxic soil (Küsel et al. 1999). Molecular analysis of sea grass and salt marsh plant roots indicates that they are colonised by diverse acetogenic genera (Küsel et al. 1999; Leaphart et al. 2003). Acetogens also occur in the rhizosphere of rice (Conrad et al. 1989). These observations suggest that acetogens that inhabit rhizopheres have strategies to deal with oxidative stress. Therefore, it is not surprising that acetogens can both tolerate and consume O_2 (Küsel et al. 2001; Drake et al. 2002; Karnholz et al. 2002). Other anaerobes, i.e. sulfate reducers, have been shown to possess similar capabilities and can also respire O_2 (Cypionka 2000).

Acetogens utilise the O_2-sensitive acetyl-CoA pathway for the conservation of energy and the biosynthesis of acetate and biomass. Recent studies indicate that acetogens have three general ways to cope with oxic conditions (Fig. 10.3).

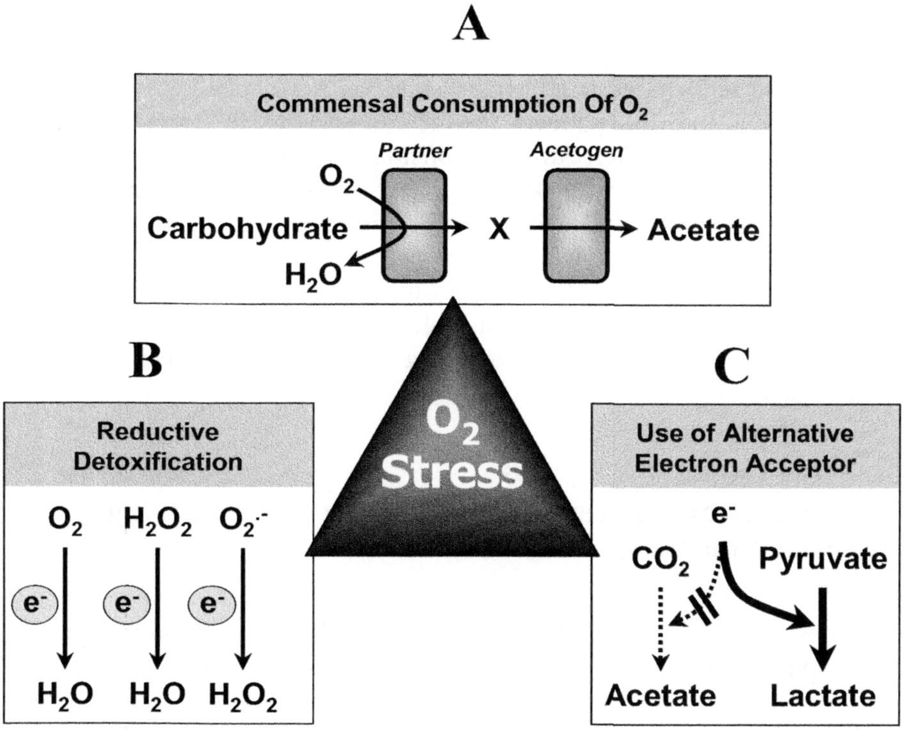

Fig. 10.3 Symbiotic and metabolic mechanisms that acetogens use for coping with unstable redox conditions: *A* commensalism, *B* reductive detoxificiation, *C* alternative electron acceptors. *X* A molecule that can serve as an interspecies substrate for an acetogen (e.g. H_2, formate, lactate), e^- electron. Modified from Müller et al. (2004), with permission

1. Trophic interaction between acetogens and either microaerophiles or aerotolerant anaerobes that consume O_2 theoretically enhances both the survival and activity of acetogens in habitats subject to transient influxes of O_2 (Fig. 10.3A). Acetogens grow commensally with microaerophiles and aerotolerant O_2-consuming anaerobes on oligosaccharides via an interspecies transfer of fermentation products (Gößner et al. 1999, unpublished data). The aerotolerant fermentative organisms not only consume O_2 (thus offering the acetogen protection from toxic O_2), but also produce products (e.g. H_2, lactate) that are utilised commensally by the acetogen for acetogenesis.
2. Acetogens have numerous oxidative stress enzymes, including NADH-oxidase, peroxidase, catalase, superoxide dismutase, rubredoxin oxidoreductase and rubrerythrin performing reductive detoxification (Fig. 10.3B; Das et al. 2001; Küsel et al. 2001; Karnholz et al. 2002; Silaghi-Dumitrescu et al. 2003). However, superoxide dismutase and catalase form O_2, while most acetogens tested to date do not have these enzymes. In general, it appears that anaerobes such as sulfate reducers and acetogens have reductive mechanisms for dealing with O_2 and its toxic side-products peroxide and superoxide (Cypionka 2000; Das et al. 2001; Kurtz 2003; Silaghi-Dumitrescu et al. 2003).
3. Acetogens can use a variety of alternative terminal electron acceptors for recycling reduced electron carriers and are not dependent upon the O_2-sensitive acetyl-CoA pathway for the conservation of energy (Fig. 10.3C; Drake et al. 2004). *Clostridium glycolicum* RD-1, an acetogen isolated from seagrass roots, tolerates 4% O_2 under agitated cultivation conditions (Küsel et al. 2001). Under these conditions, the acetogen conserves energy via lactate and ethanol fermentations, catabolic processes that are not highly sensitive to O_2. Although it is now clear that acetogens have multiple strategies for coping with the unstable redox conditions of flooded rhizospheres, the in situ activity and ecological relevance of acetogens in such habitats remain unresolved.

10.4 Functional Diversity and Activity of Free-Living N_2-Fixing Bacteria

The described actogens have to adapt to the adverse oxic conditions in the rhizosphere of wetland plants. However, the suboxic conditions can also offer a niche for aerobic microbes carrying oxygen-sensitive enzymes. For these organisms, new niches are open which can give rise to microbial diversification. Such a group of microbes are formed by diazotrophic (nitrogen-fixing) bacteria carrying the oxygen-sensitive nitrogenase enzyme (Hurek and Reinhold-Hurek 1998, 2006). Despite the crucial importance of nitrogen for essential ecological and biogeochemical processes, information on the biology of important diazotrophs in many environments (including wetlands) is still very poor, since so far apparently many if not most diazotrophs defy cultivation efforts, like is the case for the majority of prokaryotic organisms in nature. This appears to be the case also for diazotrophic prokaryotes associated with graminoid plants in subtropical wetland soils.

The biological reaction counterbalancing the loss of nitrogen from soils or ecosystems is biological nitrogen fixation (BNF), the enzymatic reduction of N_2 to ammonia. This process, unique to Bacteria and Archaea, is estimated to contribute 90–130 Tg N year^{-1} to terrestrial ecosystems (Galloway et al. 1995) and 100–200 Tg N year^{-1} to marine ecosystems (Karl et al. 2002). One of the best studied interactions between microbes and eukaryotes, in which the eukaryotic partner directly profits from N_2 fixation, is the root nodule symbiosis between rhizobia and legumes. The most important crops worldwide though, wheat, rice and maize, belong to the Poaceae, which do not naturally form these specialised symbiotic structures. However, it has been shown that some graminoid crops, such as certain Brazilian sugar cane cultivars (Boddey 1995) or flooded rice (Cassman et al. 1995), can derive a substantial part of the plant nitrogen from BNF without any inoculation in natural soils. There is increasing evidence that BNF may also be a key source for new plant nitrogen in other graminoid plants which thrive in periodically or continually flooded environments where primary production is probably nitrogen-limited. Good candidates are smooth cordgrass (*Spartina alterniflora*; Mendelssohn 1979; Lovell et al. 2000; Brown et al. 2003), Kallar grass (*Leptochloa fusca*; Reinhold et al. 1986; Hurek et al. 2002), and purple moor-grass (*Mollinia coerulea*; Hamelin et al. 2002). Just like rice, all these species are flood-tolerant plants which may grow into extensive monoculture stands in saltmarsh ecosystems of North America, in subtropical wetland soils in Asia or Africa and in nutrient-poor fens or bogs in Europe, respectively.

So far, major nitrogen-fixing diazotrophs have not yet been identified that contribute to plant growth by N_2 fixation to an extent observed in some uninoculated graminaceous plants. Molecular ecological studies based on sequence analysis (e.g. Zani et al. 2000; Hurek et al. 2002), denaturing gradient gel electrophoresis (DGGE; Lovell et al. 2000), PCR restriction length polymorphism (RFLP; Poly et al. 2001) and fluorescently labelled terminal restriction fragment length polymorphism (T-RFLP; Tan et al. 2003) of *nifH* DNA or mRNA fragments have revealed a wide range of mostly as yet uncultured bacteria occurring in natural habitats, including those dominated by graminoid plants in wetlands (Zehr et al. 2003; Hurek and Reinhold-Hurek 2006). Additionally, plants which obtain a substantial part of their plant nitrogen from BNF but do not form any symbiotic structures often cannot be readily distinguished from those which do not depend on it. Furthermore, direct or indirect nitrogenase activity measurements are not always applicable or are too interference-prone under field conditions, which also adds to the difficulty in identifying major diazotrophs, particularly in wetlands.

Phylogenetic analyses of clone libraries of 16S rDNA fragments are often applied for an inventory of prokaryotic diversity. Coupled with fluorescence in situ hybridisation (FISH) of bacteria with rRNA-targeted oligonucleotide probes, it provides a widely used tool to study different populations of uncultured microbes. However, 16S rDNA sequences do not necessarily allow assignment of functions to these bacteria. A functional analysis of a community may be achieved when key genes for certain metabolic processes are used as target genes. For example, for the process of N_2 fixation, the iron protein gene (*nifH*) of the evolutionarily conserved nitrogenase protein complex (Howard and Rees 1996) is particularly suitable for phylogenetic analysis (Hurek et al. 1997, 2002; Engelhard et al. 2000). In order to analyse the diversity of nitrogen-fixing bacteria, various sets of universal *nifH* primers are available which successfully amplify fragments of distantly related diazotrophs, but only those of Zehr and McReynolds (1989) have been shown to amplify highly diverged *nifH* genes with equal efficiencies (Tan et al. 2003), except for the oxygen-tolerant N_2-fixing system of *Streptomyces thermoautotrophicus* (Ribbe et al. 1997). Such culture-independent studies on root-associated diazotrophs have demonstrated that the majority of *nifH* fragments retrieved cannot be assigned to cultivated diazotrophs (Hurek et al. 1997, 2002; Engelhard et al. 2000; Hurek and Reinhold-Hurek 2006) Most interestingly, novel deeply branching groups of nitrogenases have been detected which cannot be assigned to any known bacterial phylum (Hurek and Reinhold-Hurek 2006).

Whereas the retrieval of *nif* DNA or other genes from the natural environment only shows the mere presence of nitrogenase genes or the diversity of diazotrophs or other groups of bacteria, for the estimation of processes in the environment it is important to assess functional diversity or activity. For example, surveys on *nifH* mRNA expression can also evaluate whether the bacteria studied are of any importance for biological nitrogen fixation in the natural environment. Studies on the expression of nitrogenase genes carried out by reverse transcription polymerase chain reaction (RT-PCR; Hurek et al. 2002) do not only evaluate the actual activity of a particular microbe within a pool of other microbes in the natural environment. They also allow identification of the primary diazotrophic bacteria by comparing sequences retrieved from the environment with sequences from cultivated organisms. When coupled to microarray analysis (see Fig. 10.4), the most active diazotrophic gene pools can be detected relatively rapidly (Zhang, Hurek and Reinhold-Hurek, unpublished data). Consequently, direct targeting and quantification of *nifH* mRNA is the tool of choice to identify primary nitrogen-fixing organisms in environments where major diazotrophs are inconspicuous.

Using these techniques, we found that the rhizospheres of several grasses in largely undisturbed wetlands may be dominated by unsuspected, highly diversified groups of globally distributed N_2-fixing bacteria which may or may not be restricted to this environment. Our data on *nifH* expression levels

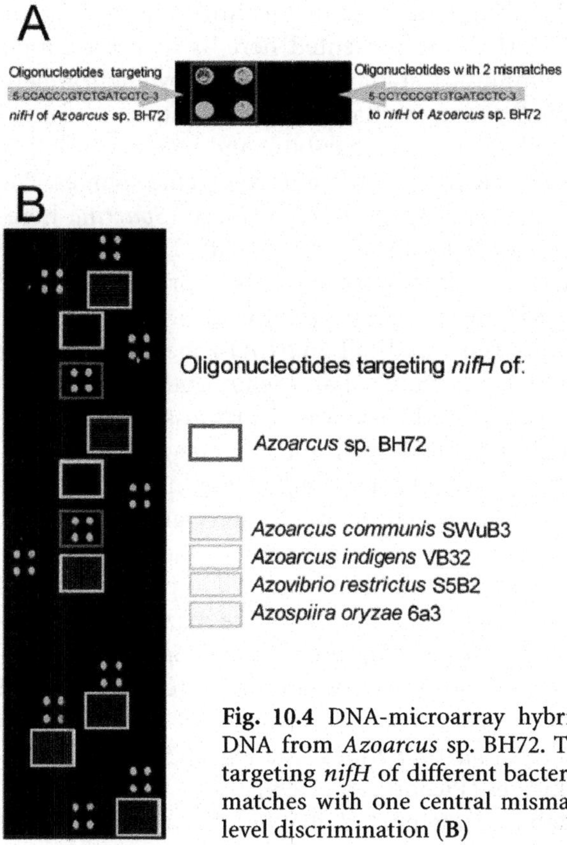

Fig. 10.4 DNA-microarray hybridisation with cy5-labeled *nifH* DNA from *Azoarcus* sp. BH72. The positions of oligonucleotides targeting *nifH* of different bacterial species are framed. Two-mismatches with one central mismatch (**A**) are suitable for species-level discrimination (**B**)

show that, similar to other plant–microbe interactions such as endomycorrhizae, very high microbial activities may be reached in the rhizosphere of uncultivated, flood-tolerant grasses by particular micro-organisms without readily discernible changes of the root system, as in nodule symbioses.

10.5 Microbial Community Stability in Response to Manipulation of the Vegetation

It is obvious that wetland plants offer many niches for microbes which, in the case presented in the previous section, can give rise to high diversity of a functionally redundant community. The question is whether all these different species contribute to the function they catalyse and how stable these communities are. The highly dynamic wetland environment may give rise to a high degree of selection or displacement of species from microbial communities.

Diazotroph diversity and assemblage stability in the North Inlet (NI) salt marsh, near Georgetown (S.C., USA), is presented here as a case example shedding light on microbial community stability as affected by environmental perturbations characteristic for many wetland environments.

The NI marsh is small (3200 ha), euhaline, bar-built and receives very limited inputs of nitrogen (or other materials) from anthropogenic sources. The dominant plant in the NI ecosystem is *Spartina alterniflora* (*Spartina* hereafter), which occurs in two growth forms (Turner 1976). The short-form plants (about 30 cm in height vs =1 m for tall-form plants) represent a stunted form produced in response to edaphic conditions (Howes et al. 1986). Edaphic variables that differ between the short and tall *Spartina* zones include pore water salinity, soluble sulfide concentration, pH and redox potential.

Among the first questions addressed to assess *Spartina*-associated diazotroph diversity was the stability of the diazotroph assemblage. These studies employed PCR to recover partial sequences of *nif*H, the gene encoding the nitrogenase iron protein (reviewed by Lovell 2002). *nif*H amplicons were resolved using DGGE, which separates DNA sequences on the basis of their melting characteristics. DGGE yields a highly reproducible pattern of *nif*H bands reflecting the composition (species richness) of the diazotroph assemblage (Fig. 10.5). DGGE banding patterns from short and tall *Spartina* rhizosphere samples were remarkably consistent across sample dates and between *Spartina* growth zones, reflecting the very strong similarity of the diazotroph assemblages in the tall and short *Spartina* zones (see Fig. 10.5) and the seasonal stability of these assemblages (Piceno et al. 1999).

Diazotroph assemblage stability was next challenged through manipulative experimentation in the field. Plots in the short *Spartina* zone were randomly assigned to treatments: (1) control, no manipulation, (2) N fertilisation, (3) N+P fertilization (Piceno and Lovell 2000a), (4) clipping of aboveground biomass, (5) shading (Piceno and Lovell 2000b). Fertilised plots were treated biweekly for two or eight weeks and clipped and shaded plots were maintained for eight weeks. DGGE profiles from control, nutrient addition and carbon resource manipulation plots were very similar. One DGGE band that was prominent in the control plots was not found in plots that had been fertilised for over ten years in a separate experiment. These results also indicate substantial stability of the diazotroph assemblage. Results of these manipulations were also examined using phospholipid fatty acid (PLFA) analysis, which provides a more inclusive examination of the rhizosphere microbiota (Lovell et al. 2001b). PLFA analysis also showed no change in the rhizosphere community in response to the experimental manipulations. These results are consistent with the seasonal study and indicate a stable diazotroph assemblage able to weather environmental variation without loss of taxa.

Changes in species evenness (as opposed to species richness) of the *Spartina*-associated diazotroph assemblage composition were examined in a

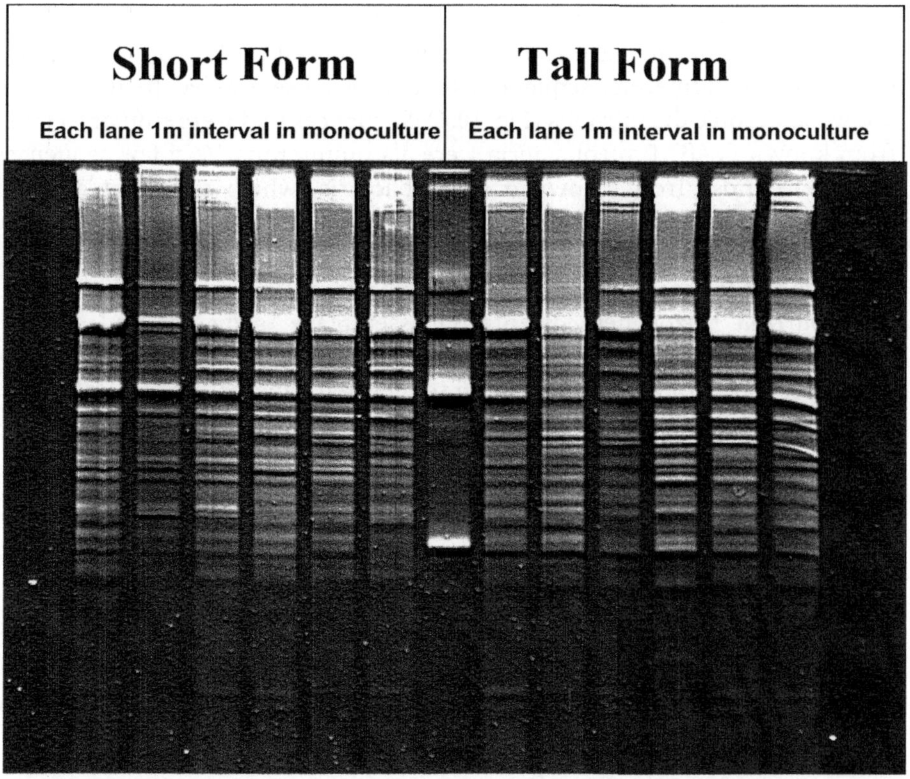

Fig. 10.5 Denaturing gradient gel showing *nif*H amplicons recovered from short growth form and tall growth form *Spartina alterniflora*. Rhizosphere samples were collected at 1 m intervals along transects in monoculture stands of short and tall *Spartina*. Gel markers are the *nif*H amplicons from *Klebsiella pneumoniae*, *Sinorhizobium meliloti* and *Azospirillum lipoferum*

series of long-term fertilisation treatments consisting of control (no nutrient additions) and N+P additions once per month for a full year (Bagwell and Lovell 2000). Fertilisation resulted in a three-fold increase in *Spartina* aboveground biomass. Changes in the relative abundances of selected diazotrophs were monitored quantitatively using reverse sample genome probing (reviewed by Bagwell and Lovell 2004). The diazotrophs were differentially responsive to environmental variability and to the fertilisation treatments, but none were displaced from the assemblage. While relative abundances of the diazotrophs varied, these organisms were persistent autochthonous flora of the *Spartina* rhizosphere.

To examine the species composition of the *Spartina* rhizosphere diazotroph assemblage in greater detail, the bands from DGGE gels run during three previous studies (see above) were sampled and amplicons were recovered, cloned and sequenced. The sequences were then used in a phylogenetic reconstruction of the diazotroph assemblage (Lovell et al. 2000). The assemblage yielded numerous sequences that were closely allied to those from known anaerobic diazotrophs (Fig. 10.6). Sequences that were similar to those from known (γ+β)-Proteobacteria were also numerous. Very few sequences similar to those from known α-Proteobacteria, which are abundant and diverse in dead aboveground *Spartina* biomass (Lovell et al. 2001a), were recovered. None of the partial *nif*H sequences recovered from *Spartina* rhizosphere was the same as any sequence from a formally described diazotroph. The *Spartina* rhizosphere diazotroph assemblage is diverse and consists entirely of organisms that are either unknown to science or whose *nif*H sequences have not been deposited in sequence databases.

The functionally important diazotrophs in the short-form *Spartina* rhizosphere were assessed by recovery of mRNA from rhizosphere cores, RT-PCR, and sequencing of the products (Brown et al. 2003). Sequences falling into the anaerobe cluster were affiliated with those from known sulfate-reducing bacteria, though again representing unknown taxa (Fig. 10.6). Most sequences falling into the (γ+β)-Proteobacteria were affiliated with *Pseudomonas*, with an additional sequence highly similar to sequences from *Vibrio*. This shows that diazotrophs using sulfate as terminal electron acceptor and those most likely using oxygen were actively expressing *nif*H in the same relatively small samples and at the same time. These results also show that there is no one dominant functional diazotroph and there are several taxa that actively participate in nitrogen fixation.

Important insights into the ecology of diazotrophs in the *Spartina* rhizosphere have been gained through these studies. The diazotroph assemblage is stable and consistent in character and is the same across different *Spartina* habitats. While there is some indication of affinity of certain diazotrophs for *Spartina* (Lovell et al. 2000; Bagwell et al. 2001), there is no evidence at present for a dominant species in this association, which means that diazotroph diversity is likely to be a parameter important to the ecological function of diazotrophy. Root and rhizosphere assemblages contain both anaerobes and oxy-

Fig. 10.6 Phylogenetic analysis of partial NifH sequences from unknown short form *Spartina alterniflora* rhizosphere diazotrophs (*Sp*), expressed NifH sequences from short form *Spartina* rhizosphere diazotrophs (*Sp mRNA*) and partial NifH sequences from various formally described diazotrophs. The dendrogram was generated by using neighbour-joining and Poisson-corrected distances. *Methanobacterium thermoautotrophicum* was used as the outgroup taxon. The percentage of 1000 bootstrap replicates that supported each node is shown. Bootstrap values <50 % are not shown. NCBI GenBank accession numbers for Sp sequences are AY091856–AY091913 and for Sp mRNA sequences are AF098493–AF098507

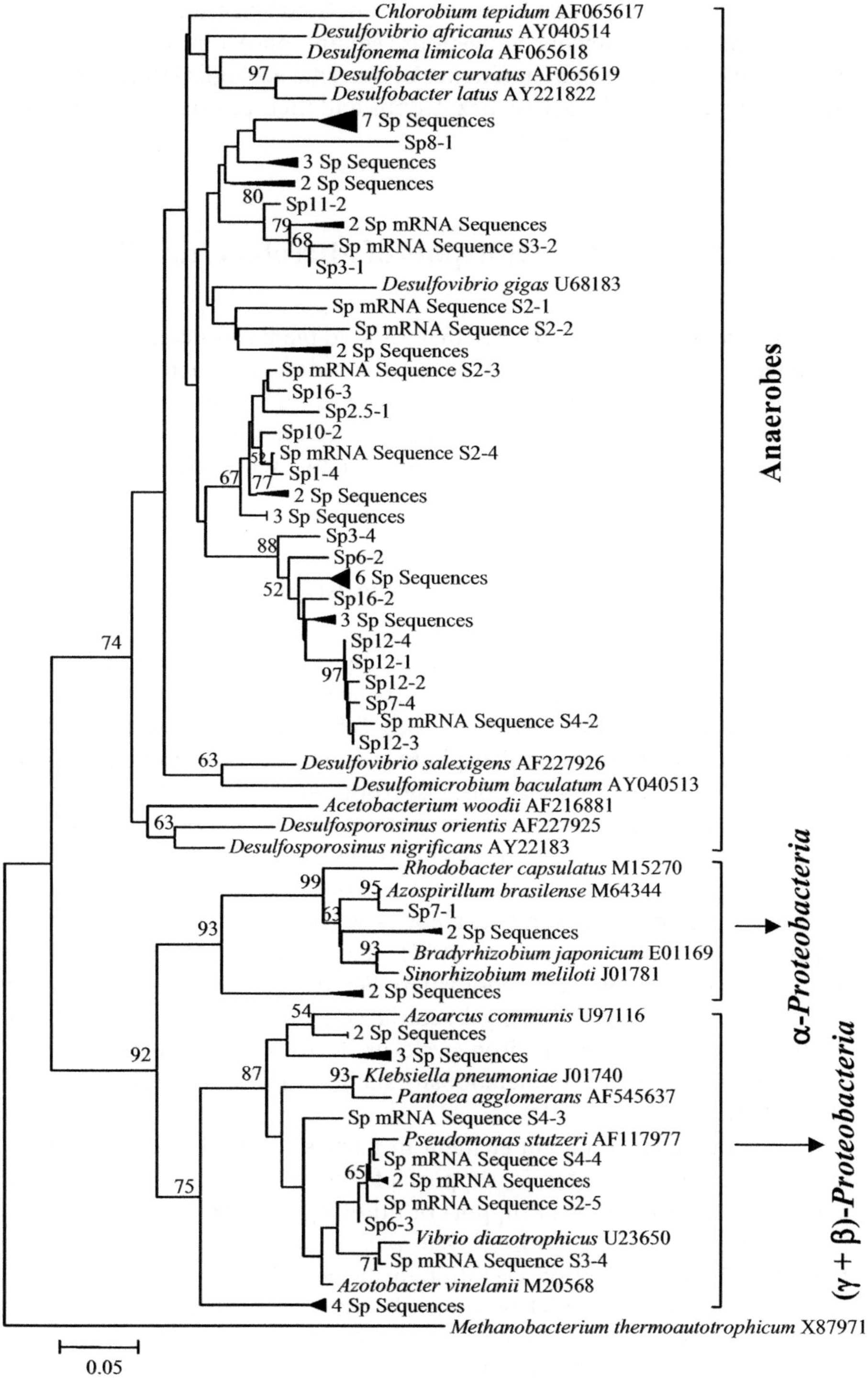
Chlorobium tepidum AF065617
Desulfovibrio africanus AY040514
Desulfonema limicola AF065618
97
Desulfobacter curvatus AF065619
Desulfobacter latus AY221822
7 Sp Sequences
Sp8-1
3 Sp Sequences
2 Sp Sequences
80
Sp11-2
79
2 Sp mRNA Sequences
68
Sp mRNA Sequence S3-2
Sp3-1
Desulfovibrio gigas U68183
Sp mRNA Sequence S2-1
Sp mRNA Sequence S2-2
2 Sp Sequences
Sp mRNA Sequence S2-3
Sp16-3
Sp2.5-1
Sp10-2
52
Sp mRNA Sequence S2-4
Sp1-4
67
77
2 Sp Sequences
3 Sp Sequences
88
Sp3-4
Sp6-2
6 Sp Sequences
52
Sp16-2
3 Sp Sequences
74
Sp12-4
Sp12-1
97
Sp12-2
Sp7-4
Sp mRNA Sequence S4-2
Sp12-3
63
Desulfovibrio salexigens AF227926
Desulfomicrobium baculatum AY040513
Acetobacterium woodii AF216881
63
Desulfosporosinus orientis AF227925
Desulfosporosinus nigrificans AY22183
Anaerobes
99
Rhodobacter capsulatus M15270
95
Azospirillum brasilense M64344
Sp7-1
63
93
2 Sp Sequences
93
Bradyrhizobium japonicum E01169
Sinorhizobium meliloti J01781
2 Sp Sequences
α-Proteobacteria
54
Azoarcus communis U97116
2 Sp Sequences
92
3 Sp Sequences
87
93
Klebsiella pneumoniae J01740
Pantoea agglomerans AF545637
Sp mRNA Sequence S4-3
Pseudomonas stutzeri AF117977
Sp mRNA Sequence S4-4
65
2 Sp mRNA Sequences
75
Sp mRNA Sequence S2-5
Sp6-3
Vibrio diazotrophicus U23650
71
Sp mRNA Sequence S3-4
Azotobacter vinelanii M20568
4 Sp Sequences
(γ + β)-Proteobacteria
Methanobacterium thermoautotrophicum X87971
0.05

gen utilising organisms and both can be functionally active in the same rhizosphere material. Many open questions remain, concerning diazotrophy and the role diversity plays in maintenance of this vital ecological function. Future studies will focus on the active diazotrophs and their responsiveness to biotic and abiotic environmental variation.

10.6 Wetland Roots as Hotspots of Microbial Iron-Cycling

The numerous habitats and niches that wetland plant roots create are not only occupied by highly diverse and functionally redundant microbial communities but also offer possibilities for non-diverse, highly specialised microbes. Some of these groups are involved in highly relevant elemental conversions which for a long time have been considered to be exclusively catalysed chemically. The cycling of iron is such an example.

10.6.1 Wetland Rhizosphere Ferrous Wheels: Introduction

Wetland plant roots are often encrusted by a rusty-red coating of Fe(III) oxyhydroxide, Fe(II) and a variety of associated elements. This *iron plaque* is colourful visual evidence of the strong influence plants exert on wetland biogeochemical cycles. The rhizosphere is a zone of rapid cycling of redox-sensitive elements such as Fe because roots are sources of both electron donors (i.e. organic carbon) and the thermodynamically dominant terminal electron acceptor O_2. Despite this general observation, iron plaque has been traditionally conceptualised as a static Fe pool that forms on wetland plant roots due to non-enzymatic (i.e. abiotic) Fe(II) oxidation driven by radial oxygen loss (Mendelssohn et al. 1995). Recent work has shown that iron plaque is actually a dynamic pool where Fe(III) rapidly accumulates or dissolves and that Fe(II)-oxidising bacteria contribute to plaque deposition. Furthermore, a rhizosphere iron cycle or 'ferrous wheel' influences other forms of anaerobic carbon metabolism, such as methanogenesis (Megonigal et al. 2004).

Iron is the most abundant element in Earth's crust that can act both as a terminal electron acceptor in support of anaerobic microbial metabolism and as an electron donor in support of lithotrophic growth. Yet, a major role for micro-organisms in the iron cycle was recognised only recently. Before the 1990s, Fe(III) reduction was thought to be either a non-enzymatic (i.e. abiotic) process or a trivial product of fermentation (Kamura et al. 1963; Lovely 2000). Similarly, enzymatic Fe(II) oxidation was thought to be negligible, except under acidic (pH=4) conditions (Emerson 2000). Although it is now clear that micro-organisms contribute to Fe(III) reduction and Fe(II) oxidation in a broad range of environments, we are just beginning to understand how these processes operate in situ, particularly in freshwater wetland ecosystems. An understanding of iron cycling in wetland ecosystems that support macrophytes requires specific consideration of rhizosphere processes.

10.6.2 Rhizosphere Fe(III) Reduction

Iron [Fe(III)]-reducing bacteria (FeRB) couple dissimilatory Fe(III) reduction to organic carbon oxidation, thereby linking the carbon and iron cycles of wetland soils (Roden and Wetzel 1996). FeRB can mediate up to 100% of anaerobic carbon oxidation in freshwater and marine environments (Thamdrup 2000), depending on many factors, the most important of which is perhaps the abundance and structure of Fe(III) minerals. The most rapidly reduced Fe(III) substrates are poorly crystalline, high surface area minerals such as ferrihydrite. The oxides in iron plaque are commonly dominated by poorly crystalline forms such as ferrihydrite, but they can contain more crystalline Fe forms, such as lepidocrocite (γ-FeOOH) and/or goethite (β-FeOOH). A second important factor that regulates Fe(III) reduction rate is organic carbon availability (Roden and Wetzel 2002).

Wetlands appear to be hotspots of iron reduction and oxidation compared to non-vegetated portions of the landscape (Kostka and Luther 1995). The earliest contributors to this field recognised that plants have the capacity to stimulate non-enzymatic and microbial Fe(III) reduction by regenerating Fe(III) oxides in the rhizosphere (Giblin and Howarth 1984; Luther et al. 1992). Plants may also enhance Fe(III) reduction through the production of labile carbon and organic ligands (Megonigal et al. 2004), but this has not been demonstrated directly and requires attention from the research community. In most cases, conclusions about the influence of plants on Fe(III) reduction have been inferred by comparing the properties of bulk soils (i.e. mixtures of rhizosphere and non-rhizosphere soil) across sites that differ in root abundance. For example, poorly crystalline Fe(III) minerals were enriched in vegetated sediments compared to unvegetated sediments in both a freshwater *Juncus effusus* marsh (Roden and Wetzel 1996) and a saltwater *Spartina alterniflora* marsh (Kostka and Luther 1995). A similar observation was made in a rice paddy by comparing soils within and below the root zone (Frenzel et al. 1999).

An important prediction of the hypothesis that plant activity promotes rapid iron cycling is that substrate availability and microbial activity will be higher in the rhizosphere than in the non-rhizosphere soil. Direct comparisons of root plaque to bulk soil support this prediction. Gribsholt et al. (2003) reported higher rates of Fe(III) reduction in the rhizosphere than non-rhizosphere soil in a *S. alterniforia*-dominated salt marsh. In a survey across a variety of wetland ecosystem types, Weiss et al. (2003) observed that FeRB were a significantly greater proportion of the microbial community in iron plaque versus non-rhizosphere soil (12 % vs 0.5 % on average, respectively). These studies demonstrated there is substantial small-scale spatial variability in factors that regulate Fe(III) reduction. This variability can be attributed in many cases to plant root activity.

Weiss et al. (2004) explored several biological and geochemical factors that could contribute to differences in FeRB abundance between the iron plaque and bulk soil. As observed in the salt marsh study of Gribsholt et al. (2003), Fe(III) reduction rates in several freshwater wetland soils were more rapid in the rhizosphere iron pool (defined as the root plaque itself) than the non-rhizosphere iron pool (Fig. 10.7A) during anaerobic laboratory incubations.

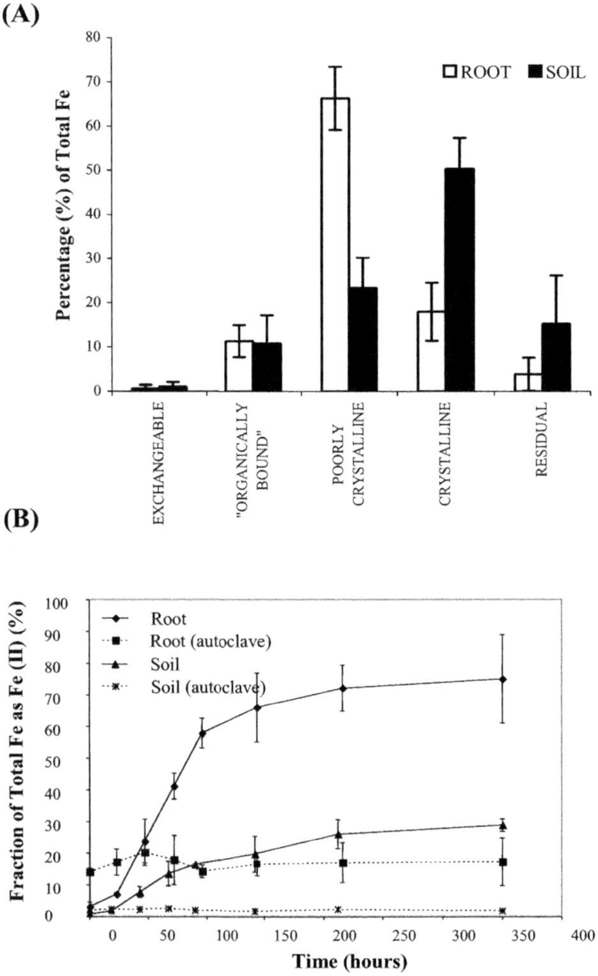

Fig. 10.7 Differences between the Fe(III) oxide pools on the root surface and non-rhizosphere soil with respect to mineralogy and reactivity. **A** Sequential chemical extraction of several Fe(III) oxide pools that differ in reactivity, particularly the poorly crystalline (0.2 M acid ammonium oxalate) and strongly crystalline (dithionate-citrate-biocarbonate) pools. **B** Time-course of Fe(III) oxide dissolution in live (*solid lines*) and autoclaved (*dotted lines*) samples. Error bars are ±1 SD. Reproduced with permission from Weiss et al. (2004)

Thus, the rhizosphere supported relatively high rates of FeRB metabolism. The most likely explanation for enhanced FeRB metabolism in the rhizosphere was a difference in the composition of Fe(III) minerals. A sequential chemical extraction revealed that the rhizosphere iron pool was enriched in poorly crystalline Fe(III) minerals compared to the non-rhizosphere iron pool (Fig. 10.7B). The variation in Fe(III) reduction rates was not due to the initial difference in FeRB abundance (Weiss et al. 2003), because the effect persisted when samples were autoclaved and amended with equal amounts of *Geobacter metallireducans*. Likewise, amending samples with organic carbon in the form of acetate or a model electron-shuttle compound (AQDS) did not diminish the difference in Fe(III) reduction rates between rhizosphere and non-rhizosphere iron pools. It is apparent that wetland plants have the potential to dramatically influence the metabolism of rhizosphere FeRB, primarily through regeneration of poorly crystalline Fe(III) minerals driven by radial O_2 loss.

Regeneration of Fe(III) oxides in wetland soils also occurs via two other mechanisms that introduce O_2, water table fluctuations and bioturbation by burrowing animals, which occurs in salt marshes. Radial O_2 loss by roots appears to be more important than bioturbation by crabs in some salt marsh sites (Gribsholt et al. 2003), but not in other sites (Kostka et al. 2002). In salt marshes, regeneration of Fe(III) oxides can be negated due to non-enzymatic Fe(III) reduction by H_2S, with subsequent declines in the abundance of FeRB (Koretsky et al. 2003).

10.6.3 Rhizosphere Fe(II) Oxidation

Dissolved or adsorbed forms of ferrous iron are oxidised to Fe(III) by Fe(II)-oxidising bacteria (FeOB) or by non-enzymatic oxidation in the presence of O_2 and then precipitate as Fe(III) oxyhydroxides. The relative contributions of enzymatic versus non-enzymatic oxidation to overall Fe(II) oxidation rates is influenced by factors such as pH, O_2 and Fe(II) concentrations and the rate of Fe(II) delivery (Neubauer et al., 2002). At pH <4, non-enzymatic oxidation of Fe(II) by O_2 is slow; and enzymatic activity increases oxidation rates by a factor $>10^6$ (Singer and Stumm, 1970). The influence of FeOB on Fe(II) oxidation has been harder to assess at circum-neutral pH because non-enzymatic oxidation can be extremely rapid. However, in the microaerophilic niche of the wetland rhizosphere, the half-life of non-enzymatic Fe^{2+} oxidation is up to 300-fold longer than at high O_2 concentration (Roden et al. 2003). This long Fe^{2+} half-life theoretically favours a role for enzymatic oxidation by FeOB.

Circum-neutral, chemolithoautotrophic FeOB have been isolated directly from iron plaque (Emerson et al. 1999), where they constitute up to 6 % of all microbial cells (Weiss et al. 2003), and from iron-rich wetland sediments

(Sobolev and Roden 2004). The later organism can also grow organotrophically with acetate. It has been difficult to assess the contribution of FeOB in situ because there are no adequate specific inhibitors or isotope methods to observe the process. However, work with pure cultures of rhizosphere FeOB suggested that they can mediate a substantial fraction (up to 60 %) of overall Fe(II) oxidation at low O_2 concentrations. Up to 90 % of Fe(II) oxidation was microbially mediated in studies of non-rhizosphere FeOB cultures (Emerson and Revsbech, 1994; Sobolev and Roden, 2001; Neubauer et al., 2002). Microbial Fe(II) oxidation was insignificant compared to non-enzymatic oxidation in a model rice system (van Bodegom et al., 2001), but these experiments involved vigorous sample agitation that may artificially favour non-enzymatic Fe(II) oxidation.

Pure culture studies also alluded to some of the factors that regulate circumneutral enzymatic Fe(II) oxidation in complex environments such as the rhizosphere. In microcosm experiments, enzymatic and non-enzymatic Fe(II) oxidation were linearly and positively related to the rate of Fe(II) addition (r^2=0.90; (Neubauer et al., 2002), suggesting that microbial Fe(II) oxidation in wetlands may be Fe(II)-limited. FeOB competed best with non-enzymatic oxidation when total iron accumulation in the chambers was low, suggesting a regulatory role for chemical surfaces that can adsorb or bond Fe(II) (see Fig. 10.8).

The presence of Fe(II)-oxidising bacteria can actually reduce the rate of overall Fe(II) oxidation in some instances (Neubauer et al. 2002). This is presumably caused by the inhibition of non-enzymatic Fe(II) oxidation due to chelation and stabilisation of aqueous Fe(II), perhaps by exopolymers or other organic extracellular molecules produced by FeOB (Emerson and

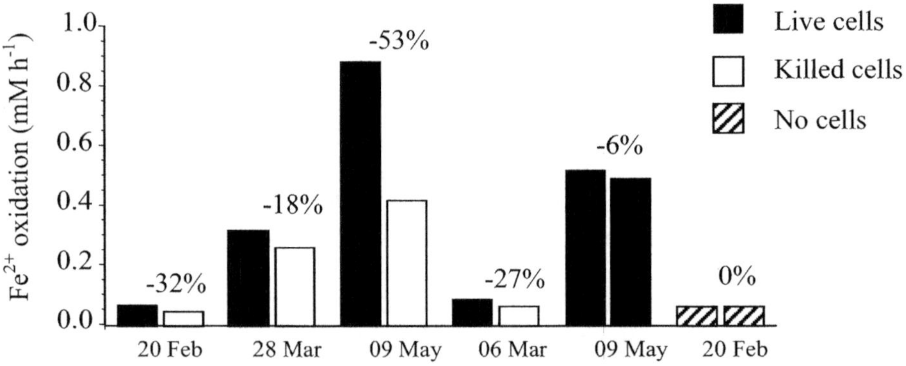

Fig. 10.8 Microcosm experiments to determine the influence of a lithotrophic Fe(II)-oxidising bacteria isolated from the rhizosphere on Fe(II) oxidation rates. Three pairs of bars show Fe(II) oxidation rates before and after killing the bacteria with sodium azide. The other experiments show successive rate determinations with live cells or no cells. Reproduced with permission from Neubauer et al. (2002)

Moyer, 1997). Similar compounds have been proposed to chelate Fe(III), allowing it to diffuse into anaerobic zones following microbial oxidation (Sobolev and Roden, 2001).

Microbial Fe(II) oxidation can also take place in the absence of O_2, coupled either to CO_2 reduction (i.e. photolithotrophy; Widdel 1993; Straub et al. 1999) or NO_3^- reduction (Straub et al., 1996). Such organisms accounted for up to 58% of the total cultivatable denitrifying bacteria in a lake sediment (Hauck et al., 2001), but not more than 0.8% of denitrifiers in other aquatic sediments (Straub and Buchholz-Cleven, 1998). Fe(II) substrates can be in the form of goethite (Weber et al., 2001) or FeS (Schippers and Jorgensen, 2002), but not FeS_2. Virtually nothing is known about anaerobic FeOB in wetlands.

10.6.4 Rhizosphere Fe(II) Oxidation Scaled to Ecosystems

The fact that the wetland rhizosphere iron pool is enriched in poorly crystalline minerals compared to bulk soil iron pool is not sufficient evidence to conclude that plant activity explains why wetlands are iron cycling hotspots in the landscape. When root mass is compared to soil mass on a volume basis, freshwater wetland soils can have nearly two orders of magnitude more poorly crystalline Fe(III) in non-rhizosphere soil than iron plaque (1.95 mg and 0.04 mg Fe cm^{-3} as poorly crystalline Fe(III); calculated from Weiss et al. 2003, 2004). Thus, the deposition and reduction of labile iron would appear to occur outside the rhizosphere. However, it is important to note that even non-rhizosphere soil in these freshwater wetlands have more poorly crystalline Fe(III) than expected for non-vegetated sediments. The mean poorly crystalline iron content of the eight non-rhizosphere soils studied by Weiss et al. (2004) was 40 mmol Fe cm^{-3}, a value 100% higher than most (37 of 40) of the non-vegetated marine sediments reported by Thamdrup (2000). Kostka and Luther (1995) reported that vegetated salt marsh sediments contained more reactive Fe(III) than other marine sediments. One interpretation of this observation is that 'non-rhizosphere' soils were once in the 'rhizosphere' category and still show the lingering influence of plant-driven iron oxide regeneration. This is a reasonable hypothesis, because root growth and turnover can be rapid in wetlands (Schubauer and Hopkinson 1984; Powell and Day 1991), and a worthwhile hypothesis to test in the interest of furthering our understanding of rhizosphere processes on microbial community metabolism.

10.7 Methane-Processing Microbes in Wetland Rhizospheres

The emission of the greenhouse gas methane from wetland environments is driven by groups of microbes which depend on the close proximity of oxic and anoxic conditions and therefore are provided with a niche by roots of wetland plants. A substantial amount of the knowledge available on the ecology of methane-consuming and methane-producing microbes comes from rice paddies, in particular from an Italian rice soil.

10.7.1 Italian (Vercelli) Rice Soil as a Model System

The heartlands of rice production are East Asia and India, with rice-growing areas in Africa and the Americas contributing to the global yield (FAO 2004). On the global scale, the European rice growing area is small. Nevertheless, rice is there a locally important crop and is part of many traditional dishes, in particular in the valley of River Po in northern Italy. The rice fields at Vercelli (northern Italy) are the first site where methane emissions were studied, together with the underlying microbial processes (Holzapfel-Pschorn et al. 1985). However, that rice paddies emit methane was first suggested for Japanese fields (Koyama 1955). Over the years, the Italian site has been studied repeatedly in field campaigns and in the laboratory, turning this paddy soil into a model system for the microbial ecology of flooded soils (for recent reviews, see e.g. Frenzel et al. 1999; Frenzel 2000; Conrad and Frenzel 2002). This work was paralleled by biogeochemical and microbiological studies in Japan (reviewed by Kimura 2000). One often tacitly assumes that all rice fields are irrigated (paddy rice) like those in Italy, but this is not true. Rice is also grown in upland fields, as rain-fed rice that is temporarily flooded by rainfall and even as deep-water rice. However, both in terms of yield and methane emission, irrigated rice is most important. Numerous field campaigns all over the world helped to develop a fairly complete picture of global source strengths (e.g. Denier van der Gon et al. 2000; Shearer and Khalil 2000; Wang and Li 2002). However, when it comes to the microscale where microbial activity takes place, many questions remain open about the actors and how they interact.

10.7.2 Microbes and Microbial Processes

The prerequisite for any methane emission is production. Methane is exclusively produced by methanogenic archaea, but not all archaea are methanogens. Methanogens are anaerobes, but some may survive oxic conditions for some time. This is essential in the Mediterranean rice fields that lay fallow during the cold season. However, the number of methanogens is not affected so much during winter, but the taxonomic composition may change (Krüger et al. 2005). When flooded, the bulk of the soil becomes anoxic within a few hours. Just a thin surface layer stays oxic, while diffusion limitation renders the bulk volume anoxic (Noll et al. 2005). Methanogenesis is the prevalent terminal process in flooded soils and freshwater sediments, while sulfate reduction dominates in marine sediments. However, during the dry period, alternative electron acceptors become re-oxidised and are then reduced in a characteristic sequence once the soil is flooded again (Ponnaperuma 1984). The reduction of electron acceptors follows a typical sequence, with reduction of oxygen, nitrate, iron and sulfate. Iron, in terms of electron flow, is the most important alternative electron acceptor (Frenzel et al. 1999; Krüger et al. 2001). Methanogenesis may start already while iron reduction goes on, albeit at a low rate (Yao et al. 1999; Krüger et al. 2001). Eventually, CO_2 and acetate-methyl is reduced and methanogenesis becomes the dominating terminal process.

Methane emission from rice fields often shows two seasonal peaks, the first one attributed to decaying straw and the second one associated with rhizodeposition. Indeed, methanogenesis in rice microcosms is tightly coupled to the plant's primary production. Pulse labeling of rice plants with $^{14}CO_2$ resulted in a transient accumulation of radioactive methanogenic precursors in the pore water, while 3–6% of the applied radioactivity was finally emitted as $^{14}CH_4$ (Dannenberg and Conrad 1999). However, methanogenesis is not confined to the soil, but can also be found on rice roots (Bosse and Frenzel 1997; Krüger et al. 2002, 2005). The populations detected on rice roots are similar to those found in the soil (Krüger et al. 2005). However, nothing is known as yet about the microsites on the roots where the methanogens are located. Methanogenesis on freshly excised rice root begins virtually without lag (Bosse and Frenzel 1997) and is driven by hydrogen at the beginning (Conrad et al. 2002). The missing lag suggests that the methanogens are active in situ, even if a final proof is still missing.

Methane oxidation is complementary to methane production: only methane that is not oxidised may be emitted from a rice field. Methane oxidation is performed by a few phylogenetically well defined groups of proteobacteria (Heyer et al. 2002). Freshwater methanotrophs are obligate aerobes that possess a methane monooxygenase to perform the first step in methane oxidation. Methanotrophs can consume more than 90% of the methane that is produced in anoxic sediments and soils (Conrad and Rothfuss 1991). Methanotrophs grow with methane as their sole source for energy and growth. The culture-dependent most probable number (MPN) assays for detecting and counting rely on this unique ability. However, some methanotrophs have at least a rudimentary ability to use substrates with C–C bonds (Roslev and King 1995). Different culture-independent tools were developed to visualise and detect methanotrophic bacteria. Microscopical techniques include specific antibodies (e.g. Gilbert et al. 1998), rRNA-directed oligonucleotide probes (e.g. Eller et al. 2001) and, more recently, mRNA-directed probes for the pmoA (Pernthaler and Amann 2004). Eventually, different functional genes (*pmo*A, *mmo*X, *mxa*F) were used in PCR-based detection (Henckel et al. 1999; Bodelier et al. 2000; Horz et al. 2000; Knief et al. 2003; Kolb et al. 2003).

Quantification of methane oxidation is best based on a mass balance:
emission = production – oxidation – accumulation

Accumulation can be ignored in most cases and, from a simple re-arrangement, it becomes obvious that either oxidation or production can be calculated, if the complementary process can be inhibited. The rate measurement reduces to simple flux (= emission) measurements, comparing flux rates with and without inhibitor. The ideal inhibitor must be a gas to pass from the headspace through the plant's aerenchyma to the soil and its effect must be highly selective. We do not yet have a gas available that selectively inhibits methanogenesis, but for the inhibition of methane oxidation we can choose between different species. Acetylene, CH_3F and dimethyl ether have all some unwanted side effects, but CH_2F_2 is a near-optimal choice (Krüger et al. 2002).

A totally different technique uses the kinetic isotope effect that imprints a strong isotopic signal on the methane that escapes biological oxidation (Krüger et al. 2001; Marik et al. 2002). However, care has to be taken not to confuse effects caused by diffusion with the isotopic discrimination caused by methanotrophs. Best results will be obtained if different techniques are combined (Krüger et al. 2002; Krüger and Frenzel 2003).

Methanotrophs are also found in tight association with rice roots, can even be cultured from surface-sterilised roots and culms and are detected microscopically inside the roots (Bosse and Frenzel 1997; Gilbert et al. 1998; Eller et al. 2005). This tight association with the roots is also found in other wetland plants (Frenzel 2000). Eller and Frenzel compared the population dynamics of methanotrophs over one season in the greenhouse and in the field (Eller et al. 2005). MPNs showed both in the rice field and in microcosms that the growth of methanotrophs occurred predominantly at the root surface and within the root. On an aerial base, however, most methanotrophs were counted in the bulk soil that functions like a seed bank. Methanotrophs survive well the fallow period with oxic but low-methane conditions and also the anoxic porewater during flooding. Only if a root grows near by will the methanotrophs experience favourable conditions, with both oxygen and high methane concentrations, at least for some hours to a few days (Flessa and Fischer 1992). The rhizosphere is the microsite that compensates for losses in abundance and diversity. With culture-independent molecular assays, only a few seasonal changes could be detected within the Methylocystaceae, whereas a higher variability was found for the Methylococcaceae. The diversity detected was comparable between field and greenhouse, even though the MPN counts differed slightly (Eller et al. 2005).

10.7.3 The Controls

Once the soil is anoxic, methanogenesis is controlled by competition for substrates with energetically more favourable processes (see above). Methane oxidation is controlled by the availability of oxygen. However, nutrients may be equally important: it was found in a series of experiments with microcosms and in the field that methane oxidation may be limited not only by oxygen, but also by the availability of nitrogen (Bodelier et al. 2000; Krüger et al. 2001, 2002; Krüger and Frenzel 2003; Eller et al. 2005). This finding was surprising, because the ability to utilise dinitrogen as a nitrogen source is an important phenotypic trait in many currently known methanotrophic bacteria (c.f. Dedysh et al. 2004). In contrast, NH_4^+ and all NH_4^+-based nitrogen fertilisers are thought to inhibit methane oxidation (c.f. Bodelier and Laanbroek 2004). However, the high methane concentrations in the pore water of flooded rice fields (Krüger et al. 2001) lead to a NH_4^+:CH_4 ratio that prevents an inhibitory effect. A recent review shows that indications for nitrogen limitation of methane consumption have been reported in a variety of lowland soils, upland soils and sediments (Bodelier and Laanbroek 2004). Obviously, depriving methane-oxidising bacteria of a suitable source of nitrogen hampers their growth and activity. However, the methane-oxidising activity in rice soils reacts instantaneously on the addition of mineral nitrogen, calling for an alternative explanation (Bodelier et al. 2000). Switching from mineral N assimilation to the fixation of molecular nitrogen may explain this phenomenon, but has not been verified experimentally as yet (Bodelier and Laanbroek 2004).

10.8 Summary and Prospects

Wetland ecosystems have a great impact on the life on this planet. They offer a habitat to a vast number of plant and animal species, as well as human populations. The high productivity and diversity of species largely depends on the efficient recycling of nutrients mainly catalysed by microbes. The latter are also responsible for the fact that wetlands are among the most prominent sources of the greenhouse gas methane.

Despite the impact of microbes on wetland ecosystems, little is known about the species present and about their ecology and functioning in situ, especially in the rhizosphere. On the one hand, this is caused by the general dilemma of contemporary microbial ecology studies which is the inability to cultivate the majority of species in the environment. Molecular biological techniques adopted in the past decennia brought insight into the diversity and abundance of particular groups of microbes, but we still have little information on the eco-physiology and in situ functioning of these microbes. The latter still strongly depends on having cultured organisms in the laboratory.

On the other hand, the lack of knowledge is also due to the complexity of the wetland rhizosphere ecosystem. Temporal, spatial and quantitative aspects of oxic–anoxic interfaces depend on numerous factors. The anatomy of wetland plant shoots and roots is a determining yet unpredictable variable, from a microbial point of view. Nevertheless, despite temporary substrate deprivation and exposure to inhibiting agents, microbes manage to adapt. The latter not only includes physiological mechanisms but also commensalistic interactions with other microbes.

The high degree of functional redundancy, like in the case of diazotrophic microbes, may also be regarded as a 'community adaptation'. The high number of different phylogenetic microbial groups performing N_2 fixation ensures a broad eco-physiological potential, sustaining this function under various environmental conditions. The high degree of stability of these diazotrophic communities under environmental disturbance indicates that no members of these communities are removed from the assemblage, preserving the functional potential.

However, important ecosystem functions like the cycling of iron and the emission of trace gases is mediated by relative non-diverse and phylogenetically restricted groups of microbes. Work in recent years established the importance of iron-cycling microbes in the wetland rhizosphere, a phenomenon assumed to be chemically mediated for many years. The investigations in an Italian rice field revealed comprehensive and novel insights into the controls of methane cycling in wetlands. Also in this case, as in all others discussed, the ecology of the microbes involved turns out to be of crucial importance.

Understanding the real importance of microbes in ecosystem functioning in wetlands will require insight into the diversity of the microbes, their functioning or activity in situ and their eco-physiology. Molecular biological techniques reveal great insight into diversity, but we have still no clue to the actual extent of microbial diversity. Hence, diversity assessment will continue, using the latest micro-array technology for high-throughput screening of samples. The latter can both yield diversity data and can also be used for detecting the presence of functional genes (Tiquia et al. 2004). The possibility to screen the presence of genes in large sample numbers offers possibilities to link gene-diversity to environmental parameters through multivariate statistical techniques.

More information about functions will also be gained through genomic techniques like BAC libraries (Liles et al. 2003), possibly revealing the presence of novel genes elucidating the functioning and survival of some of the poorly understood microbial groups in the wetland rhizosphere. However, real ecologically valuable data can only be obtained when function can be coupled to identity. This is still the Holy Grail in microbial ecology. Techniques like stable isotope probing of PLFA (Boschker et al. 1998) and DNA/RNA (Lüders et al. 2003), CARD-FISH (Pernthaler and Amann 2004) and FISH-MAR (Lee et al. 1999) may serve this purpose. However, in soil environments, the application of these techniques is still problematic.

The biggest challenge in wetland research, however, will be scaling-up and linking data obtained from the micrometer scale to ecosystem fluxes. Stable isotopes, modeling and multidisciplinary projects will be important ingredients in tackling this challenge.

Acknowledgements The authors want to acknowledge Jan LaRock for providing the DGGE gel image used in Fig. 10.1. This research was supported by U.S. National Science Foundation awards DEB-9903623, DEB-0084526 and MCB-0237854. This publication is publication number 3576 of the NIOO-KNAW Netherlands Institute of Ecology.

References

Armstrong W, Beckett PM (1987) Internal aeration and the development of stelar anoxia in submerged roots. A multishelled mathematical model combining axial diffusion of oxygen in the cortex with radial losses to the stele, the wall layers and the rhizosphere. New Phytol 105:221–245

Armstrong J, Armstrong W, Beckett PM (1992) Phragmites australis: venturi- and humidity-induced pressure flows enhance rhizome aeration and rhizosphere oxidation. New Phytol 120:197–207

Armstrong W, Justin SHFW, Beckett PM, Lythe S (1991) Root adaptation to soil waterlogging. Aquat Bot 39:57–73

Armstrong W, Brändle R, Jackson MB (1994) Mechanisms of flood tolerance in plants. Acta Bot Neerl 43:307–358

Armstrong W, Cousins D, Armstrong J, Turner DW, Beckett PM (2000) Oxygen distribution in wetland plant roots and permeability barriers to gas-exchange with the rhizosphere: a microelectrode and modelling study with *Phragmites australis*. Ann Bot 86:687–703

Bagwell CE, Lovell CR (2000) Persistence of selected *Spartina alterniflora* rhizoplane diazotrophs exposed to natural and manipulated environmental variability. Appl Environ Microbiol 66:4625–4633

Bagwell CE, Lovell CR (2004) A DNA–DNA hybridization method for the detection and quantification of specific bacterial taxa in natural environments. In: Spencer JFT, Ragout de Spencer AL (eds) Environmental microbiology, Humana, Totowa, pp 169–174

Bagwell CE, Dantzler M, Bergholz, PW, Lovell CR (2001) Host specific ecotypic diversity of rhizoplane diazotrophs of the perennial glasswort, *Salicornia virginica* and selected salt marsh grasses. Aquat Microb Ecol 23:293–300

Blom CWPM, Voesenek LACJ (1996) Flooding: the survival strategies of plants Trends Evol Ecol 11:290–295

Boddey R M (1995) Biological nitrogen fixation in sugar cane: a key to energetically viable biofuel production. Critical Rev Plant Sci 14:263–279

Bodelier PLE, Laanbroek HJ (2004) Nitrogen as a regulatory factor of methane oxidation in soils and sediments. FEMS Microbiol Ecol 47:265–277

Bodelier PLE, Roslev P, Henckel T, Frenzel P (2000) Stimulation by ammonium-based fertilizers of methane oxidation in soil around rice roots. Nature 403:421–424

Boschker HTS, Nold SC, Wellsbury P, Bos D, DeGraaf W, Pel R, Parkes RJ, Cappenberg TE (1998) Direct linking of microbial populations to specific biogeochemical processes by ^{13}C-labelling of biomarkers. Nature 392:801–805

Bosse U, Frenzel P (1997) Activity and distribution of methane-oxidizing bacteria in flooded rice soil microcosms and in rice plants (*Oryza sativa*). Appl Environ Microbiol 63:1199–1207

Bouma TJ, Nielsen KL, Hal B van, Koudstaal B (2001) Root system topology and diameter distribution of species from habitats differing in inundation frequency. Funct Ecol 15: 360–369

Brown MM, Friez MJ, Lovell CR (2003) Expression of nifH genes by diazotrophic bacteria in the rhizosphere of short form *Spartina alterniflora*. FEMS Microbiol Ecol 43:411–417

Cassman KG, De Datta SK, Olk DC, Alcantara JM, Samson MI, Descalsota JP, Dizon MA (1995) Yield decline and the nitrogen economy of long-term experiments on continuous, irrigated rice systems in the tropics. In: Lal R, Stewart BA (eds) Soil management: experimental basis for sustainability and environmental quality, Lewis/CRC, Boca Raton, pp 181–222

Conrad R, Frenzel P (2002) Flooded soils. In: Britton G (ed) Encyclopedia of environmental microbiology. Wiley, New York, pp 1316–1333

Conrad R, Rothfuss F (1991) Methane oxidation in the soil surface layer of a flooded ride field and the effect of ammonium. Biol Fertil Soils 12:28–32

Conrad R, Bak F, Seitz H.J Thebrath B, Mayer HP Schütz H (1989) Hydrogen turnover by psychrotrophic homoacetogenic and mesophilic methanogenic bacteria in anoxic paddy soil and lake sediment. FEMS Microbiol Ecol 62:285–294

Conrad R, Klose M, Claus P (2002) Pathway of CH_4 formation in anoxic rice field soil and rice roots determined by ^{13}C-stable isotope fractionation. Chemosphere 47:797–806

Cypionka, H (2000) Oxygen respiration by *Desulfovibrio* species. Annu Rev Microbiol 54:827–848

Dannenberg S, Conrad R (1999) Effect of rice plants on methane production and rhizospheric metabolism in paddy soil. Biogeochemistry 45:53–71

Das A, Coulter ED, Kurtz DM Jr, Ljungdahl LG (2001) Five-gene cluster in *Clostridium thermoaceticum* consisting of two divergent operons encoding rubredoxin oxidoreductase – rubredoxin and rubrerythrin-type flavodoxin – high-molecular-weight rubredoxin. J Bacteriol 183:1560–1567

Dedysh SN, Ricke P, Liesack W (2004) NifH and NifD phylogenies: an evolutionary basis for understanding nitrogen fixation capabilities of methanotrophic bacteria. Mikrobiologia 48:592–598

Denier van der Gon HAC, Bodegom PM van, Houweling S, Verburg PH, Breemen N van (2000) Combining upscaling and downscaling of methane emissions from rice fields: Methodologies and preliminary results. Nutr Cycl Agroecosyst 58:285–301

Drake HL, Küsel K, Matthies C (2002) Ecological consequences of the phylogenetic and physiological diversities of acetogens. Antonie van Leeuwenhoek 81:203–213

Drake HL, Küsel K, Matthies C (2004) Acetogenic prokaryotes. In: Dworkin M, et al. (eds) The prokaryotes, 3rd edn: an evolving electronic resource for the microbiological community, rel 3.17. Springer, New York, http://springeronline.com, accessed August 2004

Eller G, Stubner S, Frenzel P (2001) Group specific 16S rRNA targeted probes for the detection of type I and type II methanotrophs by fluorescence in situ hybridisation. FEMS Microbiol Lett 198:91–97

Eller G, Krüger M, Frenzel P (2005) Comparing field and microcosm experiments: A case study on methano- and methylotrophic bacteria in paddy soil. FEMS Microbiol Ecol 51:279–291

Emerson D (2000) Microbial oxidation of Fe(II) and Mn(II) at circumneutral pH. In: Lovley DR (ed) Environmental microbe–metal interactions. ASM, Washington, D.C., pp 31–52

Emerson D, Moyer C (1997) Isolation and characterization of novel iron-oxidizing bacteria that grow at circumneutral pH. Appl Environ Microbiol 63:4784–4792

Emerson D, Weiss JV, Megonigal JP (1999) Iron-oxidizing bacteria are associated with ferric hydroxide precipitates (Fe-plaque) on the roots of wetland plants. Appl Environ Microbiol 65:2758–2761

Engelhard M, Hurek T, Reinhold-Hurek B (2000) Preferential occurrence of diazotrophic endophytes, *Azoarcus* spp, in wild rice species and land races of *Oryza sativa* in comparison with modern races. Environ Microbiol 2:131–141

FAO (2004) FAOstat data. FAO, Rome

Flessa H, Fischer WR (1992) Redox processes in the rhizosphere of terrestrial and paludal plants. Z Pflanzenernaehr Bodenk 155:373–378

Frenzel P (2000) Plant-associated methane oxidation in ricefields and wetlands. Adv Microb Ecol 16:85–114

Frenzel P, Rothfuss F, Conrad R (1992) Oxygen profiles and methane turnover in a flooded rice microcosm. Biol Fertil Soils 14:84–89

Frenzel P, Bosse U, Janssen PH (1999) Rice roots and methanogenesis in a paddy soil: ferric iron as an alternative electron acceptor in the rooted soil. Soil Biol Biochem 31:421–430

Galloway JN, Schlesinger WH, Levy HI, Michaels AF, Schnoor JL (1995) Nitrogen fixation: anthropogenic enhancement–environmental response. Global Biogeochem Cycles 9:235–252

Giblin AE, Howarth RW (1984) Pore water evidence for a dynamic sedimentary iron cycle in salt marshes. Limnol Oceanogr 29:47–63

Gilbert B, Aßmus B, Hartmann A, Frenzel P (1998) In situ localization of two methanotrophic strains in the rhizosphere of rice plants by combined use of fluorescently labeled antibodies and 16S rRNA signature probes. FEMS Microbiol Ecol 25:117–128

Gößner A, Devereux R, Ohnemüller N, Acker G, Stackebrandt E, Drake HL (1999) *Thermicanus aegyptius* gen. nov., sp. nov., isolated from oxic soil, a facultative microaerophile that grows commensally with the thermophilic acetogen *Moorella thermoacetica*. Appl Environ Microbiol 65:5124–5133

Gribsholt B, Kostka JE, Kristensen E (2003) Impact of fiddler crabs and plant roots on sediment biogeochemistry in a Georgia saltmarsh. Mar Ecol Prog Ser 259:237–251

Hamelin J, Fromin N, Tarnawski S, Teyssier-Cuvelle S, Aragno M (2002) *NifH* gene diversity in the bacterial community associated with the rhizosphere of *Molinia coerulea*, an oligonitrophilic perennial grass. Environ Microbiol 4:477–481

Hauck S, Benz M, Brune A, Schink B (2001) Ferrous iron oxidation by denitrifying bacteria in profundal sediments of a deep lake (Lake Constance). FEMS Microbiol Ecol 37:127–134

Henckel T, Friedrich M, Conrad R (1999) Molecular analyses of the methane-oxidizing microbial community in rice field soil by targeting the genes of the 16S rRNA, particulate methane monooxygenase, and methanol dehydrogenase. Appl Environ Microbiol 65:1980–1990

Heyer J, Galchenko VF, Dunfield PF (2002) Molecular phylogeny of type II methane-oxidizing bacteria isolated from various environments. Microbiol SGM 48:592-598; 148:2831–2846

Hines ME, Evans RS, Sharak Genthner BR, Willis SG, Friedman S, Rooney-Varga JN, Devereux R (1999) Molecular phylogenetic and biogeochemical studies of sulfate-reducing bacteria in the rhizosphere of *Spartina alterniflora*. Appl Environ Microbiol 65:2209–2216

Holzapfel-Pschorn A, Conrad R, Seiler W (1985) Production, oxidation and emission of methane in rice paddies. FEMS Microbiol Ecol 31:343–351

Horz HP, Rotthauwe JH, Lukow T, Liesack W (2000) Identification of major subgroups of ammonia-oxidizing bacteria in environmental samples by T-RFLP analysis of amoA PCR products. J Microbiol Methods 39:197–204

Howard JB, Rees DC (1996) Structural basis of biological nitrogen fixation. Chem Rev 96:2965–2982

Howes BL, Dacey JWH, Goehringer DD (1986) Factors controlling the growth form of *Spartina alterniflora*: feedbacks between above-ground production, sediment oxidation, nitrogen and salinity. J Ecol 74:881–898

Hurek T, Reinhold-Hurek B (1998) Interactions of *Azoarcus* sp. with rhizosphere fungi. In: Varma A, Hock B (eds) Mycorrhiza, 2nd edn. Springer, Berlin Heidelberg New York, pp 595–614

Hurek T, Reinhold-Hurek B (2006) Molecular ecology of N_2-fixing microbes associated with graminaceous plants. In: Werner D, Newton WE (eds) Agriculture, forestry, ecology and the environment. Kluwer, Dordrecht, pp 173–198

Hurek T, Egener T, Reinhold-Hurek B (1997) Divergence in nitrogenases of *Azoarcus* spp, Proteobacteria of the α-subclass. J Bacteriol 179:4172–4178

Hurek T, Handley L, Reinhold-Hurek B, Piché Y (2002) *Azoarcus* grass endophytes contribute fixed nitrogen to the plant in an unculturable state. Mol Plant Microb Interact 15:233–242

Karl D, Bergman B, Capone D, Carpenter E, Letelier R, Lipschultz F, Paerl H, Sigman D, Stal L (2002). Dinitrogen fixation in the world's oceans. Biogeochemistry 57/58:47–98

Kamura T, Takai Y, Ishikawa K (1963) Microbial reduction mechanism of ferric iron in paddy soils (part 1). Soil Sci Plant Nutr 9:171–175

Karnholz A, Küsel K, Gößner A, Schramm A, Drake HL (2002) Tolerance and metabolic response of acetogenic bacteria toward oxygen. Appl Environ Microbiol 68:1005–1009

Kimura M (2000) Anaerobic microbiology in waterlogged rice fields. In: Bollag JM, Stotzky G (eds) Soil biochemistry. Dekker, New York, pp 35–138

Knief C, Lipski A, Dunfield PF (2003) Diversity and activity of methanotrophic bacteria in different upland soils. Appl Environ Microbiol 69:6703–6714

Kolb S, Knief C, Stubner S, Conrad R (2003) Quantitative detection of methanotrophs in soil by novel pmoA-targeted real-time PCR assays. Appl Environ Microbiol 69:2423–2429

Koretsky CM, Moore CM, Lowe KL, Meile C, DiChristina TJ, Van Cappellen P (2003) Seasonal oscillation of microbial iron and sulfate reduction in saltmarsh sediments (Sapelo Island, GA, USA). Biogeochemistry 64:179–203

Kostka JE, Luther GWI (1995) Seasonal cycling of Fe in saltmarsh sediments. Biogeochemistry 29:159–181

Kostka JE, Gribsholt B, Petrie E, Dalton D, Skelton H, Kristensen E (2002) The rates and pathways of carbon oxidation in bioturbated saltmarsh sediments. Limnol Oceanogr 47:230–240

Koyama T (1955) Gaseous metabolism in lake muds and paddy soils. J Earth Sci 3:65–76

Krüger M, Frenzel P (2003) Effects of N-fertilisation on CH_4 oxidation and production, and consequences for CH_4 emissions from microcosms and rice fields. Global Change Biol 9:773–784

Krüger M, Frenzel P, Conrad R (2001) Microbial processes influencing methane emission from rice fields. Global Change Biol 7:49–63

Krüger M, Eller G, Conrad R, Frenzel P (2002) Seasonal variation in pathways of CH_4 production and in CH_4 oxidation in rice fields determined by stable carbon isotopes and specific inhibitors. Global Change Biol 8:265–280

Krüger M, Frenzel P, Kemnitz D, Conrad R (2005) Activity, structure and dynamics of the methanogenic archaeal community in a flooded Italian rice field. FEMS Microbiol Ecol 51:323–331

Kurtz DM (2003) Oxygen and anaerobes. In: Ljungdahl LG, Adams M, Barton L, Ferry JG, Johnson M (eds) Biochemistry and physiology of anaerobic bacteria. Springer, Berlin Heidelberg New York, pp 128–142

Küsel K, Pinkart HC, Drake HL, Devereux R (1999) Acetogenic and sulfate-reducing bacteria inhabiting the rhizoplane and deep cortex cells of the sea grass *Halodule wrightii*. Appl Environ Microbiol 65:5117–5123

Küsel K, Karnholz A, Trinkwalter T, Devereux R, Acker G, Drake HL (2001) Physiological ecology of *Clostridium glycolicum* RD-1, an aerotolerant acetogen isolated from sea grass roots. Appl Environ Microbiol 67:4734–4741

Leaphart AB, Friez MJ, Lovell CR (2003) Formyltetrahydrofolate synthetase sequences from salt marsh plant roots reveal a diversity of acetogenic bacteria and other bacterial functional groups. Appl Environ Microbiol 69:693–696.

Lee N, Nielsen PH, Andreasen KH, Juretschenko S, Nielsen JP, Schleifer KH, Wagner M (1999) Combination of fluorescent in situ hybridization and micro-autoradiography – a new tool for structure-function analyses in microbial ecology. Appl Environ Microbiol 65:1289–1297

Lenssen JPM, Menting FBJ, Putten WH van der, Blom CWPM (1999) Effects of sediment type and water level on biomass production of wetland plant species. Aquat Bot 64:151–165

Liles MR, Manske BF, Bintrim SB, Handelsman J, Goodman RM (2003) A census of rRNA genes linked genomic sequences within a soil metagenomic library. Appl Environ Microbiol 69:2684–2691

Lovell CR (2002) Plant–microbe interactions in the marine environment. In: Bitton G (ed) Encyclopedia of environmental microbiology, vol 5. Wiley, New York, pp 2539–2554

Lovell CR, Piceno YM, Quattro JM, Bagwell CE (2000) Molecular analysis of diazotrophic diversity in the rhizosphere of the smooth cordgrass, *Spartina alterniflora*. Appl Environ Microbiol 66:3814–3822

Lovell CR, Friez MJ, Longshore JW, Bagwell CE (2001a) Recovery and phylogenetic analysis of *nifH* sequences from diazotrophic bacteria associated with dead aboveground biomass of *Spartina alterniflora*. Appl Environ Microbiol 67:5308–5314

Lovell CR, Bagwell CE, Czákó M, Márton L, Piceno YM, Ringelberg DB (2001b) Stability of a rhizosphere microbial community exposed to natural and manipulated environmental variability. FEMS Microbiol Ecol 38:69–76

Lovley DR (2000) Fe(III) and Mn(IV) reduction. In: Lovley DR (ed) Environmental microbe–metal interactions. ASM, Washington, D.C., pp 3–30

Lüders T, Manefield M, Friedrich MW (2003) Enhanced sensitivity of DNA- and rRNA-based stable isotope probing by fractionation and quantitative analysis of isopycnic centrifugation gradients. Environ Microbiol 6:73–78

Luther GW III, Kostka JE, Church TM, Sulzberger B, Stumm W (1992) Seasonal iron cycling in a salt marsh sedimentary environment: the importance of ligand complexes with Fe(II) and Fe(III) in the dissolution of Fe(III) minerals and pyrite, respectively. Mar Chem 40:81–103

Marik T, Fischer H, Conen F, Smith K (2002) Seasonal variations in stable carbon and hydrogen isotopes ratios in methane from rice fields. Global Biogeochem Cycles 16:1094

Megonigal JP, Hines ME, Visscher PT (2004) Anaerobic metabolism: linkages to trace gases and aerobic processes. In: Schlesinger WH (ed) Biogeochemistry. Elsevier-Pergamon, Oxford, pp 317–424

Mendelssohn IA (1979) The influence of nitrogen level, form, and application method on the growth response of *Spartina alterniflora* in North Carolina. Estuaries 2:106–112

Mendelssohn IA, Kleiss BA, Wakeley JS (1995) Factors controlling the formation of oxidized root channels: a review. Wetlands 15:37–46

Müller V, Inkamp F, Rauwolf A, Küsel K, Drake HL (2004) Molecular and cellular biology of acetogenic bacteria. In: Nakano MM, Zuber P (eds) Strict and facultative anaerobes: medical and environmental aspects. Horizon Bioscience, Norfolk, pp 251–281

Neubauer SC, Emerson D, Megonigal JP (2002) Life at the energetic edge: kinetics of circumneutral iron oxidation by lithotrophic iron-oxidizing bacteria isolated from the wetland-plant rhizophere. Appl Environ Microbiol 68:3988–3995

Noll M, Matthies D, Frenzel P, Derakshani M, Liesack W (2005) Succession of bacterial community structure and diversity in a paddy soil oxygen gradient. Environ Microbiol 7:382–395

Pernthaler A, Amann R (2004) Simultaneous fluorescence in situ hybridization of mRNA and rRNA in environmental bacteria. Appl Environ Microbiol 70:5426–5433

PicenoYM, Lovell CR (2000a) Stability in natural bacterial communities. I. Nutrient addition effects on rhizosphere diazotroph assemblage composition. Microb Ecol 39:32–40

Piceno YM, Lovell CR (2000b) Stability in natural bacterial communities. II. Plant resource allocation effects on rhizosphere diazotroph assemblage composition. Microb Ecol 39:41–48

Piceno YM, Noble PA, Lovell CR (1999) Spatial and temporal assessment of diazotroph assemblage composition in vegetated salt marsh sediments using denaturing gradient gel electrophoresis analysis. Microb Ecol 38:157–167

Poly F, Ranjard L, Nazaret S, Gourbière F, Monrozier LJ (2001) Comparison of gene pools in soils and soil microenvironments with contrasting properties. Appl Environ Microbiol 67:2255–2262

Ponnamperuma FN (1984) Effects of flooding on soils. In: Kozlowski TT (ed) Flooding and plant growth. Academic, New York, pp 9–45

Powell SW, Day FP (1991) Root production in four communities in the Great Dismal Swamp. Am J Bot 78:288–297

Reinhold B, Hurek T, Niemann E-G, Fendrik I (1986) Close association of *Azospirillum* and diazotrophic rods with different root zones of Kallar grass. Appl Environ Microbiol 52:520–526

Ribbe M, Gadkari D, Meyer O (1997) N_{-2} fixation by *Streptomyces thermoautotrophicus* involves a molybdenum-dinitrogenase and a manganese-superoxide oxidoreductase that couple N_{-2} reduction to the oxidation of superoxide produced from O_{-2} by a molybdenum-CO dehydrogenase. J Biol Chem 272:26627–26633

Roden EE (2003) Fe(III) oxide reactivity toward biological versus chemical reduction. Environ Sci Technol 37:1319–1324

Roden EE, Wetzel RG (1996) Organic carbon oxidation and suppression of methane production by microbial Fe(III) oxide reduction in vegetated freshwater wetland sediments. Limnol Oceanogr 41:1733–1748

Roden EE, Wetzel RG (2002) Kinetics of microbial Fe (III) oxide reduction in freshwater wetland sediments. Limnol Oceanogr 47:198–211

Roslev P, King GM (1995) Aerobic and anaerobic starvation metabolism in methanotrophic bacteria. Appl Environ Microbiol 61:1563–1570

Schippers A, Jorgensen BB (2002) Biogeochemistry of pyrite and iron sulfide oxidation in marine sediments. Geochim Cosmochim Acta 66:85–92

Schubauer JP, Hopkinson CS (1984) Above- and belowground emergent macrophyte production and turnover in a coastal marsh ecosystem, Georgia. Limnol Oceanogr 29:1052–1065

Shearer MJ, Khalil MAK (2000) Rice agriculture: emissions. In: Atmospheric methane: its role in the global environment. Springer, Berlin Heidelberg New York, pp 170–189

Silaghi-Dumitrescu R, Coulter ED, Das A, Ljungdahl LG, Jameson GNL, Huynh BH, Kurtz DM Jr (2003) A flavodiiron protein and high molecular weight rubredoxin from *Moorella thermoacetica* with nitric oxide reductase activity. Biochemistry 42:2806–2815

Singer PC, Stumm W (1970) Acid mine drainage: the rate-determining step. Science 167:1121–1123

Sobolev D, Roden EE (2001) Suboxic deposition of ferric iron by bacteria in oppposing gradients of Fe(II) and oxygen at circumneutral pH. Appl Environ Microbiol 67:1328–1334

Sobolev D, Roden EE (2004) Characterization of a neutrophilic, chemolithotrophic Fe(II)-oxidizing B-proteobacterium from freshwater wetland sediments. Geomicrobiol J 21:1–10

Sorrell BK (1994) Airspace structure and mathematical modelling of oxygen diffusion, aeration and anoxia in *Eleocharis sphacelata* R. Br. roots. Aust J Mar Freshwater Res 45:1529–1541

Sorrell BK, Mendelssohn IA, McKee KL, Woods RA (2000) Ecophysiology of wetland plant roots: a modeling comparison of aeration in relation to species distribution. Ann Bot 86:675–685

Straub KL, Buchholz-Cleven BEE (1998) Enumeration and detection of anaerobic ferrous iron-oxidizing, nitrate-reducing bacteria from diverse European sediments. Appl Environ Microbiol 64:4846–4856

Straub KL, Benz M, Schink B, Widdel F (1996) Anaerobic, nitrate-dependent microbial oxidation of ferrous iron. Appl Environ Microbiol 62:1458–1460

Straub KL, Rainey FA, Widdel F (1999) *Rhodovulum iodosum* sp. nov. and *Rhodovulum robiginosum* sp. nov., two new marine phototrophic ferrous-iron-oxidizing purple bacteria. Int J Syst Bacteriol 49:729–735

Summers JE, Ratcliffe RG, Jackson MB (2000) Anoxia tolerance in the aquatic monocot *Potamogeton pectinatus*: absence of oxygen stimulates elongation in association with an unusually large Pasteur effect. J Exp Bot 51:1413–1422

Tan Z, Hurek T, Reinhold-Hurek B (2003) Effect of N-fertilization, plant genotype and environmental conditions on *nifH* gene pools in roots of rice. Environ Microbiol 2:1009–1015

Thamdrup B (2000) Bacterial manganese and iron reduction in aquatic sediments. Adv Microb Ecol 16:41–84

Tiquia SM, Liyou W, Chong SC, Passovets S, Xu D, Xu Y, Zhou J (2004) Evaluation of 50-mer oligonucleotide arrays for detecting microbial populations in environmental samples. BioTechniques 36:664–675

Turner RE (1976) Geographic variations in salt marsh macrophyte production: a review. Contrib Mar Sci 20:47–68

Van Bodegom P, Stams F, Mollema L, Boeke S, Leffelaar P (2001) Methane oxidation and the competition for oxygen in the rice rhizosphere. Appl Environ Microbiol 67:3586–3597

Wang MX, Li J (2002) CH_4 emission and oxidation in Chinese rice paddies. Nutr Cycl Agroecosyst 64:43–55

Weber KA, Picardal FW, Roden EE (2001) Microbially catalyzed nitrate-dependent oxidation of biogenic solid-phase Fe(II) compounds. Environ Sci Technol 35:1644–1650

Weiss JV, Emerson D, Backer SM, Megonigal JP (2003) Enumeration of Fe(II)-oxidizing and Fe(III)-reducing bacteria in the root zone of wetland plants: Implications for a rhizosphere iron cycle. Biogeochemistry 64:77–96

Weiss JV, Emerson D, Megonigal JP (2004) Geochemical control of microbial Fe(III) reduction potential in wetlands: comparison of the rhizosphere to non-rhizosphere soil. FEMS Microbiol Ecol 48:89–100

Widdel F, Schnell S, Heising S, Ehrenreich A, Assmus B, Schink B (1993) Ferrous iron oxidation by anoxygenic phototrophic bacteria. Nature 362:834–836

Yao H, Conrad R, Wassmann R, Neue HU (1999) Effect of soil characteristics on sequential reduction and methane production in sixteen rice paddy soils from China, the Philippines, and Italy. Biogeochemistry 47:269–295

Zani S, Mellon MT, Collier JL, Zehr JP (2000). Expression of nifH genes in natural microbial assemblages in Lake George, New York, detected by reverse transcriptase PCR. Appl Environ Microbiol 66:3119–3124

Zehr JP, McReynolds LA (1989) Use of degenerate oligonucleotides for amplification of the *nifH* gene from the marine cyanobacterium *Trichodesmium thiebautii*. Appl Environ Microbiol 55:2522–2526

Zehr JP, Jenkins BD, Short SM, Steward GF (2003) Nitrogenase gene diversity and microbial community structure: a cross-system comparison. Environ Microbiol 5:539–554

11 Linkages Between Microbial Community Composition and Biogeochemical Processes Across Scales

A. Ogram, S. Bridgham, R. Corstanje, H. Drake, K. Küsel, A. Mills, S. Newman, K. Portier, and R. Wetzel (deceased)

11.1 Introduction

Much of the biogeochemical cycling that is critical to the functioning of wetlands is controlled by complex communities of microorganisms (Fig. 11.1). These communities form the basis of detrital food webs which mineralize nutrients bound up in plant biomass and are therefore a critical link in the nutrient cycles that drive wetland ecosystems. Considerable research has been conducted in recent years to define energetic controls on biogeochemical cycles in wetlands, and a basic knowledge of the ecology and physiology of individual groups of many of the microorganisms responsible for biogeochemical cycling has existed for many years. Much previous research regarded wetland microbial communities as merely biomass, however, with little regard for the activities or numbers of individual groups that control these functions. Individual components of these communities respond differently to changes in their environment, which in turn may affect the rates and pathways for biogeochemical cycles in wetlands. Anthropogenic impacts resulting from air and nutrient pollution and increased atmospheric CO_2 concentrations effect biogeochemical processes in wetlands, but the mechanisms by which these changes occur cannot be fully understood until the ecology of the key microorganisms in these systems is understood. Linking microbial community structure and activities with biogeochemical cycles will provide greater insight into the mechanisms that drive these cycles and how they respond to anthropogenic impacts.

The development and application of new tools and new approaches to the characterization of microbial communities and their activities leads to increased appreciation of the roles of individual functional groups of bacteria

Ecological Studies, Vol. 190
J.T.A. Verhoeven, B. Beltman, R. Bobbink, and D.F. Whigham (Eds.)
Wetlands and Natural Resource Management

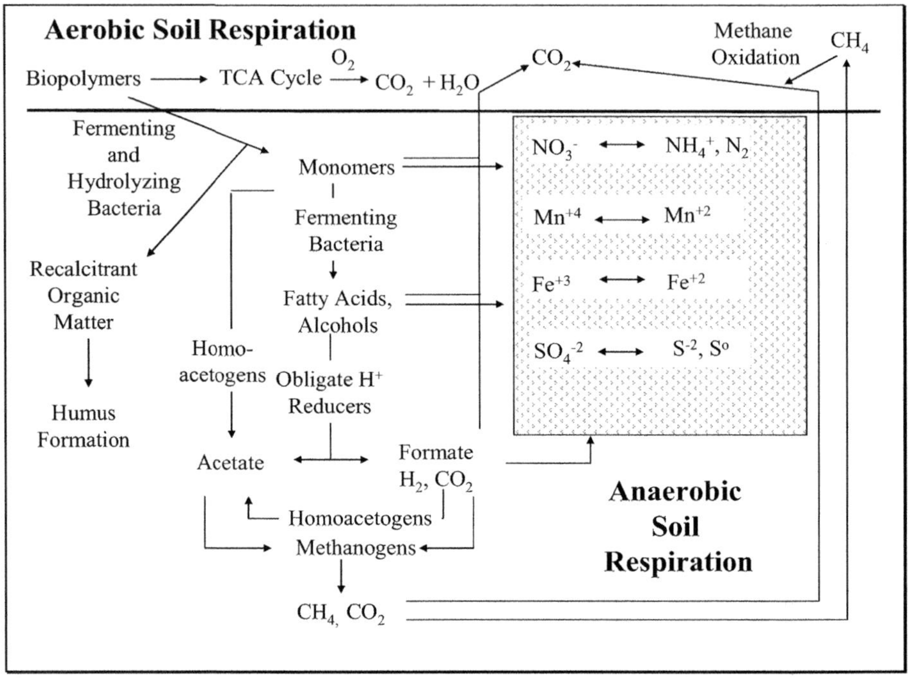

Fig. 11.1 Aerobic and anaerobic microbial processes (adapted from Conrad 1989)

in biogeochemical cycles and of the specific mechanisms that control their activities in wetlands. Recent advances in a wide array of analytical approaches, including advances in instrumental analysis, analysis of enzyme function (Carriero et al. 2000), and application of molecular genetics (Castro et al. 2002, 2004) are providing significant insights into linkages between microorganisms and the functions they control. Recent advances in molecular ecology include the specific characterization of functional groups of microorganisms (Castro et al. 2005) and the application of genomic tools to rapid analysis of microbial communities through DNA fingerprinting (Blum et al. 2004) and the use of microarrays (Loy et al. 2004).

Most of these studies were conducted on limited geographic scales; extensive sampling and analysis of microbial community structure and function can be time-consuming and costly, precluding extensive sampling schemes across landscapes. New methods allow detailed analysis of specific components of community structure to be efficiently conducted on hundreds of samples, such that linking microbial community structure and activity may be linked with processes that extend to landscape scales. The degree of specialization required for studies that transcend broad scales of research, such as those that extend from the molecular to the landscape scale, require the integration of interdisciplinary teams of wetland biogeochemists, microbial ecologists, and statisticians. Interdisciplinary projects such as these can result in very complex data sets that require new approaches toward data analysis, such that individual biogeochemical parameters collected on small may be linked with specific aspects of microbial community structure and activity across scales (Corstanje 2003).

This chapter reviews some ongoing studies that seek to link microorganisms or their activities with their environment and biogeochemical processes and to use this information to better define controls on carbon decomposition in wetlands and the specific pathways carbon flows through different wetlands. The chapter will begin at the beginning: decomposition of plant matter in wetlands, and it will then proceed through various scales and systems, culminating with an approach for statistical analysis linking microbial community composition with complex biogeochemical data in wetlands.

11.2 Microbial Controls on Decomposition

11.2.1 Decomposition of Plant Matter in Wetlands

The initial decomposition of plant matter in wetlands is largely controlled by predominately aerobic microbial communities attached to living and dead plants (periphyton). The physical structure of the periphyton is such that gaseous and ionic exchange rates with the surrounding aqueous environment occur at orders of magnitude lower than occur in the surrounding water. As a result, much of the metabolism of microbial groups within periphyton communities is mutualistic, involving the passage of metabolites between different functional groups of microorganisms (Wetzel 1993, 1996). Because of this intimate physiological interdependence, internal nutrient and energetic recycling is intense within this relatively closed community. Metabolism, growth, and productivity within periphyton communities rely heavily on internal recycling and the conservation of resources; the result is an unusually high efficiency of utilization and retention of captured external resources and a high productivity (Wetzel 1990).

Most organic carbon fixed by eukaryotic algae and cyanobacteria within the periphyton is ultimately decomposed to CO_2 by bacteria and eukaryotic microorganisms within the periphyton. Many of the nutrients released during decomposition are actively sequestered and retained by the viable components of the periphyton community. The fate of organic matter from living or dead microbes of the periphytic communities that is not utilized and respired is very complex and often depends on the changing chemical and physical conditions of the supporting macrophyte substrata. Most of the organic production is metabolized by heterotrophic microorganisms; and relatively little of the carbon fixed by the periphyton passes to higher trophic levels outside of the periphyton (Wetzel 2005).

The major sources of organic matter produced by the wetland–littoral complex in most freshwater ecosystems are often augmented by inputs from terrestrial vegetation. In wetlands surrounding lakes or in wetlands of river flood plains, the loading of organic matter as particulate terrestrial vegetation directly to the water may be small in comparison to the continual inputs of dissolved organic matter. In most wetlands, the vegetation is dominated by herbaceous perennials that exhibit multiple population cohorts or more or less continuous growth, senescence, and turnover. These plant tissues in various stages of fungal and bacterial degradation release large quantities of dissolved organic matter (DOM). As that DOM moves within the wetlands, a selective degradation occurs in removal of the more labile and energy-rich components.

Most particulate organic matter is decomposed within the land–water interface of wetlands and littoral regions. Losses result from a combination of decomposition of plant organic matter with release of respiratory CO_2 and methane to the atmosphere, fragmentation of particulate organic matter by turbulence, macroinvertebrates and vertebrates, and movement of dissolved and fine particulate organic matter. Fragmentation of coarse particulate organic matter by feeding activities of immature macroinvertebrates can markedly increase the rates of microbial degradation. Much greater amounts of CO_2 are released from benthic respiration than the amount of oxygen consumed, which emphasizes the prevalence of anaerobic metabolism utilizing alternative electron acceptors (nitrate, ferrous iron, sulfate, organic compounds) rather than oxygen (e.g., Reddy et al. 2005). Export of organic matter is predominantly as dissolved organic matter of relatively recalcitrant chemical compounds, often originating from lignin and cellulose structural tissues of higher plants and various bacterial degradation products of degradation of these higher plant tissues.

During the senescence of macrophytes, appreciable nutrient content is translocated to rooting tissues, particularly among herbaceous perennials that dominate among aquatic plants. Some senescence and degradation of leaf tissues, however, results in the loss of cellular integrity and leaching of nutrients and dissolved organic compounds from the leaves. Labile components of these leachates, particularly amino acids, for example, are readily utilized and sequestered by the periphytic microflora (e.g., Cunningham and Wetzel 1989; Bicudo et al. 1998). Similarly, as leaching of dissolved organic matter from senescent macrophyte vegetation under various stages of degradation occurs, a selective utilization of more labile constituents by attached microflora occurs as the dissolved organic matter moves through the wetland communities. The result is an increasing recalcitrance of the dissolved organic compounds, particularly the humic substances. These phenolic substances have a high aromaticity, particularly from lignin of structural tissues, and are difficult for microorganisms to hydrolyze.

11.2.2 Microbial Enzyme Activities as Indicators of Controls on Decomposition

Partially decomposed plant matter eventually enters the soil, and much of the decomposition is effected by soil microbial communities. Regardless of the location, a key feature of microbial decomposition is the hydrolysis of macromolecules into low molecular weight compounds that can be assimilated to support microbial metabolism. This process is catalyzed by extracellular enzymes and is considered the rate-limiting step in organic matter decomposition (Meyer-Reil 1991; Sinsabaugh et al. 1993).

Due to the complexity of organic matter, multiple enzymes acting in concert are required for the effective degradation of different compounds. As with other aspects of decomposition, enzyme activity is influenced by a variety of factors, including nutrient availability, substrate quality (including availability and complexity), metabolic activities of the microbial community, and inhibition by organic acids and lignin.

A model with some success in linking plant litter decomposition to enzymatic hydrolysis is MARCIE (Microbial Allocation of Resources Among Community Indicator Enzymes; Sinsabaugh and Moorhead 1994). The foundation for this model is that lignocellulose degradation by enzymes is controlled by environmental nutrient concentrations. Decomposition or bacterial productivity is linked to the availability of C, N and P, as evidenced by enzymatic activity. The ratios of C-acquiring enzymes to both N and P can then be used as indicators of relative nutrient availability (Sinsabaugh et al. 2002).

Microbial decomposition is generally expected to increase in response to nutrient enrichment, although this is dependent on the interaction of substrate quality with the specific enzymes that facilitate decomposition. Carreiro et al. (2000) demonstrated that high concentrations of N could either stimulate or inhibit decomposition, dependent on the substrate quality. In low-lignin material, N addition stimulates decomposition through increased activity of cellulase enzymes. In contrast, decomposition rates in high-lignin material may be negatively related to N addition due to suppression of phenol oxidase activity, a key enzyme responsible for the degradation of lignin.

Phosphorus metabolism of the microbial community is often linked to the activity of phosphomonoesterases that are induced in response to P limitation and inhibited under high concentrations of inorganic P (Jansson et al. 1988; Newman et al. 2003). More recently, significant attention has been directed towards the importance of phosphodiesters, and in association, phosphodiesterases in P-cycling in wetlands (Turner and Newman 2005).

Using ^{31}P NMR, Turner and Newman (2005) analyzed P species in surficial soils in the Florida Everglades. In general, the diester to monoester ratio was >1. Further, in a constructed wetland designed for P removal, >70 % of the soluble organic phosphorus in the surface water was hydrolyzed by phosphodiesterase, with <10 % hydrolyzed by phosphomonoesterase (Pant et al. 2002). However, our understanding of the importance of phosphodiesterase activity in P metabolism and microbial decomposition is confounded by the potential use of phosphodiesters as a source of energy and nitrogen rather than phosphorus (Heath 2004).

It is apparent from our studies of enzyme-mediated decomposition that small shifts in activity of critical enzymes will significantly alter decomposition rates (Sinsabaugh et al. 2002), with ecosystem and global significance. For example, a study at the ecosystem level of a suite of enzyme activities in open-water slough and densely vegetated emergent marsh in the Everglades indicated greater carbon quality in the open-water environment. This in turn suggests that enzyme-mediated decomposition of the resultant peat is an important factor in creating topographical features in the Everglades landscape (Penton 2004). At a global level, Freeman et al. (2001) examined the enzyme activity of peat under anoxic and oxic conditions, with and without inhibition by phenolic compounds. They concluded that oxygen limitation of phenol oxidase activity in peat wetlands is the key factor controlling the current storage of carbon in peat (Freeman et al. 2001).

Establishment of linkages between microbial activity and decomposition processes is brought closer through the measurement of enzyme activities that hydrolyze specific compounds. With an increasing availability of model substrates to measure enzyme activity and increased knowledge of microbial community structure, future research will provide greater links between structure and function to assess microbially mediated decomposition and provide insights into the biological bases for controls on decomposition.

11.3 Linking Decomposition with Microbial Community Composition

11.3.1 Anaerobic Carbon Cycle

As described above, much of the initial breakdown of large organic molecules is accomplished through the activities of extracellular enzymes to produce smaller molecules that can be taken up by microorganisms and metabolized (Conrad 1989). Oxygen is rapidly depleted in the first few centimeters of a flooded soil and much of the succeeding decomposition occurs in oxygen-limited or anoxic conditions. Once labile carbon enters anoxic soil, it is metabolized through the passage of metabolites between different microbial groups via different pathways and is eventually mineralized to CO_2 and methane (CH_4). If mineral terminal electron acceptors (TEAs) such as NO_3^-, Fe (III), Mn (IV), and SO_4^{2-} are exhausted, the 'cooperation' of four groups of bacteria are necessary for the effective degradation of organic matter under anoxic conditions (Fig. 11.1): (a) hydrolytic and fermenting bacteria, (b) H^+-reducing bacteria, (c) methanogens, and (d) acetogenic bacteria. Hydrolytic and fermenting bacteria break down complex organic molecules into monomers and subsequently ferment them to simple alcohols, short-chain fatty acids, H_2, and CO_2. H^+-reducing bacteria (so-called 'syntrophs') further reduce low molecular weight organic end-products of fermenters, utilizing H^+ as the terminal electron acceptor and forming H_2, acetate, and CO_2. Natural populations of methanogens are largely dependent on H_2 and acetate for energy substrates in freshwater wetlands (Conrad 1996). Homoacetogens are a versatile group of bacteria capable of utilizing a range of organic substrates, including H_2–CO_2, CO, and methanol, to produce acetate (Dolfing 1988; Conrad 1996) and can compete with the previous three microbial groups for these substrates. This is a delicately balanced syntrophic microbial consortium, with H^+ reduction only thermodynamically favorable at very low partial pressures of H_2 that are achieved through its consumption by methanogens, acetogens, and sulfate reducers (Dolfing 1988; Conrad 1996). This has been termed interspecies H_2 transfer. If the H_2 partial pressure becomes too high, the system will become unbalanced and lead to an accumulation of fermentation products with a subsequent lowering of the pH.

Recent work suggests that humic acids may be important electron acceptors in many wetlands, particularly in high-organic-matter soils such as peats (Lovley et al. 1996; Coates et al. 1998; Segers and Kengen 1998; van Hulzen et al. 1999). Humic acids may also serve as an intermediary in iron reduction, where electrons from fermentation reduction reactions are shuttled via humics to Fe(III), which is reduced to Fe(III) (Lovley et al. 1996). The active electron-accepting portion of humic acids appears to be quinones (Lovley et al. 1996; Cervantes et al. 2000).

11.3.2 Controls over CO_2:CH_4 Ratios in Anaerobic Respiration in Wetlands

Factors that determine the relative importance of different pathways for anaerobic carbon mineralization result in large differences in the ratios of CH_4 and CO_2 produced in wetlands. This question is important both in determining the carbon budget of a wetland and in assessing the potential impacts on climate change; both CO_2 and CH_4 are important greenhouse gases, with wetlands (including rice fields) being responsible for approximately one-third of the global CH_4 flux (Schlesinger 1997). Additionally, the global warming potential of CH_4 is approximately between three and 22 times that of CO_2, depending on the time-frame considered (Rodhe 1990).

Under oxic conditions, microorganisms utilize the tricarboxylic acid cycle for respiration, with CO_2 as the final end-product. Under anoxic conditions, however, the dominant pathway may lead to either CO_2 or CH_4 as the end-product of microbial energy generation and requires the cooperation of several groups of prokaryotes. The CO_2:CH_4 ratio produced during anaerobic metabolism varies over several orders of magnitude among different types of wetlands. For example, almost no carbon was mineralized to CH_4 during long-term anoxic incubations of soils taken from bogs and acidic fens in the northern United States and several peatlands in the southeastern United States, but the CO_2:CH_4 ratio was as low as eight to nine in minerotrophic peatlands and wetlands with mineral soils (Fig. 11.2A).

Three possible hypotheses can be proposed that may explain the wide range of anaerobic respiratory CO_2:CH_4 ratios observed in wetlands:

1. CO_2:CH_4 ratios in wetlands are controlled by environmental factors that directly control methanogens (e.g., temperature, pH, nutrient availability, trace metal availability).
2. CO_2:CH_4 ratios in wetlands are controlled by substrate competition between bacterial groups that use different inorganic TEAs. Because of the

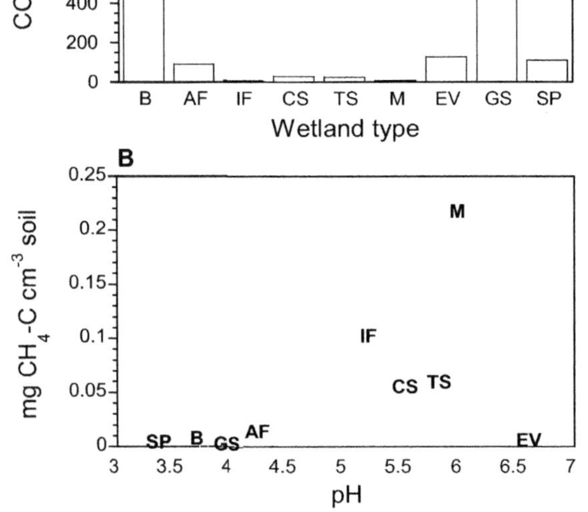

Fig. 11.2 A The ratio of the cumulative amount of CO_2 and CH_4 produced during continuous anoxic incubations of wetland soils for 59 weeks. The following sites are in northern Minnesota, USA: bog (*B*, *n*=5), acidic fen (*AF*, *n*=2), intermediate fens (*IF*, *n*=2), cedar swamp (*CS*, *n*=3), tamarack swamp (*TS*, *n*=2), and beaver-created marsh (*M*, *n*=2). A single peatland was sampled for a gum swamp (*GS*) and short pocosin (*SP*) in North Carolina, USA, and in the Everglades (*EV*), Florida, USA. Five replicate cores were incubated per site. Data for the Minnesota sites were previously reported in Bridgham et al. (1998). **B** Scatter plot of cumulative methane production versus soil pH for these same sites

low energy yield of methanogenic pathways, methanogens are less competitive for substrates than are other anaerobic bacteria.

3. CO_2:CH_4 ratios in wetlands are controlled by substrate limitation of methanogenesis that reflects inefficient syntrophic fermentation reactions in anaerobic carbon mineralization. Factors such as low pH, low temperature, and poor soil carbon quality cause: (a) a slowing of the overall fermentation process; and (b) a build up of fermentation products (short-chain fatty acids and alcohols). Thus, environmental controls over methanogenesis occur primarily from effects on 'upstream' microbial syntrophic pathways.

In reality, these scenarios are not mutually exclusive and probably all operate to some extent in most wetlands. However, we believe that they form a set of hierarchical controls that vary between different types of wetlands and along climatic gradients. Thus, they may form a fundamental microbial constraint on CH_4 production and flux in wetlands in response to climate change.

Regarding the first hypothesis, the wide range of CO_2:CH_4 ratios found in wetlands may result in part from a wide range of soil pHs, although important outlier sites such as the Everglades are evident (Fig. 11.2B). Most methanogens that have been isolated prefer a circumneutral pH; raising the pH of acidic wetland soils has increased CH_4 production in some studies (Dunfield et al. 1993;Valentine et al. 1994), but not in others (Bridgham and Richardson 1992). In a study comparing pathways for methanogenesis in various wetlands, Hines and Duddleston (2001) found that acetoclastic methanogenesis was insignificant in peatlands when the pH was <4.6 or temperature was <15 °C, but acetate was readily consumed in temperate fens. All methanogenesis in bogs and northern fens in this study proceeded via a H_2–CO_2 pathway. It may be that acetate accumulated in pore water until it was consumed through aerobic respiration when the water table dropped, effectively limiting potential methanogenesis in these peatlands. In contrast to these results, accumulation of acetate was observed in acidic bogs in Michigan during the winter and spring, but the acetate was quickly consumed by methanogens as the growing season ensued (Shannon and White 1996).

Regarding the second hypothesis, a competitive hierarchy exists among potential anaerobic electron acceptors based on factors such as Eh, pH, concentrations of other TEAs, and thermodynamic yields (Fig. 11.1). In general, TEAs with higher thermodynamic yields are used first, in the order: denitrification, Mn(IV) reduction, Fe(III) reduction, sulfate reduction, and methanogenesis (Megonigal et al. 2004). Humic acids as a TEA appear to have thermodynamic yields between denitrification and sulfate reduction (Cervantes et al. 2000). There is strong evidence to support this hypothesis from many studies, including sequential use of TEAs from higher to lower thermodynamic yields over time in batch reactors with wetland sediments that are closed to the atmosphere and with depth in oxic–anoxic interfaces in sediments (Megonigal et al. 2004). Numerous studies have shown that higher concentrations of more thermodynamically favorable TEAs depress methanogenesis (e.g., Gauci et al. 2004). However, denitrification is limited in many wetlands by low nitrate concentrations, caused by acidic and/or anoxic conditions inhibiting nitrification (Bowden 1987; Bridgham et al. 2001). Low soil mineral content limits Mn(IV) and Fe(III) reduction in peatlands. Porewater sulfate concentrations are quite low in most freshwater wetlands, although the few studies that have quantified sulfate reduction rates in these wetlands suggest that it can be surprisingly important in anaerobic carbon cycling (Giblin and Wieder 1992; Castro et al. 2002; Vile et al. 2003). However, modeling and empirical evidence suggests that humic acids may be important TEAs in peatlands and lead to the high CO_2:CH_4 ratios and high temperature responses of CH_4 observed in these systems (Segers and Kengen 1998; van Hulzen et al. 1999).

Regarding the third hypothesis, methanogenesis requires the tightly linked transfer of metabolic intermediates between different functional groups of prokaryotes, as described above. If the passage of electron donors or electron acceptors is disrupted, either by competition with groups such as homoacetogens, or by environmental factors, methanogenesis will be affected.

While extensive research has focused on understanding fermentation dynamics in rice fields (e.g., Krylova and Conrad 1998; Chidthaisong et al. 1999) and lake sediments (e.g., Phelps and Zeikus 1984; Krumböck and Conrad 1991), little research has been conducted in natural wetlands. Several studies have found strong substrate-induced limitations of methanogenesis in peatlands (Bridgham and Richardson 1992; Valentine et al. 1994; Updegraff et al. 1995; Bridgham et al. 1998). Despite the fact that peat is often >90 % organic carbon, most of that carbon is highly recalcitrant, and microorganisms can be considered to exist in a sea of inedible organic carbon. It has been suggested that much of the organic carbon that ultimately fuels methanogenesis is derived from root exudates (Happell et al. 1993; Megonigal et al. 1999), and plant production and CH_4 fluxes are often positively correlated in wetlands (Whiting and Chanton 1993; Updegraff et al. 2001; Vann and Megonigal 2003).

Denitrification and Fe(III) and Mn(IV) reduction are insignificant in many bogs and acidic fens, such that that fermentation is typically considered the dominant anaerobic pathway in sites with low atmospheric sulfur deposition. As will be discussed in detail below, sulfate reduction can be an important pathway for carbon mineralization, however, and methanogenesis a minor pathway (Table 11.1). Under optimal conditions, the fermentative pathway should result in a CO_2:CH_4 ratio of approximately 1:1 (Megonigal et al. 2004), such that the results presented in Table 11.1 strongly suggest that the ratio of anaerobic CO_2:CH_4 in different peatlands reflects substrate availability for methanogens that is mediated through other microbial groups partici-

Table 11.1 The percentage of total carbon mineralized anaerobically that can be accounted for by sulfate reduction (assuming two moles of carbon mineralized for every mole of sulfate consumed), methanogenesis, and the unexplained fraction in five peatlands. To the extent that denitrification, Fe(III) reduction, and Mn(IV) reduction are insignificant, the unexplained fraction of carbon mineralization is due to fermentation and humic acids acting as electron acceptor

Locale	Sulfate reduction	Methane production	Unexplained/ fermentation	Reference
Bleak Lake Bog, Alberta	1.2	0.05	98.8	Vile et al. (2003)
Cervené Blato, Czech Republic	35.9	0.01	64.1	Vile et al. (2003)
Oceán Bog, Czech Republic	26.7	0.004	73.3	Vile et al. (2003)
Big Run Bog, W.Va.	64.4	11.7	23.9	Wieder et al. (1990)
Buckle's Bog, Md.	38.1	2.8	59.1	Wieder et al. (1990)

pating in anaerobic carbon cycling. Thus, we suggest that inefficiencies and competition within this complex anaerobic microbial consortium provide an important limitation on CH_4 production in peatlands. Evidence indicating inefficiency of anaerobic C degradation in peatlands is the substantial fraction (~10%) of annual net primary production that is stored as soil C (Gorham 1995). Additionally, high concentrations of organic acids within DOC in peatlands result from incomplete anaerobic carbon decomposition, which lead to the characteristically low pHs observed in these soils (Hemond 1980; Gorham et al. 1985). The DOC is primarily composed of fulvic acids and humic acids that also appear to be important TEAs in peatlands (Segers and Kengen 1998; van Hulzen et al. 1999). Therefore, we hypothesize direct linkages between substrate quality, anaerobic carbon flow, pH, TEAs, and CO_2:CH_4 ratios in anaerobic respiration in peatlands.

All three possible pathways proposed above are likely to be responsible for specific anaerobic CO_2:CH_4 ratios in wetlands; and the relative importance of each pathway is likely to be different in different types of wetlands. Recent evidence suggests that humic acids may be an important TEA in many wetlands and their importance should be investigated more thoroughly. When alternative TEAs are not present in significant concentrations, fermentation reactions provide a strong 'upstream' control over methanogenesis in peatlands, and possibly in other wetlands, by controlling the availability of acetate and H_2. A better understanding should be developed regarding controls over fermentation rates in wetlands, given their importance in the atmospheric fluxes of the greenhouse gases CH_4 and CO_2. Given the importance of wetlands as a soil carbon store (largely the result of slow anaerobic carbon degradation) and as a CH_4 source, it is essential that a better understanding of controls over anaerobic carbon catabolism be developed.

11.3.3 Sulfate and Iron Reduction as Important Routes for Mineralization in Fens

Input of sulfate and iron can be significant in some peatlands, resulting in significant mineralization of carbon by dissimilatory sulfate- and iron-reducing prokaryotes. Anaerobic respiration via sulfate and iron reduction can be important competitors for carbon in these peatlands, accounting for much of the relatively high CO_2:CH_4 ratios observed in some of these environments. Minerotrophic fens in northern Bavaria emit only 0.02–15.0 mmol CH_4 m^{-2} day^{-1} (Horn et al. 2003). Significant sulfur enters these fens as a result of atmospheric deposition from combustion of soft coal in Eastern Europe (Berge et al. 1999). Sulfate runs off from upland soils to fens at lower elevations (Alewell and Giesemann 1996; Fig. 11.3), and $\delta^{34}S$ values and ^{35}S-labeling patterns indicate

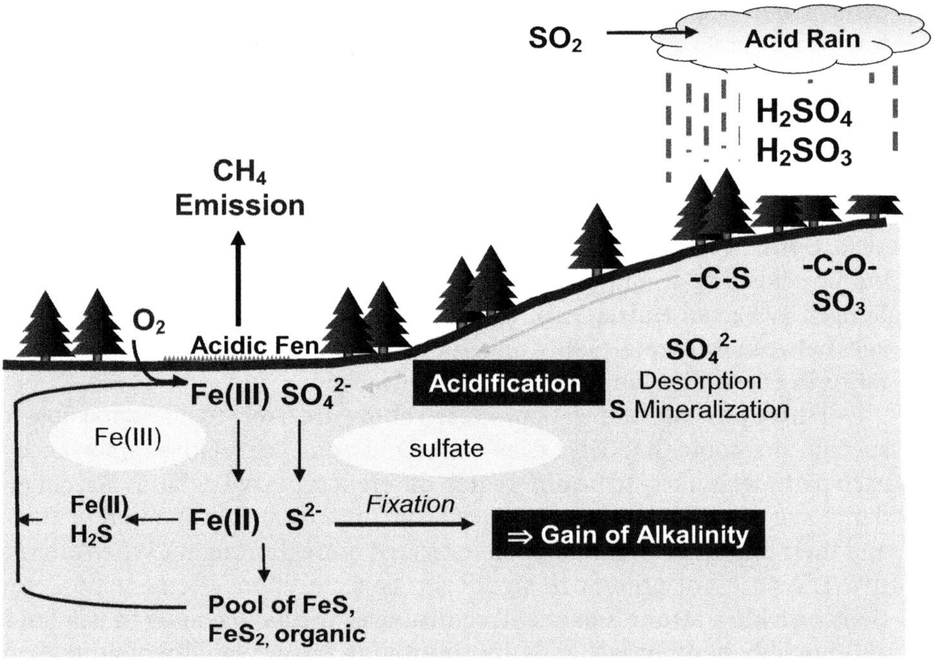

Fig. 11.3 Hypothetical model illustrating the turnover of sulfur and iron in a minerotrophic fen in a catchment that has been subjected to atmospheric deposition of sulfur. Republished with permission from Küsel and Alewell (2004)

that dissimilatory sulfate reduction is an ongoing process (Alewell and Gehre 1999; Alewell and Novak 2001). Maximum concentrations of sulfate in the soil solution of these fens never exceed 400 μM, and the small concentrations (20 μM) of sulfate in the soil solution obtained during late summer and autumn (Küsel and Alewell 2004) might be insufficient for dissimilatory sulfate reduction (Lovley and Klug 1983). Interestingly, the numbers of methanogens [10^6–10^7 cells g^{-1} (fresh weight soil)] cultured at pH 5 are only 10-fold higher than those of sulfate-reducers cultured at pH 5. Addition of H_2 to microcosms containing soil from this fen resulted in increased methanogenesis, and supplemental acetate has no effect on methanogenesis. It is likely that hydrogenotrophic members of the methanogen families Methanobacteriaceae, Methanomicrobiales, and Methanosarcinaceae are responsible for most of the methane formed (Horn et al. 2003). Rates for the consumption of supplemental sulfate in soil microcosms are twice as high as rates for the formation of methane in microcosms lacking sulfate (Loy et al. 2004) and are comparable to those observed in eutrophic Everglades soils (Castro et al. 2002). Taken together, these observations indicate that: (1) sulfate, when available, is subject to rapid utilization by dissimilatory sulfate-reducing prokaryotes; and (2) low concentrations of sulfate are a seasonal limiting factor for the sulfate-reducing activity of sulfate-reducing bacteria in these fens.

Knowledge of the distribution of individual phylogenetic groups of sulfate-reducing prokaryotes (SRP) and methanogens in these fens would provide important information regarding the mechanisms of carbon mineralization and the stability of these functions with time and changing environmental conditions. 16S rRNA gene-based oligonucleotide microarray analyses reveal stable diversity patterns of known SRP in the upper 30 cm of the fen, including members of the family Syntrophobacteraceae and of the genus *Desulfomonile* (Loy et al. 2004). Comparative analyses of dissimilatory (bi)sulfite reductase (*dsrAB*) gene sequences obtained from these soils include sequences from *Desulfobacca*-like sulfate reducers. Due to the low concentration of sulfate in these fens and the likelihood that H_2 is an important substrate for moderately acid-tolerant methanogens in this ecosystem, it is tempting to speculate that, in the absence of sulfate, some sulfate-reducing bacteria in these fens function as syntrophs in consortia with methanogens.

In addition to sulfate, these Bavarian fens receive Fe(II) from anoxic groundwater (Fig. 11.3). Fe(II) is rapidly oxidized near the soil surface by both chemical and biological routes (Sobolev and Roden 2002), such that the upper 10 cm of these soils are highly enriched with both dithionate-extractable pedogenic iron oxides [up to 19 g kg^{-1} soil; representing both crystaline and poorly crystalized Fe(II)] and oxalate-extractable poorly crystalized, or amorphous, iron oxides (up to 14 g kg^{-1} soil; Küsel and Alewell 2004). Amorphous iron oxides are the favored forms of Fe(II) for microbial reduction due to the greater availability of reducible Fe(II). Average concentrations of dissolved

Fe(II) in these fens range over 100–200 μM, although maximum concentrations can reach 7 mM (Küsel and Alewell 2004). High concentrations of Fe(II) in the soil water are positively correlated with high concentrations of acetate, and anoxic incubation experiments with samples from the surface layer indicate high Fe(III)-reducing, but negligible sulfate-reducing or methanogenic activities. Rates of Fe(III) reduction approximate 2.6 μmol Fe(II) g^{-1} (fresh weight soil) day^{-1}. The predicted hierarchy of terminal electron-accepting processes is observed in these microcosms, when Fe(III)-reducing activities decrease in microcosms, sulfate is consumed, and finally CH_4 formation is observed. The sequential temporal formation of Fe(II) followed by consumption of sulfate and the late increase of CH_4 production in these studies indicate that dissimilatory Fe(III) reduction can supress both sulfate reduction and methanogenesis in these fens, thereby oxidizing a considerable amount of organic carbon (Roden and Wetzel 1996).

A greater understanding of the genomic and ecophysiological characteristics of anaerobic microbial community members in peatlands will provide significant insight into the precise pathways through which carbon is mineralized in these environments.

11.3.4 Linking Community Composition with Nutrient Status in Wetlands

As presented in Fig. 11.1, methanogens and SRP form the top level of complex chains of microorganisms in anaerobic carbon cycling, utilizing H_2, CO_2, and simple fermentation products formed during the decomposition of complex plant matter. The activities of methanogens and SRP depend on the activities of groups responsible for production of their electron donors and carbon sources, such as fermenters, and the activities of competitors, such as homoacetogens. As a result of dependence on the activities of other groups, SRP and methanogens may be regarded as integrators of environmental effects on those groups. The activities and composition of methanogens and SRP are therefore likely to reflect environmental conditions, including nutrient status, that affect the activities of all groups of microorganisms that participate in carbon mineralization in anoxoic soils. Linking the composition of assemblages of methanogens and SRP with biogeochemical parameters and nutrient status (such as degree of eutrophication) is likely to provide insight into environmental controls on microbial functioning and on the pathways through which carbon is mineralized in a given environment.

SRP are a very diverse group of microorganisms and include representatives among the Proteobacteria, gram-positive bacteria, and Archaea (Castro et al. 2000). Many of these species can be classified in two very general metabolic groups: those capable of completely oxidizing acetate to CO_2, and those incapable of oxidizing acetate to CO_2. Similarly, methanogens may be broadly grouped into those which can use H_2 and CO_2 as sole source of electrons and carbon, respectively (hydrogenotrophs), and those that can use acetate as sole source of electrons and carbon (acetotrophs). Specific environmental conditions may be likely to favor one metabolic group of either SRP or methanogens over the other.

Nutrient inputs into the northern Florida Everglades provide a unique opportunity to study the impacts of nutrient pollution on biogeochemical cycling and to establish linkages between nutrient status, the composition and activities of microbial groups, and biogeochemical processes. The phosphate gradient in the northern Everglades is well defined, and an array of environmental parameters is routinely monitored along the gradient by the South Florida Water Management District (DeBusk et al 1994).

Among the most obvious indicators of eutrophication in this region is the displacement of sparse stands of sawgrass (*Cladium* sp.), an indigenous grass that is well adapted to the extremely low phosphate concentrations that naturally occur in the Everglades, with dense stands of cattail (*Typha* sp.). As one moves from the cattail-dominated eutrophic regions, through transition regions where oligotrophic zones are beginning to be impacted by elevated phosphorus concentrations, to the oligotrophic regions where the moving phosphorus front has not yet reached, a number of ecosystem level changes are apparent. As a result of the increased biomass and phosphate in the eutrophic regions, carbon cycling occurs at a more rapid rate than observed in the oligotrophic regions (DeBusk and Reddy 1998). As might be expected, microbial processes such as respiration, methanogenesis, and sulfate reduction rates are higher in the eutrophic regions than in the oligotrophic regions.

With increased amounts of carbon passing through the eutrophic regions relative to the oligotrophic regions, it might be expected that the species that comprise resident assemblages of SRP and methanogens would differ with respect to position along the nutrient gradient. If so, an understanding of potential differences in the composition of these groups might shed light on the impacts of eutrophication on the pathways through which carbon is mineralized. Genes characteristic of SRP (dissimilatory sulfite reductase, *dsrAB*) and methanogens (methyl coenzyme M reductase, *mcrA*) were cloned from DNA extracted directly from soil samples taken from eutrophic (site F1) and oligotrophic (site U3) regions of WCA-2A during spring and summer sampling times (Castro et al. 2002, 2004).

Sequence analysis of *dsrAB* clone libraries revealed a clear dominance by incomplete acetate oxidizing species of *Desulfotomaculum* in the oligotrophic region, and *Desulfotomaculum* species in the eutrophic region were dominated by complete acetate oxidizing species, a difference which may provide clues to the functions of these bacteria in their respective regions (Castro et al. 2002).

The robustness of these observations was tested by the application of terminal restriction fragment length polymorphism (T-RFLP; Liu et al. 1997) methods for the analysis of *dsrAB* and *mcrA* distribution in these regions over the course of one year (Castro et al. 2005). T-RFLP profiles reflect the distribution of the numerically dominant genotypes in each sample, based on analysis of the terminal end of the target gene. In this analysis, genotypes are represented as a peak (terminal restriction fragment, T-RF) on an electropherogram generated by an automated nucleic acid sequencer (equivalent to a band on an electrophoresis gel). T-RFLP analyses were conducted on DNA extracted from soil samples taken monthly from the oligotrophic (U3), transition (F4), and eutrophic (F1) regions, and encompassed one hydroperiod. Individual T-RFLP profiles (representing assemblages of the respective genotypes from each sample) were analyzed by principal components analysis (Figs. 11.4, 11.5). Assemblages of both *dsrAB* (Fig. 11.4) and *mcrA* (Fig. 11.5) genotypes were relatively stable throughout the hydroperiod, and the compositions of these assemblages were clearly characteristic of their position along the nutrient gradient. Assemblages of SRP from the three sites separated more clearly by principal components analysis than did methanogens, which may be a function of the greater metabolic diversity of SRP than of methanogens.

The clear dominance of complete oxidizing SRP in eutrophic and transition regions may be an indication of a broader array of available electron donors in these regions than in the oligotrophic region. Complete oxidizing SRP are typically capable of utilizing a broader range of electron donors than are incomplete oxidizers (Widdel 1992). It is speculated that incomplete oxidizers are more efficient at taking up low concentrations of a few substrates, such as lactate, than are complete oxidizers. This differentiation might be expected between eutrophic regions with higher microbial activities and higher carbon fluxes than are present in oligotrophic regions. Higher activities, and greater amounts of carbon put into the soil, may select for generalists capable of taking advantage of a greater diversity of carbon substrates. Oligotrophic regions, characterized by little available carbon, may select for SRP that are highly efficient at using a limited range of available substrates.

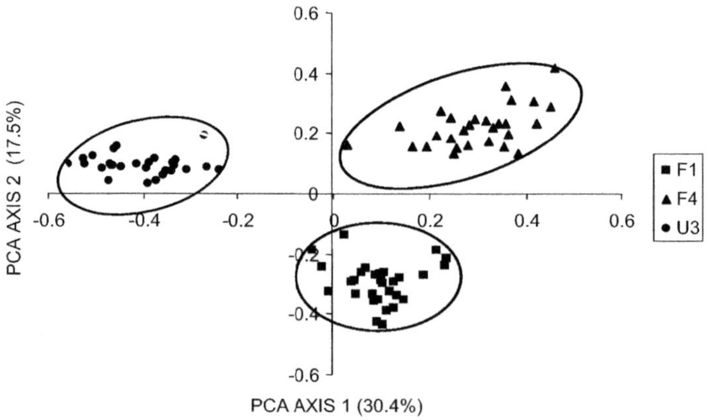

Fig. 11.4 PCA ordering generated from T-RFLP profiles for the *dsrAB* gene of eutrophic (*F1*), transition (*F4*), and oligotrophic (*U3*) soils. Figure reprinted from Castro et al. (2005)

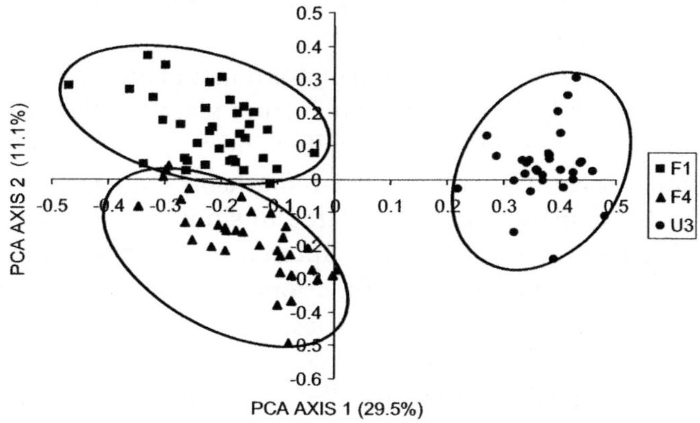

Fig. 11.5 PCA ordering generated from T-RFLP profiles for *mcrA* of eutrophic (*F1*), transition (*F4*), and oligotrophic (*U3*) soils. Figure reprinted from Castro et al. (2005)

No clear separation of methanogens on the basis of broad functional divisions, such as hydrogenotrophs versus acetotrophs, was observed; all regions were dominated by hydrogenotrophs. A clear division was observed, however, between *mcrA* genotypes in the oligotrophic and the transition and eutrophic regions (Castro et al. 2005). Methanogens are far less metabolically versatile than are SRP, and hydrogenotrophs that dominate the Everglades soil may be responding to differences between the sites in available H_2. One might expect that higher H_2 concentrations would be available in the eutrophic and transition regions than in the oligotrophic region and that different hydrogenotrophic species would be selected by the different H_2 concentrations. Speculations regarding factors that drive species diversity and selection in various environments might be supported by developing and defining clearer links between microbial community composition and their environment.

11.3.5 Plant-Associated Microbial Communities Across Lanscapes

Observations that the microbial world is hyperdiverse (Curtis et al. 2002; Roberts et al. 2004) and that microbial communities are characterized by functional redundancy (Edelman and Galley 2001; Roberts et al. 2004) lead to questions related to the distribution of microbial taxa across sites that are similar in their functional capabilities. For example, salt marshes contain a limited variety of plants that have a broad range of climatic conditions within which they flourish. Common marsh grasses include species of the genus *Spartina* and *Juncus* with several others intermingled. In the eastern United States, *S. alterniflora* is commonly found from the central Florida coast to Maine, with inclusions of *S. patens* in the cooler regions and *S. bakeri* in the warmer south. In the warmest part of the south, *S. bakeri* is commingled with *Paspalum* and *Distichlis*. Rushes like *Juncus* are commonly found interspersed along the entire coastline. In essence, a single plant species may be found in widely differing climatic regions, and several different grasses may be closely juxtaposed at a single site.

Despite this distribution of dominant plants, microbial communities function in a highly similar fashion from place to place. Decay of organic matter occurs, cycling of nitrogen, sulfur, and other mineral elements goes on, and in most cases, the processes are remarkably similar. Is the functional similarity in the disparate habitats related to similarity in the microbial communities that are present, or are there differences in composition among the communities that is masked in the functional signature, presumably due to the degenerate nature of the communities?

Microbial processes and community diversity (evenness and richness) and composition are controlled by many factors including substrate quality (e.g., Godshalk and Wetzel 1978; Stuart et al. 1981; Rice 1982; Tenore 1983; Valiela et al. 1984), substrate particle size (Swift et al. 1979), edaphic conditions (e.g., Haines and Hanson 1979; Valiela et al. 1984), and climate (Meentenmeyer 1978; Bunnell and Tait 1977; Bunnell et al. 1977; Jansson and Berg 1985). Regardless of the ultimate effects of these factors on decomposition, little information is available on the structure and similarity of the microbial communities that carry out the process in different environmental settings.

An examination of community similarity in sediment samples collected from the high marsh environment at ten sites on the east coast of the United States suggested that the microbial communities varied from site to site and that there was a gradient in community structure that roughly corresponded to the location of the site sampled along a latitudinal gradient (Blum et al. 2004). The gradient included sites from the Florida Everglades (estuarine portion), mid-Florida (Upper Indian River Lagoon), Georgia, South Carolina, North Carolina, Virginia, Delaware, Connecticut, Massachusetts, and Maine. Average temperatures varied systematically (–0.9 °C to 24 °C mean daily high temperature in January) along the north to south gradient. Other variables, e.g., rainfall, tide range, salinity, organic content, and bulk density, also varied, but the variance was non-systematic.

The community structure measure was based on similarities in T-RFLP profiles of 16S rRNA genes amplified from DNA extracted from the samples. This gene is thought to be highly conserved taxonomically and is commonly used to examine relationships among both isolated and uncultured strains. Each T-RF obtained was treated as a unit character, and the results were analyzed using principal components analysis (PCA; Franklin et al. 1999; Wikstrom et al. 1999; Franklin and Mills 2003). The value of the second principal component (PC) score increased with increasing latitude (Fig. 11.6). It is interesting to note that no difference was observed among the ten sites for total microbial abundance. The authors implied that these data would be consistent with functionally similar communities that vary (along the gradient) in community composition.

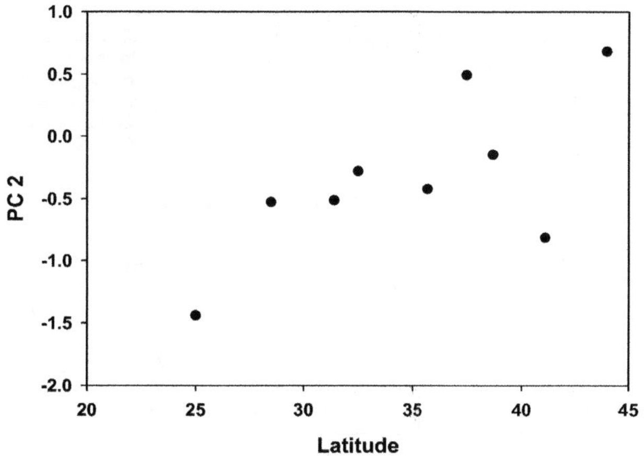

Fig. 11.6 Distribution of Principal Component PC2 values for bacterial sediment community composition with latitude to display latitudinal gradient in community composition of the sediment. The correlation between the variables was r=0.583. Figure reprinted from Blum et al. (2004)

Standing dead plants at each of the sites were also examined for compositional similarities among their resident microbial communities (Blum et al. 2004). Overall, sediment- and plant-associated microbial communities clearly differed from one another; however, the plant associated microbes did not show the same kind of large-scale biogeographic distribution as did the sediment communities. Six plants were examined, one plant from each site that represented the dominant high-marsh vegetation. The plant species were *Spartina alterniflora* (Maine and South Carolina), *S. patens* (Massachusetts and Delaware), *S. bakeri* (Indian River Lagoon, Florida), *Juncus roemerianus* (Virginia, North Carolina, and Georgia), *Phragmites australis* (Connecticut), and *Rhizophora mangle* (red mangrove, Florida Everglades). The abundance of both bacteria and fungi showed a latitudinal distribution (Fig. 11.7), but the community structures of the two domains were not correlated with latitude. Bacterial counts were highest in mangrove leaves from south Florida and lowest in the *S. alterniflora* leaves from Maine. Fungal counts were lowest in the more southern samples and tended to increase in a northerly direction. The authors speculated that the results might indicate an antagonistic or competitive relation between the bacteria and fungi on the standing dead plant material that was somehow influenced by the temperature. They also noted that two of the plant types (that were also from the two most southern sites) did not contain fungi above the minimum detection limit for the assays used to quantify them.

While there were no differences in the microbial communities associated with standing dead plants, differences among plant types were clear for both bacteria and fungi (Fig. 11.8). For both fungi and bacteria, there were major differences in the communities associated with *J. roemerianus* as compared with all the other plants. Generally, the samples from all the *Spartina* species clustered well together, but communities from *Rhizophora* and *Phragmites* also clustered closely with those from *Spartina*.

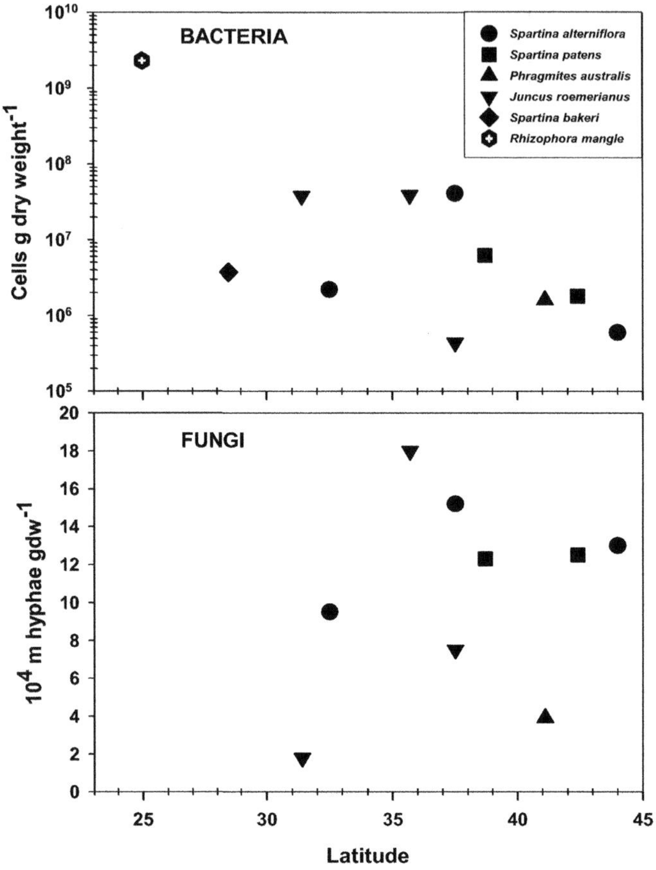

Fig. 11.7 Abundance of bacteria (*top*) and fungi (*bottom*) in the plant material collected from the ten sites. Fungi were below the limit of detection (170 m hyphae g^{-1} dry weight plant) for *S. bakeri* or *R. mangle*. Figure reprinted from Blum et al. (2004)

The observed differences in sediment community composition across these large scales suggests that different microbial inocula are available to colonize standing-dead plant material after it falls to the sediment surface and continues the decay process. If the developing microbial decomposition community is a combination of the resident standing-dead and sediment communities, the mature decay communities may be substantially different between sites, even though the fundamental decay process is similar between sites. The balance of natural selection and recruitment in these mature decomposition communities may be strongly influenced by the biogeography of the environment. It will be interesting to observe if the biogeographical component of sediment microbial composition is maintained within the decomposition communities or if its signature is lost through time. Decomposition experiments currently in progress at the ten sites will enable resolution between these potential ecological outcomes.

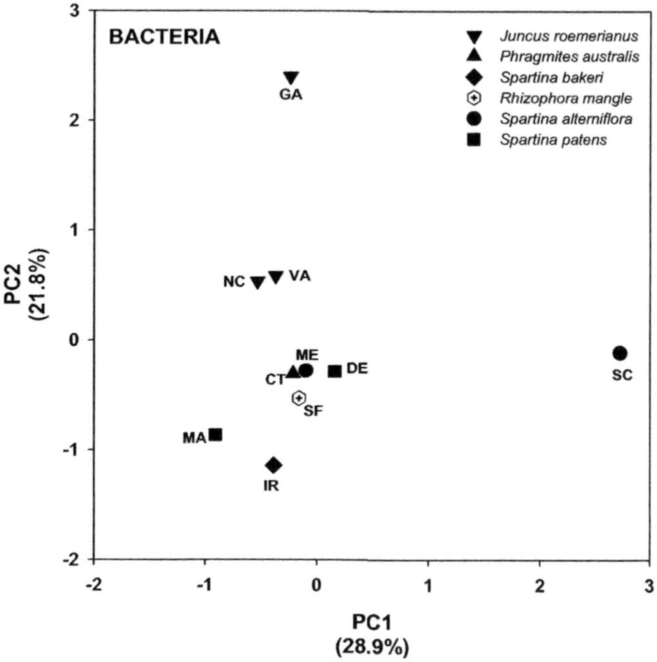

Fig. 11.8 Principal Components plots of bacterial and fungal communities associated with standing dead materials from different plants. Note that no data were recovered from amplifications of extracts of *R. mangle* using fungal primers. Figure reprinted from Blum et al. (2004)

11.4 Linking Microbial Community Structure and Function with Environmental Parameters

As described above in Section 11.3, soil microbial communities control nutrient cycling and organic matter decomposition in wetlands, and conversely, environmental factors form the fundamental forces that control the structures and activities of microbial communities. Analysis of quantitative linkages between microbial community structure, function, and environmental parameters will provide a greater understanding of the factors that control nutrient cycling in wetlands and of the effects of anthropogenic impacts on these cycles. Analysis of these linkages requires an integrative approach that includes the different components, resulting in a relatively holistic analysis of combinations of biogeochemical and microbial measures (Fig. 11.9).

A variety of tools are available to characterize microbial community composition (Ogram and Sharma 2001), although the means to integrate these measures to obtain a meaningful assessment are often not adequate (Kennedy and Papendick 1995). Microbial community composition data that lists the presence or absence of particular operational taxonomic units (i.e., microbial measures) must be integrated with nutrient cycling data, which tends to be process-oriented. Furthermore, underlying relationships are often complex and can be non-linear. Multivariate analysis of microbial measures results in

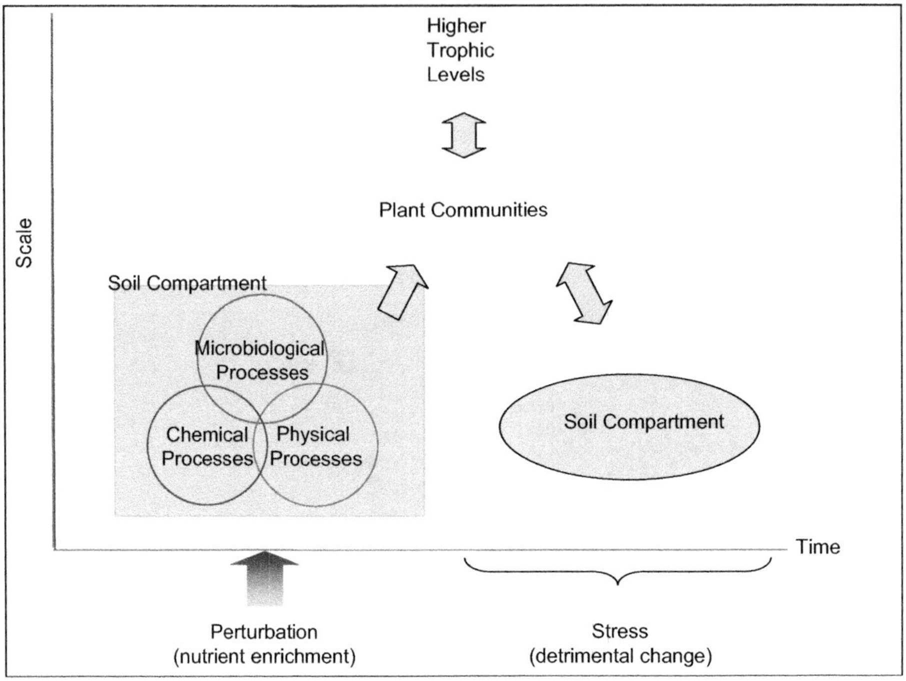

Fig. 11.9 Schematic illustration of a system response to a perturbation as contrasted with a stress. A perturbation such as nutrient enrichment would result in measurable changes in soil biogeochemical properties. These can consequently lead to gross-scale ecosystem structural changes which, when seen as detrimental, are classified as stress. These larger-scale changes will consequently affect soil biogeochemical properties

a form of integration that is significant both in terms of the effects of the microbial communities on nutrient cycling and in terms of environmental parameters on microbial community composition. Attempts at integrating process oriented measures and microbial compositional measures are numerous, encompassing data reduction approaches such principal component analysis (Castro et al. 2005) to association approaches as neural networks (Palumbo et al. 2004). Results of these analyses are interpretative and meaningful in that they synthesize overall processes and reduce these to a limited number of variables.

An understanding of scaling factors is also important when investigating biological interactions among or between microbial communities, via interpretation of distribution patterns in the landscape (Fig. 11.10; Blanchard 1990). An inappropriate sampling scale may underestimate spatial heterogeneity, thus leading to erroneous conclusions. Moreover, different ecosystem processes occur at different scales, each differentially affecting microbial communities. Nutrient movement through soil columns has been identified

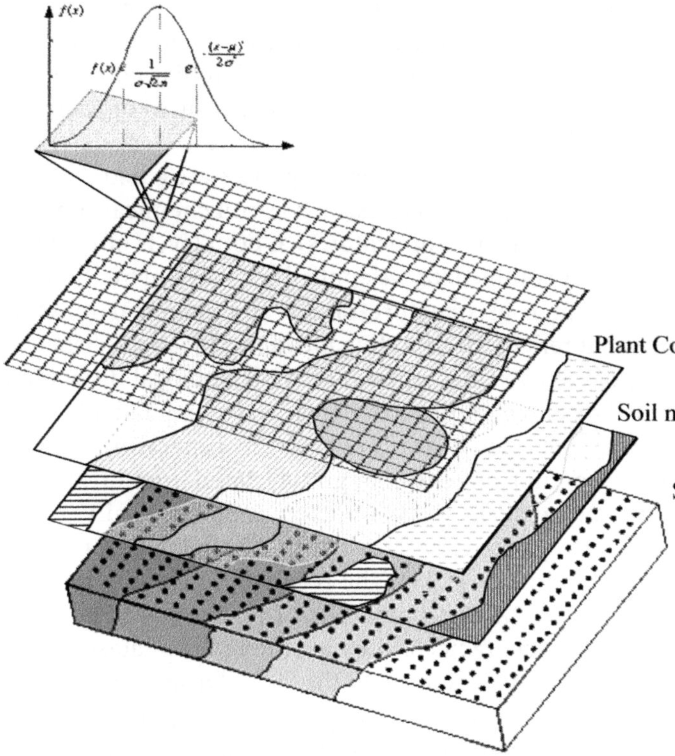

Fig. 11.10 Stylized representation of the estimated information layers as a function of ecosystem components

as a possible structuring agent for short-scale (<1 mm) spatial patterns observed in the subsoil (Nunan et al. 2003). At a larger scale, spatial patterns of plant communities, soil characteristics, and landscape position will affect nutrient dynamics and microbial communities.

Understanding interactions between nutrient cycling and microbial community structure and composition requires the application of an array of quantitative analytical tools that are capable of distinguishing spatial (co)variation structures over different scales and are capable of unraveling complex associations that are often non-linear. These efforts will result in valuable insights in the drivers of microbial diversity and how microbial communities and community structure influence nutrient cycling.

11.4.1 Case Study: a Northern Everglades Marsh System

A dataset was analyzed at three sites in the Northern Everglades (see Section 11.3.4), representing eutrophic (F1), transition (F4), and oligotrophic (U3) areas. This study encompassed a comprehensive multivariate analysis of soil chemical and microbiological characteristics at these sites with the objective of determining the relationships between resident microbial communities and their eco-physiologies, and the response of these measures to ecological perturbations (Corstanje 2003). Specifically, our objective was to determine whether integrator measures such as microbial responses are a better indicator of nutrient impact than primary measures such as soil chemical composition, and to identify the most sensitive indicator(s) at both levels of response. To preclude pre-classification of the data, we applied a clustering method with the expectation that the internal structure of the soil chemical measures would result in groups that reflect the sampling location. Once established, we applied a combination of stepwise discrimination and canonical discrimination (stepwise canonical discrimination) to determine which particular combinations of chemical characteristics are influential in generating (abiotic) and predicting (biotic) the multidimensional groups. Validation of the stepwise canonical discriminant analysis was conducted using jack-knifing.

Application of the biotic parameters resulted in a smaller group of discriminatory variables (table in Fig. 11.11). The resultant linear contrasts can then be used to describe WCA-2A in general terms (quadrants in Fig. 11.11). The oligotrophic area (U3) is characterized by alkaline phosphatase activity (APA), which represents the biotic response to the lack of available phosphorus (P). The eutrophic area (F1) generally has higher levels of potentially mineralizable P (PMP), contrasted with low levels of potentially mineralizable N (PMN). The transition area, F4, is characterized by a greater amount of micro-

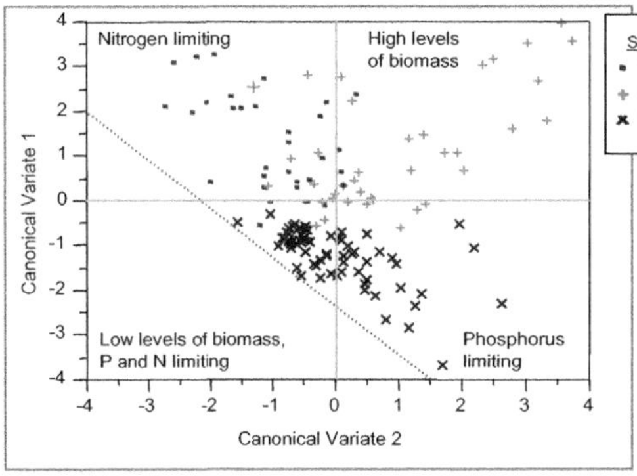

Selected Variable	Standardized Canonical Scores	
	Canonical Variate 1	*Canonical Variate 2*
PMP	1.22	-0.43
MBP	0.016	1.38
APA	-0.16	0.22
PMN	-0.31	-0.45

Fig. 11.11 The ability of the canonical variates to distinguish sites using soil biotic characteristics, level II indices. The *right panel* depicts the results of the canonical discrimination; each quadrant is given a certain designation based on the contrast represented by canonical variates in the table. *MBP* Microbial biomass phosphorus (mg kg^{-1}), *PMP* potential mineralizable phosphorus (mg kg^{-1} day^{-1}), *PMN* potential mineralizable nitrogen (mg kg^{-1} day^{-1}), *APA* alkaline phosphatase activity (μg g^{-1} h^{-1})

bial biomass P (MBP).Our study suggested that microbial variables are responsive.

The results of cross-validation of discriminant analysis of the microbial measures in the Everglades established the integrative function of these measures. The same information was conveyed in fewer variables, and jackboot cross-validation indicated that these variables were significantly more stable than the soil chemical discriminant variables. Multivariate analysis of the combination of biotic and abiotic variables resulted in a relatively successful site characterization of this ecosystem. The most effective linear combinations of variables were constructed, which de facto can be employed as indicators. The application of these linear combinations of variables as indicators of phosphorus enrichment could be further validated by a sampling approach that takes into account the spatial variation of these indicators' relationships.

References

Alewell C, Gehre M (1999) Patterns of stable S isotopes in a forested catchment as indicators for biological S turnover. Biogeochemistry 47:319–333

Alewell C, Giesemann A (1996) Sulfate reduction in a forested catchment as indicated by $\delta^{34}S$ values of sulfate in soil solutions and runoff. Isotopes Environ Health Stud 32:203–210

Alewell C, Novak M (2001) Spotting zones of dissimilatory sulfate reduction in a forested catchment: the ^{34}S-^{35}S approach. Environ Pollut 112:369–377

Berge E, Bartnicki J, Olendrzynski K, Tsyro SG (1999) Long-term trends in emissions and transboundary transport of acidifying air pollution in Europe. J Environ Manage 57:31–50

Bicudo DC, Ward AK, Wetzel RG (1998) Fluxes of dissolved organic carbon within attached aquatic microbiota. Verhand Int Ver Limnol 26:1608–1613

Blanchard GF (1990) Overlapping microscale dispersion patterns of meiofauna and microphytobenthos. Mar Ecol Prog Ser 68:101–111

Blum LK, Roberts MS, Garland JL, Mills AL (2004) Microbial communities among the dominant high marsh plants and associated sediments of the United States east coast. Microb Ecol 48:375–388

Bowden WB (1987) The biogeochemistry of nitrogen in freshwater wetlands. Biogeochemistry 4:313–348

Bridgham SD, Richardson CJ (1992) Mechanisms controlling soil respiration (CO_2 and CH_4) in southern peatlands. Soil Biol Biochem 24:1089–1099

Bridgham SD, Updegraff K, Pastor J (1998) Carbon, nitrogen, and phosphorus mineralization in northern wetlands. Ecology 79:1545–1561

Bridgham SD, Updegraff K, Pastor J (2001) A comparison of nutrient availability indices along an ombrotrophic–minerotrophic gradient in Minnesota wetlands. Soil Sci Soc Am J 65:259–269

Bunnell FL, Tait DEN (1977) Microbial respiration and substrate loss. II. A model of the influences of chemical composition. Soil Biol Biochem 9:41–47

Bunnell FL, Tait DEN, Flanagan PW, Cleave K van (1977) Microbial respiration and substrate loss. I. A general model of the influences of abiotic factors. Soil Biol Biochem 9:33–40

Carreiro MM, Sinsabaugh RL, Repert DA, Parkhurst DF (2000) Microbial enzyme shifts explain litter decay responses to simulated nitrogen deposition. Ecology 81:2359–2365

Castro H, Williams NH, Ogram A (2000) Phylogeny of sulfate-reducing bacteria. FEMS Microbiol Ecol 31:1–9

Castro H, Reddy KR, Ogram A (2002) Composition and function of sulfate-reducing prokaryotes in eutrophic and pristine areas of the Florida Everglades. Appl Environ Microbiol 68:6129–6137

Castro H, Reddy KR, Ogram AV (2004) Phylogenetic characterization of methanogenic assemblages in eutrophic and oligotrophic areas of the Florida Everglades. Appl Environ Microbiol 70:6559–6568

Castro H, Newman S, Reddy KR, Ogram AV (2005) Distribution and stability of sulfate reducing prokaryotic and hydrogenotrophic methanogenic assemblages in nutrient-impacted regions of the Florida Everglades. Appl Environ Microbiol 71:2695–2704

Cervantes FJ, Velde S van der, Lettinga G, Field JA (2000) Competition between methanogenesis and quinone respiration for ecologically important substrates in anaerobic consortia. FEMS Microbiol Ecol 34:161–171

Chidthaisong A, Rosenstock B, Conrad R (1999) Measurement of monosaccharides and conversion of glucose to acetate in anoxic rice field soil. Appl Environ Microbiol 65:2350–2355

Coates JD, Ellis DJ, Blunt-Harris EL, Gaw CV, Roden EE, Lovley DR (1998) Recovery of humic-reducing bacteria from a diversity of environments. Appl Environ Microbiol 64:1504–1509

Conrad R (1989) Control of methane production in terrestrial ecosystems. In: Andreae MO, Schimel DS (eds) Exchange of trace gases between terrestrial ecosystems and the atmosphere. Wiley, Chichester, pp 39–58

Conrad R (1996) Soil microorganisms as controllers of atmospheric trace gases (H_2, CO, CH_4, OCS, N_2O, and NO). Microbiol Rev 60:609–640

Corstanje R (2003) Experimental and multivariate analysis of biogeochemical indicators of change in wetland ecosystems. PhD thesis, University of Florida, Gainesville, 226 pp

Cunningham HW, Wetzel RG (1989) Kinetic analysis of protein degradation by a freshwater wetland sediment community. Appl Environ Microbiol 55:1963–1967

Curtis TP, Sloan WT, Scannel JW (2002) Estimating prokaryotic diversity and its limits. Proc Natl Acad Sci USA 99:104914–10499

DeBusk WF, Reddy KR (1998) Turnover of detrital organic carbon in a nutrient-impacted Everglades marsh. Soil Sci Soc Am J 62:1460–1468

DeBusk WF, Reddy KR, Koch MS, Wang Y (1994) Spatial distribution of soil nutrients in a northern Everglades marsh. Soil Sci Soc Am J 62:1460–1468

Dolfing J (1988) Acetogenesis. In: Zehnder AJB (ed), Biology of anaerobic microorganisms. Wiley, New York, pp 417–468

Dunfield P, Knowles R, Dumont R, Moore TR (1993) Methane production and consumption in temperate and subarctic peat soils: response to temperature and pH. Soil Biol Biochem 25:321–326

Edelman GM, Galley JA (2001) Degeneracy and complexity in biological systems. Proc Natl Acad Sci USA 98:13763–13768

Franklin RB, Mills AL (2003) Multi-scale variation in spatial heterogeneity for microbial community structure in an eastern Virginia agricultural field. FEMS Microbiol Ecol 44:335–346

Franklin RB, Taylor DR, Mills AL (1999) Characterization of microbial communities using randomly amplified polymorphic DNA (RAPD). J Microbiol Methods 35:225–235

Freeman C, Ostle N, Kang H (2001) An enzymic 'latch' on a global carbon store. Nature 409:149

Gauci V, Matthews E, Dise N, Walter B, Koch D, Granberg G, Vile M (2004) Sulfur pollution suppression of the wetland methane source in the 20th and 21st centuries. Proc Natl Acad Sci USA 101:12583–12587

Giblin AE, Wieder RK (1992) Sulphur cycling in marine and freshwater wetlands. In: Howarth RW, Stewart JWB, Ivanov MV (eds) Sulphur cycling on the continents. Wiley, New York, pp 85–117

Godshalk GL, Wetzel RG (1978) Decomposition of aquatic angiosperms. II. Particulate components. Aquat Bot 5:301–327

Gorham E (1995) The biogeochemistry of northern peatlands and its possible responses to global warming. In: Woodwell GM, Mackenzie FT (eds) Biotic feedbacks in the global climatic system. Oxford University Press, New York, pp 169–187

Gorham E, Eisenreich SJ, Ford J, Santelman MV (1985) The chemistry of bog waters. In: Stumm W (ed) Chemical processes in lakes. Wiley, New York, pp 339–363

Haines EB, Hanson RB (1979) Experimental degradation of detritus made from the salt marsh plants *Spartina alterniflora* Loisel, *Salicornia virginica* L., and *Juncus roemerianus* Scheele. J Exp Mar Biol Ecol 40:27–40

Happell JD, Chanton JP, Whiting GJ, Showers WJ (1993) Stable isotopes as tracers of methane dynamics in Everglades marshes with and without active populations of methane oxidizing bacteria. J Geophys Res 98:14771–14782

Heath RT (2004) Microbial turnover of organic phosphorus in aquatic systems. In: Turner BL, Frossard E, Baldwin DS (eds) Organic phosphorus in the environment. CAB International, Wallingford, pp 185–204

Hemond HF (1980) Biogeochemistry of Thoreau's Bog, Concord, Massachusetts. Ecol Monogr 50:507–526

Hines ME, Duddleston KN (2001) Carbon flow to acetate and C_1 compounds in northern wetlands. Geophys Res Lett 28:4251–4254

Horn MA, Matthies C, Küsel K, Schramm A, Drake HL (2003) Hydrogenotrophic methanogenesis by moderately acid-tolerant methanogens of a methane-emitting acidic peat. Appl Environ Microbiol 69:74–83

Hulzen J van, Segers R, Bodegom PM van, Leffelaar PA (1999) Temperature effects on soil methane production: an explanation for observed variability. Soil Biol Biochem 31:1919–1929

Jansson PE, Berg B (1985) Temporal variation of litter decomposition in relation to simulated soil climate: long-term decopositon in a Scots pine forest. Can J Bot 63:1008–1016

Jansson M, Olsson H, Pettersson K (1988) Phosphatases; origin, characteristics and function in lakes. Hydrobiologia 170:157–175

Kennedy AC, Papendick RI (1995) Microbial characteristics of soil quality. J Soil Water Conserv 50:243–248

Krumböck M, Conrad R (1991) Metabolism of position-labeled glucose in anoxic methanogenic paddy soil and lake sediment. Microb Ecol 85:247–256

Krylova NI, Conrad R (1998) Thermodynamics of propionate degradation in methanogenic paddy soil. FEMS Microbiol Ecol 26:281–288

Küsel K, Alewell C (2004) Riparian zones in a forested catchment: hot spots for microbial reductive processes. In: Matzner E (ed) Biogeochemistry of two German forested catchments in a changing environment. Springer, Berlin Heidelberg New York, pp 377–395

Liu WT, Marsh TL, Cheng H, Forney LJ (1997) Characterization of microbial diversity by determining terminal restriction fragment length polymorphisms of genes encoding 16S rRNA. Appl Environ Microbiol 63:4516–4522

Lovley DR, Klug MJ (1983) Sulfate reducers can outcompete methanogens at freshwater sulfate concentrations. Appl Environ Microbiol 45:187–192

Lovley DR, Coates JD, Blunt-Harris EL, Phillips EJP, Woodward JC (1996) Humic substances as electron acceptors for microbial respiration. Nature 382:445–448

Loy A, Küsel K, Lehner A, Klein M, Drake HL, Wagner M (2004) Microarray and functional gene analyses of sulfate-reducing prokaryotes in low sulfate, acidic fens reveal co-occurrence of recognized genera and novel lineages. Appl Environ Microbiol 70:6998–7009

Meentenmeyer V (1978) Macroclimate and lignin control of litter decomposition rates. Ecology 59:465–472

Megonigal JP, Whalen SC, Tissue DT, Bovard BD, Albert DB, Allen AS (1999) Radiocarbon cycling from photosynthesis through methanogenesis in a wetland plant–soil–atmosphere microcosm. Soil Sci Soc Am J 63:665–671

Megonigal JP, Hines ME, Visscher PT (2004) Anaerobic metabolism: linkages to trace gases and aerobic processes. In: Schlesinger WH (ed) Biogeochemistry. Elsevier-Pergamon, Oxford, pp 317–424

Meyer-Reil L-A (1991) Ecological aspects of enzymatic activity in marine sediments. In: Chróst RJ (ed) Microbial enzymes in aquatic environments. Springer, Berlin Heidelberg New York, pp 84–95

Newman S, McCormick PV, Backus JG (2003) Phosphatase activity as an early warning indicator of wetland eutrophication: problems and prospects. J Appl Phycol 15:45–59

Nunan N, Kejian W, Young IM, Crawford JW, Ritz K (2003) Spatial distribution of bacterial communities and their relationships with the micro-architecture of soil. FEMS Microbiol Ecol 44:203–215

Ogram A, Sharma K (2001) Analysis of soil microbial community structure. In: Hurst CJ, Crawford RL, Knudsen GR, McInerney MJ, Stetzenbach LD (eds) Manual of environmental microbiology, 2nd edn. Am Soc Microbiol, Washington, D.C., pp 554–563

Palumbo AV, Schryver JC, Fields MW, Bagwell CE, Zhou J-Z, Yan T, Liu X, Brandt CC (2004) Coupling of functional gene diversity and geochemical data from environmental samples. Appl Environ Microbiol 70:6525–6534

Pant HK, Reddy KR, Dierberg FE (2002) Bioavailability of organic phosphorus in a submerged aquatic vegetation-dominated treatment wetland. J Environ Qual 31:1748–1756

Penton CR (2004) Influences of nutrient loading, vegetative habitats and simulated drought on microbial enzyme activities in the Everglades. MS thesis, University of Florida, Gainesville

Phelps TJ, Zeikus JG (1984) Influence of pH on terminal carbon metabolism in anoxic sediments from a mildly acidic lake. Appl Environ Microbiol 48:1088–1095

Reddy KR, Wetzel RG, Kadlec R (2005) Biogeochemistry of phosphorus in wetlands. In: Sims JT, Sharpley AN (eds) Phosphorus: agriculture and the Environment. Soil Sci Soc Am, Madison, pp 263–316

Rice DL (1982) The detritus nitrogen problem: new observations and perspectives from organic geochemistry. Mar Ecol Prog Ser 9:153–162

Roberts MS, Garland JL, Mills AL (2004) Microbial astronauts: assembling microbial communities for advanced life support systems. Microb Ecol 47:137–149

Roden ER, Wetzel RG (1996) Organic carbon oxidation and suppression of methane production by microbial Fe(III) oxide reduction in vegetated and unvegetated freshwater wetland sediments. Limnol Oceanogr 41:1733–1748

Rodhe H (1990) A comparison of the contribution of various greenhouse gases to the greenhouse effect. Science 248:1217–1219

Schlesinger WH (1997) Biogeochemistry: an analysis of global change. Academic, San Diego

Segers R, Kengen SWM (1998) Methane production as a function of anaerobic carbon mineralization: a process model. Soil Biol Biochem 30:1107–1117

Shannon RD, White JR (1996) The effects of spatial and temporal variations in acetate and sulfate on methane cycling in two Michigan peatlands. Limnol Oceanog 41:435–443

Sinsabaugh RL, Moorhead DL (1994) Resource allocation to extracellular enzyme production: a model for nitrogen and phosphorus control of litter decomposition. Soil Biol Biochem 26:1305–1311

Sinsabaugh RL, Antibus RK, Linkins AE, McClaugherty CA, Rayburn L, Repert D, Weiland T (1993) Wood decomposition: nitrogen and phosphorus dynamics in relation to extracellular enzyme activity. Ecology 74:1586–1593

Sinsabaugh RL, Carriero MM, Repert DA (2002) Allocation of extracellular enzymatic activity in relation to litter composition, N deposition, and mass loss. Biogeochemistry 60:1–24

Sobolev D, Roden E (2002) Evidence for rapid microscale bacteria redox cycling of iron in circumneutral environments. Antonie van Leeuwenhoek 81:587–597

Stuart V, Lucas MI, Newell RC (1981) Heterotrophic utilization of particulate matter from the kelp *Laminaria pallida*. Mar Ecol Prog Ser 4:337–348

Swift MJ, Heal OW, Anderson JM (1979) Decomposition in terrestrial ecosystems. University of California Press, Los Angeles

Tenore KR (1983) What controls the availability to animals of detritus derived from vascular plants: organic nitrogen enrichment or caloric availability? Mar Ecol Prog Ser 10:307–309

Turner B, Newman S (2005) Phosphorus cycling in wetland soils: the importance of phosphate diesters. J Environ Qual 34:1921–1929

Updegraff K, Bridgham SD, Pastor J (1995) Temperature influences on CO_2 and CH_4 fluxes in intact cores from Minnesota wetlands. Bull Ecol Soc Am 76:396

Updegraff K, Bridgham SD, Pastor J, Weishampel P, Harth C (2001) Response of CO_2 and CH_4 emissions in peatlands to warming and water-table manipulation. Ecol Appl 11:311–326

Valentine DW, Holland EA, Schimel DS (1994) Ecosystem and physiological controls over methane production in northern wetlands. J Geophys Res 99:1563–1571

Valiela I, Wilson J, Buschbaum R, Rietsma C, Bryant D, Foreman K, Teal J (1984) Importance of chemical composition of salt marsh litter on decay rates and feeding by detritivores. Bull Mar Sci 35:261–269

Vann CD, Megonigal JP (2003) Elevated CO_2 and water depth regulation of methane emissions: comparison of woody and non-woody plant species. Biogeochemistry 63:117–134

Vile MA, Bridgham SD, Wieder RK, Novák M (2003) Atmospheric sulfur deposition alters pathways of gaseous carbon production in peatlands. Global Biogeochem Cycles 17:1058–1064

Wetzel RG (1990) Land–water interfaces: Metabolic and limnological regulators. (Edgardo Baldi memorial lecture, 24th Congress Societas Internationalis Limnologiae) Verhand Int Ver Limnol 24:6–24

Wetzel RG (1993) Microcommunities and microgradients: linking nutrient regeneration and high sustained aquatic primary production. Neth J Aquatic Ecol 27:3–9

Wetzel RG (1996) Benthic algae and nutrient cycling in standing freshwater ecosystems. In: Stevenson RJ, Bothwell M, Lowe R (eds) Algal ecology: benthic algae in freshwater ecosystems. Academic, New York, pp 641–667

Wetzel RG (2005) Periphyton in the aquatic ecosystem and food webs. In: Azim E, Verdegem M, Dam A van, Beveridge M (eds) Periphyton: ecology, exploitation, and management. CABI, London, pp 51–69

Whiting GJ, Chanton JP (1993) Primary production control of methane emissions from wetlands. Nature 364:794–795

Widdel F (1992) The genus *Desulfotomaculum*. In: Balows A, Truper HG, Dworkin M, Harder W, Schleifer KH (eds) The prokaryotes. Springer, Berlin Heidelberg, New York, pp 1792–1799

Wieder RK, Yavitt JB, Lang GE (1990) Methane production and sulfate reduction in two Appalachian peatlands. Biogeochemistry 10:81–104

Wikstrom P, Andersson A-C, Forsman M (1999) Biomonitoring complex microbial communities using random amplified polymorphic DNA and principal component analysis. FEMS Microbiol Ecol 28:131

Section IV

Wetlands and Climate Change Worldwide

12 Coastal Wetland Vulnerability to Relative Sea-Level Rise: Wetland Elevation Trends and Process Controls

Donald R. Cahoon, Philippe F. Hensel, Tom Spencer, Denise J. Reed, Karen L. McKee, Neil Saintilan

12.1 Introduction

The distribution of tidal saline wetlands (e.g., salt marshes and mangroves) is increasingly impacted by global environmental change, including human alteration of the world's coasts and sea-level rise (Kennedy et al. 2002; Poff et al. 2002). Rates of salt marsh and mangrove loss appear to be accelerating (Nicholls et al. 1999). A better understanding of wetland accretionary dynamics, controls and constraints, and in particular responses to sea-level rise is required to inform the maintenance and restoration of these systems. Differences in wetland form and function result from a range of continental- and regional-scale processes (Mitsch and Gosselink 2000). Local geomorphology, climate regime, and hydrology result in differences in sediment supply, primary production and decomposition, subsidence, and autocompaction, resulting in variations in elevation among both salt marsh and mangrove sites. Combinations of these controls mean that individual wetland sites show different degrees of vulnerability to current and near-future environmental change.

Several vulnerability assessment approaches have been developed during the past two decades, including the Global Vulnerability Assessment, first developed by Hoozemans et al. (1993) and recently refined in DINAS_COAST (McFadden et al. 2003), and the Coastal Vulnerability Index (Gornitz 1991; Hammar-Klose and Thieler 2001). These approaches offer valuable planning tools for coastal resource managers at the broad (i.e., global, continental, macro-regional) scale as they allow the identification of both patterns of relative vulnerability and vulnerability 'hotspots'. However, the numerical scores that they produce cannot be easily assigned a precise physical meaning and

Ecological Studies, Vol. 190
J.T.A. Verhoeven, B. Beltman, R. Bobbink, and D.F. Whigham (Eds.)
Wetlands and Natural Resource Management

they suffer from downscaling problems, yielding predictions that are not always correct at the local scale. A more thorough understanding of coastal wetland vulnerability requires data from specific wetland sites on not only the physical and biotic parameters described above, but also: (1) the relationship among sedimentation, organic soil formation, vertical accretion, and surface elevation change and (2) the interaction of these processes, themselves subject to considerable spatial and temporal variability, with local relative sea-level rise (Day et al. 1999; Fig. 12.1).

To provide site-specific information on these key processes and interactions, the Surface Elevation Table–Marker Horizon method was developed for use in marshes (Cahoon et al. 1995) and shallow water environments (Cahoon et al. 2000a, b). The method incorporates simultaneous measurements of wetland elevation from a surface elevation table (SET; Boumans and Day 1993; Cahoon et al. 2002a, b) and vertical accretion from artificial soil marker horizons (MH; e.g., Cahoon and Turner 1989). The SET-MH methodology allows determination of not only the current accretion and elevation trajectories of the wetland relative to local sea-level rise, but also the influence of subsurface processes on elevation change. The latter has been termed 'shallow subsi-

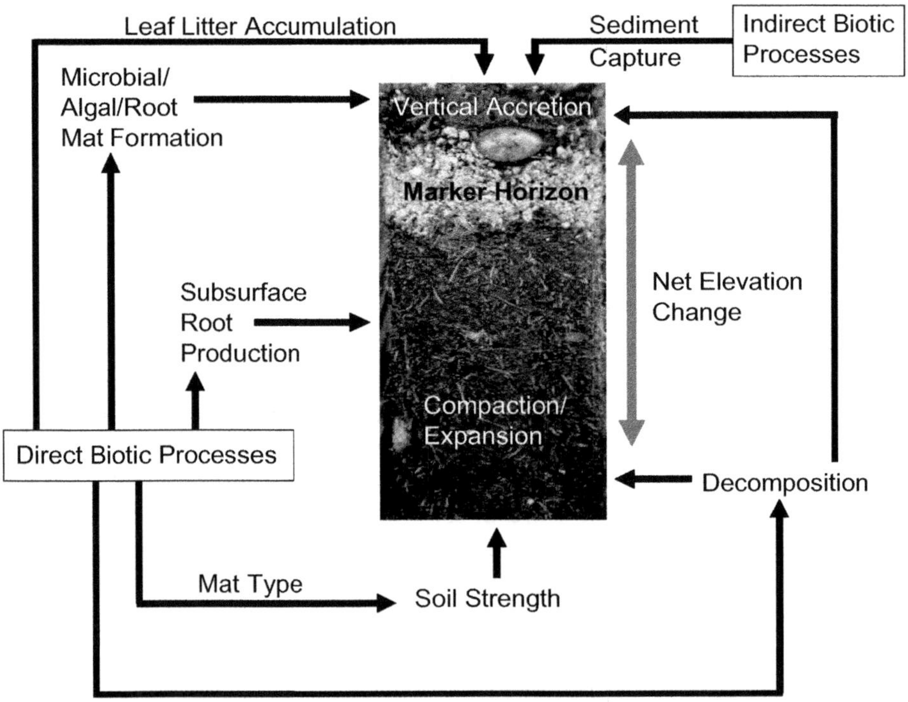

Fig. 12.1 Diagram showing direct and indirect biotic controls on vertical accretion and elevation change

dence' (Cahoon et al. 1995) to distinguish it from long-term, deep subsidence on a geological scale.

An informal network of SET-MH monitoring stations was developed during the past 15 years as 65 scientists from 18 countries applied this approach in nearly 200 coastal wetland settings. The stations are concentrated in Europe, North America, the Caribbean, and western Pacific. A few data records are now >10 years duration. The informal SET-MH network, and associated measurements (e.g., RSLR, tidal range, soil properties), provides a nascent global network of elevation and accretion monitoring stations using a common methodology that spans a range of coastal wetland types and settings. This methodology allows a quantitative comparative assessment of wetland response to current sea-level rise across sites and begins to identify the key processes that control the rate and style of wetland submergence or survival. Although this paper focuses on near-term responses to sea-level rise, this network database also provides site-specific data required for modelling longer-term elevation trajectories (e.g., Day et al. 1999; Rybczyk and Cahoon 2002), where feedbacks occurring on varying timescales influence wetland elevation change (e.g., relative elevation, productivity, sedimentation).

In this paper, we describe biotic and hydrologic controls on coastal wetland elevation dynamics. Specifically, we review the limited amount of literature that directly compares measures of hydrologic and biotic processes with direct measures of wetland elevation change. In addition, we present a first and preliminary analysis of near-term trajectories of the global SET-MH data set in order to improve our understanding of salt marsh and mangrove responses to current sea-level rise and to identify critical factors and processes controlling coastal wetland elevation dynamics across a diversity of wetland settings.

12.2 Biotic Process Controls

Habitat stability of coastal wetlands in relation to sea-level rise is dependent on a feedback relationship between hydro-edaphic conditions and the biotic community that characterizes the ecosystem (Fig. 12.1). Within certain environmental bounds, as yet not well defined, the feedback relationship between biotic and abiotic components allows the wetland to self-adjust to prevailing water levels and, consequently, to keep pace with rising sea-level (Stevenson et al. 1986; Nyman et al. 1993, 1995). Although organic matter accumulation and the processes controlling it are widely recognized and well studied in some wetland habitats, little empirical information exists directly linking biotic processes to vertical accretion and surface elevation change, despite wide recognition that biotic contributions to soil elevation may vary from low (mineral) through intermediate (organic) to very high (peat).

A review of soil organic matter dynamics is beyond the scope of this paper. Instead, this section describes several biotic processes potentially contributing to elevation change and provides data linking these processes where they exist. Biotic processes can be divided into two categories: (1) indirect, which modify mineral deposition or erosion, and (2) direct, which directly alter soil volume and/or soil strength and resistance to erosion or compaction.

12.2.1 Indirect Biotic Processes

A number of studies have demonstrated the effects of vegetation on sediment deposition (e.g., Gleason et al. 1979; Leonard et al. 1995; Furukawa et al. 1997; Rooth et al. 2003). The shoots or aerial roots of wetland plants can baffle flows (Furukawa et al. 1997; Leonard and Reed 2002; Morris et al. 2002), which may increase rates of sediment deposition. The presence of wrack or plant litter on the soil surface may also trap mineral sediment, thus contributing to vertical accretion. In a manipulative experiment in which the amount of *Phragmites australis* litter was varied from 250 g DW m^{-2} to 2000 g DW m^{-2}, the amount of trapped sediment increased exponentially (Rooth et al. 2003). The combined effect of reducing flows and stabilising the substrate may lead to a positive relationship between vegetative aboveground biomass and marsh elevation change, although few studies have shown this link experimentally. Morris et al. (2002), for example, found that nutrient enrichment increased production and standing biomass of salt marsh plants, which accelerated elevation gain by increased sediment trapping from 5.1 mm $year^{-1}$ to 7.1 mm $year^{-1}$. In a study of the effects of mangrove aerial roots on mineral sedimentation, Krauss et al. (2003) found that elevation change rates varied with root type from −0.2 mm $year^{-1}$ to 3.4 mm $year^{-1}$.

12.2.2 Direct Biotic Processes

Although mineral sedimentation may be affected by vegetation characteristics, biotic processes that contribute directly to soil volume have the greatest potential to influence vertical accretion and elevation change. These biological processes can be divided into surface and subsurface processes. Surface processes include the accumulation of decaying organic matter such as leaf litter, or the formation of living benthic mats (e.g., microbial, algal, root), which contribute to vertical accretion and also influence the resistance of the deposit to compaction or erosion. Subsurface processes such as root production, root mortality, and decomposition influence soil volume, contributing either to expansion or to subsidence.

Litter Deposits

The relative contribution of plant litter deposition to vertical accretion and elevation change in coastal wetlands likely varies spatially and temporally. However, data linking these processes is limited. We examined vertical accretion and leaf litter standing stocks in mangrove forests where the accumulation of litter varies depending upon aboveground production, consumption by detritivores, microbial decomposition, and tidal flushing (McKee and Faulkner 2000; Middleton and McKee 2001). Basin mangrove forests can accumulate thick deposits of leaf litter that contribute to vertical accretion. For example, vertical accretion rates measured in a basin forest in Southwest Florida were positively related to litter standing stocks (Fig. 12.2A). These types of forests occur in depressional areas where tidal action is less than

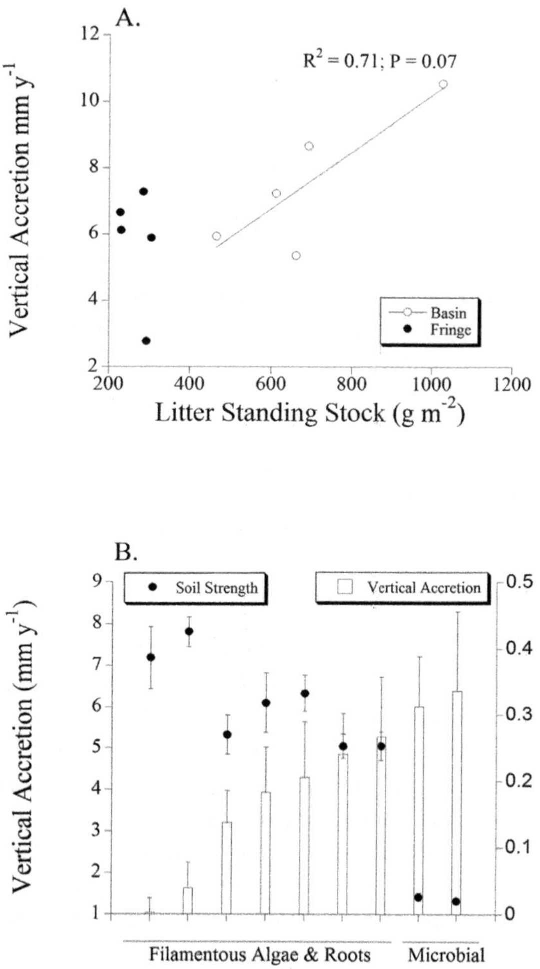

Fig. 12.2 Surface biotic processes directly contributing to vertical accretion in mangrove forests. **A** Litter accumulation is significantly correlated with vertical accretion in basin forests, but not in fringe forests where tidal action and other factors prevent it. **B** Variation in vertical accretion rates of benthic mats (above mesh screens pinned to the soil surface) composed of turf algae/mangrove roots and microbial communities in relation to their shear strength

along exposed shorelines, but where the soil and litter layer typically remain saturated, slowing decomposition. Over time, many layers of decaying leaves accumulate on the forest floor. In comparison, less litter accumulates in fringe forests with more frequent tidal flushing and faster degradation rates. In fringe forests, no relationship was found between accumulated litter and vertical accretion (Fig. 12.2A). Further work is needed to determine how these differences in litter deposition influence elevation change in mangrove forests and if surface accumulations of litter are important in maintaining elevations in other wetland types.

Benthic Algal/Microbial/Root Mats

Benthic mats composed of filamentous algae, roots of emergent plants, microbial communities, or a combination of these may form on the soil surface of coastal wetlands. For example, turf algae may form thick, living mats in red mangrove (*Rhizophora mangle*) forests and have been observed throughout the Caribbean Region (e.g., Florida, Belize, Honduras, Panama; McKee, personal communication). Another type of benthic mat is formed by growth of microbial communities consisting of cyanobacteria (Cyanophyta), diatoms (Chrysophyta), and other microorganisms (Joye and Lee 2004). Rates of vertical mat growth, which were determined on screens (1 mm mesh) pinned to the soil surface in a mangrove forest in Belize (Fig. 12.3B), varied from 1 mm year^{-1} along the shoreline (turf algal mats) to 6 mm year^{-1} in interior dwarf mangrove stands (microbial mats). The slower growing mats composed of filamentous algae and mangrove roots had high shear strength, whereas the microbial mats had almost no resistance to shearing (Fig. 12.3B). Variation in vertical accretion and shear strength of these benthic mats is intriguing and suggests a potential role in elevation change. Microbial mats have also been observed in some salt marshes where they form mounds of 10 cm or more in thickness (S. Joye, personal communication.). Further investigation is required, however, to determine whether benthic mats can influence elevation change in these and other coastal wetlands.

Subsurface Root Production

Accumulation of soil organic matter occurs when production exceeds decomposition rates. In coastal wetlands, slow decomposition, particularly of plant roots, occurs under the predominately anaerobic conditions in flooded soils. Thus, variation in root production may be the primary influence on elevation change through root inputs to soil volume. For example, reductions in soil organic matter accumulation following mass plant mortality (i.e., lack of root input) can lead to elevation declines through peat collapse (Cahoon et al. 2003, 2004). Despite general recognition of these relationships, supported by modelling studies, no direct measure of root contributions (or losses through decomposition) to elevation change have previously been

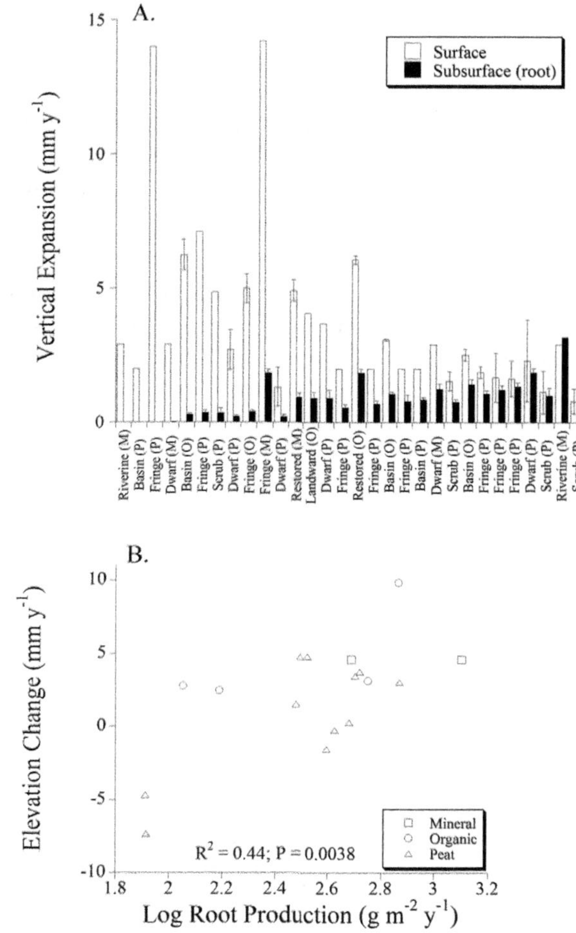

Fig. 12.3 A Comparison of surface (deposition above a marker horizon) and subsurface (root production) contributions to vertical soil expansion across different mangrove forest types and soils (*M* mineral, *O* organic, *P* peat). Root volume produced (cm^3 m^{-2} year^{-1}) was calculated from root production rates using an average root density of 1.1 g cm^{-3}. Values are the mean and SE of one to nine replicate plots. **B** Relationship between root production rates (in-growth bags) and elevation change (SET) measured in 18 mangrove forests in different geographic locations and three soil types (mineral, organic, peat). Values are the mean and SE of one to nine replicate plots

made in coastal wetlands. We consequently assessed subsurface contributions of root volume to vertical change in comparison with measured vertical accretion rates to examine surface versus subsurface inputs to soil volume. In addition, we examined the relationship between root production and elevation change.

Data on root production and vertical accretion were collected from 30 mangrove sites; surface elevation change was also measured at 18 of these sites, although the length of the records differed. Contribution of root volume to vertical expansion was calculated by converting mass to volume using mangrove root density (g cm^{-3}) determined with a pycnometer. The relative contribution of root volume to subsurface expansion is compared to surface vertical accretion in Fig. 12.3A. Depending on the site characteristics, root contribution to vertical expansion varied from 0 mm year^{-1} to 3 mm year^{-1}, indicating a potentially substantial effect on elevation in some locations. For those sites (undisturbed only) instrumented with SETs, there was a significant correlation between elevation change and root production (Fig. 12.3B).

The relative importance of biotic processes in the sustainability of coastal wetlands obviously depends upon the physical setting and their role relative to that of abiotic factors in marsh vertical development. In this section, we provide a few examples in which biotic processes influence soil elevation change by contributing directly or indirectly to vertical accretion and/or soil volume. However, additional studies are needed to fully evaluate the role of biotic processes in maintaining soil elevations relative to sea-level in coastal wetlands.

12.3 Hydrologic Process Controls

There is an extensive literature on wetland hydrology and its indirect influence on wetland elevation by: (1) delivery of suspended sediments to wetland surfaces and (2) regulation of plant growth, decomposition, and soil organic matter accumulation through control of the soil oxidative state. However, data on the direct relationship between water flux/storage and wetland elevation change is limited. In this section, we review the literature on the direct effects of hydrology on wetland elevation by the processes of compression and dilation storage (e.g., shrink–swell).

There are two mechanisms of water storage in wetland soils. Saturation storage is associated with the displacement of gases from pore spaces by water and is limited to the sediment above the water table where pressures in the 'free' (e.g., unbound) water are low enough to allow atmospheric gases to enter the sediment (Nuttle et al. 1990). Dilation storage is associated with changes in the bulk volume of the sediment and can occur throughout the sediment. As more water is incorporated into the sediment below the water table, the sediment must dilate (e.g., swelling clays) and the sediment volume change is equal to the volume of water added to storage (Nuttle et al. 1990). This water storage mechanism displaces the wetland surface. Evidence for the mechanics of dilation storage in wetland sediments can be found in the high degree of compression under a load and the significant shrinkage when dried. The amount of dilation storage and degree of surface displacement will vary with soil type (i.e., organic vs mineral). It is important to understand such water-storage controls on wetland elevation in order to properly evaluate sources of variation in long-term elevation trends. The role of wetland hydrology should be considered both at the surface and within the subsurface.

12.3.1 Surface Wetland Hydrology

As a result of surface topography, water floods a marsh to different depths, resulting in larger vertical loading in areas of deeper flooding. Tidal flooding of 10 cm depth is sufficient to deform a salt marsh surface by both compression and lateral movement, resulting in uplift (Nuttle et al. 1990). Storm tides, which can be several meters deep, can also deform the salt marsh surface (Cahoon 2003). The storm surge from Hurricane Andrew compressed the surface of a rapidly deteriorating salt marsh in Louisiana by 33 mm (Cahoon et al. 1995) and the deformed marsh surface had not rebounded 8 years later (Rybczyk and Cahoon 2002). In North Carolina, the surface of a salt marsh with 60 % soil organic matter content was compressed by 5–6 mm after each of two storms to strike the marsh in consecutive years. The deformed surface rebounded >10 mm the following year, only to be compressed 20 mm by another storm (Cahoon 2003). At these sites, the influence of water table fluctuations, organic substrates, and pre-existing marsh deterioration all appear to have influenced the compression of the substrate. However, the mechanism driving the potential rebound of such deformed surfaces is not known.

12.3.2 Subsurface Wetland Hydrology

Infiltration during flooding by astronomical tides can cause shrink–swell of the marsh substrate by dilation storage (Harrison 1975; Nuttle and Hemond 1988; Nuttle et al. 1990), although tidal flooding does not cause shrink–swell in all marshes (Cahoon et al. 1995). In addition, given that the tidal effects in some marshes decline exponentially with distance from the creek bank (Nuttle 1988), loss of water through evapotranspiration (ET) by vegetation can become an important control on water storage in interior portions of the marsh. Indeed, up to 3.0 mm of diurnal variation in surface elevation was related to ET in salt and brackish marshes in the Bay of Fundy (Paquette et al. 2004) and a salt marsh in Louisiana (Perez et al. 2003; Fig. 12.4).

Longer-term trends (e.g., weekly to seasonal) in groundwater levels can influence wetland surface elevation. Incremental changes in ground water level explained 40 % of the variation in incremental surface elevation change in mangrove forests along the Shark River in Everglades National Park (Smith and Cahoon 2003), with surface elevation greatest during the wet season. Similarly, lowering of the marsh water table by drought resulted in a decrease in surface elevation in marshes in east Texas (10–15 mm; Perez and Cahoon 2004), in south Louisiana (20 mm; Perez et al. 2003; Swarzenski et al. 2006; Fig. 12.4), and in southeast Australia (Rogers et al. 2005).

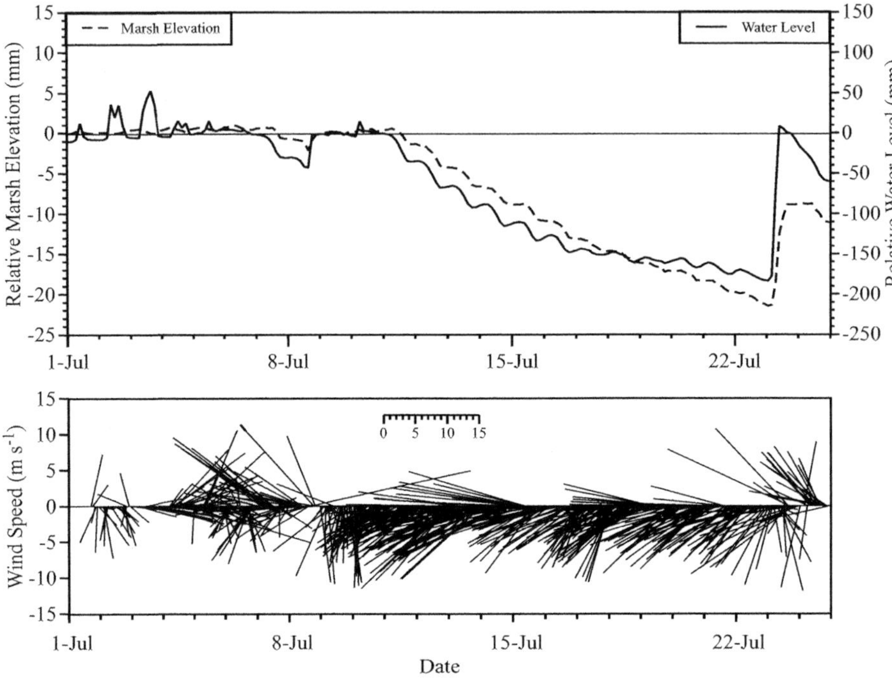

Fig. 12.4 *Top* Change in relative marsh elevation and marsh water level during a two-week wind-driven water-level drawdown event (note difference in scales). *Bottom* Wind speed and direction during the two-week period. Source: Perez et al. (2003)

12.4 Findings from the SET Network

The combination of biotic and hydrologic processes described above, acting at different rates and over different timescales in the wide array of wetland types and settings, exerts an aggregate influence on surface elevation change. The SET-MH approach can provide site-specific information on elevation trends, the degree of subsurface influence on elevation; and when used in combination with process measures, the relative contribution of different processes to elevation change. In this section, we present the first and preliminary analysis of the SET-MH monitoring network.

12.4.1 Data Analysis

Linear accretion and elevation trajectory estimates were made for each wetland setting. Shallow subsidence was calculated as the difference between the linear accretion and elevation trajectories (Cahoon et al. 1995). Average trajectories were compared to relative sea-level rise (RSLR) and *t*-tests were used to test for significant difference. A simple linear regression was used to test for associations between accretion and elevation change trajectories and RSLR, tide range, soil bulk density, and soil organic matter.

Temporal and spatial scaling constraints influence data analysis and interpretations. The present analysis compares short-term accretion and SET measurements to longer-term, historical records of sea-level rise. Further difficulties arise where sea-level records are gathered from tide gauges at some distance from, and in different geotectonic settings to, the wetland of interest. It is also difficult to relate individual site behavior to more regional trends in sea-level change and marsh morphological change. These data constraints limit the direct comparison by site. Nevertheless, it is still possible to establish a broad-based assessment of relationships between aspects of marsh sedimentation, elevation change and sea-level change. The SET network provides a starting point from which questions of wetland vulnerability and sustainability can be assessed. The data analysis is presented for salt marshes and mangroves separately because of their distinct vegetation structures. Furthermore, salt marsh and mangrove settings occupy different climatological (i.e., latitudinal) regions, although in some locales they can co-exist (e.g., southeastern Australia).

Since both salt marsh and mangrove systems influence and are influenced by the dominant biogeomorphic processes occurring within the coastal zone, trends in accretion and elevation change are analyzed according to geomorphic settings. Of the many classification schemes available, the one developed by Woodroffe (2002) was chosen as the starting point for this analysis because it applies to both salt marsh and mangrove systems and builds on many previous classifications. The geomorphic classes include: back-barrier, embayment, estuarine, deltaic, open coast, and drowned valley. We added 'oceanic islands' as an additional class to more adequately reflect the geographic diversity of the network sites. Settings such as Twin Cays (Belize), Roatan (Honduras), and several islands of the Federated States of Micronesia fit into this class. We further adapted the classification by including our drowned valley sites (located only in Australia) in the estuarine class because Roy et al. (2001) included drowned river valleys of Australia in the estuarine class. Thus, our analyses included the following classes: back-barrier, embayment, estuarine (including drowned valley), deltaic, open coast, and oceanic island. Although salt marsh and mangrove data in the network are distributed across these classes differently (e.g., oceanic islands are only represented by mangrove settings, and deltas are only represented by salt marshes), this classification scheme, modelled after Woodroffe (2002), is a useful way of examining salt marsh and mangrove data on a global scale.

12.4.2 The Salt Marsh SET Network

The SET salt marsh network comprises 65 geographical locations and 108 coastal wetland settings in North America, Europe and Australia. The 'sample network' reported here, where data availability is of sufficient quality for analytical purposes, covers 34 of these locations and 78 wetland settings. As expected in a developing methodology, short record lengths predominate; one-third of all records are less than three years in length and two-thirds are of less than five years' duration. The analytical problems caused by short record lengths are discussed further below.

Four geomorphic classes are represented in the salt marsh dataset (back-barrier, coastal bay, deltaic, estuarine) from seven main geographic regions: NE North America, SE North America, Northern Gulf of Mexico, SW United States (Pacific), SE Australia, NW Europe, and Northern Mediterranean. Unlike the Mediterranean, NW Europe is geologically stable and has a legacy of sediment supply, sea-level history, and coastal configurations strongly influenced by the Pleistocene glaciations. NW Europe is meso- to macro-tidal, and much reduced contemporary sediment supply from early Holocene levels and extensive reclamation in historical times has meant that many remaining marsh areas still show high rates of areal loss and internal marsh break-up (Carpenter and Pye 1996). The Mediterranean is geologically active and micro-tidal, with river processes in combination with wave-driven processes producing barriers to coastal lagoons and their associated wetlands. Historically, these systems have been sustained by episodic forcings, which have injected pulses of sediment onto wetland surfaces. However, coastal retreat, saltwater intrusions, and reductions in riverine sediment supply (e.g., disconnection of river channels from delta plains) have led to low accretion rates and poor vegetation productivity in many Mediterranean deltas (e.g. Rhone: Pont et al. 2002; Ebro: Curco et al. 2002; Po/Venice lagoon: Abrami and Day 2003; Day et al. 1999).

The Atlantic shoreline of eastern North America is 80–90 % estuaries and lagoons. The northeast coast has undergone cycles of glaciation, is meso- to macrotidal and has a moderate to high sediment supply. The southeastern coast is characterized by extensive barrier islands, lagoons, and drowned river valley estuaries. Two main geomorphic settings present in our sample from both the northeast and southeast coasts are back-barrier and coastal bay marshes. The northern Gulf of Mexico shoreline is 80–90 % estuaries and lagoons dominated by micro-tidal settings, including the Mississippi River delta, which is characterized by high rates of subsidence and historically high sedimentation rates. Much of the recent sediment load of the Mississippi River is currently deposited on the continental slope and is thus not available to coastal marshes. The Pacific coast of the United States is geologically active, has a narrow continental shelf, and has a limited extent of mesotidal coastal marshes (10–20 %). In the meso-tidal, swell wave environment of SE Australia, wetlands are associated with low-energy locations in bays and gulfs, in estuaries, and between barriers on multiple barrier coastlines.

Results

Average accretion rates were greater than corresponding rates of RSLR ($P<0.001$), and accretion rates showed a significant positive linear relationship to RSLR ($P=0.002$; Fig. 12.5). Average elevation rates were also greater than RSLR, ($P=0.06$), but no linear trend was evident between elevation and RSLR rates ($P=0.20$). The close correspondence between rates of surface accretion and sea-level rise suggests that, in general, salt marsh surface processes are exhibiting 'keep up' behavior as the sea level rises.

The four sites in Fig. 12.5 with strong accretionary responses are characterized by high sediment supply. The one site that lags behind sea-level rise for both accretion and surface elevation change is Old Oyster Bayou, a *Spartina alterniflora* marsh in a rapidly subsiding part of the Mississippi River delta. Although a riverine sediment source is nearby, the severely altered local hydrology apparently limits sedimentation such that vertical accretion and elevation change are essentially insignificant except during hurricane storm events (Rybczyk and Cahoon 2002). Model results indicate that this marsh will become submerged within the next 30 years because of deep subsidence rates, even if there is a 12-year return frequency for major hurricanes (Rybczyk and Cahoon 2002).

Average elevation rates were significantly less than accretion rates ($P<0.001$), indicating that shallow subsidence occurs at many sites. However,

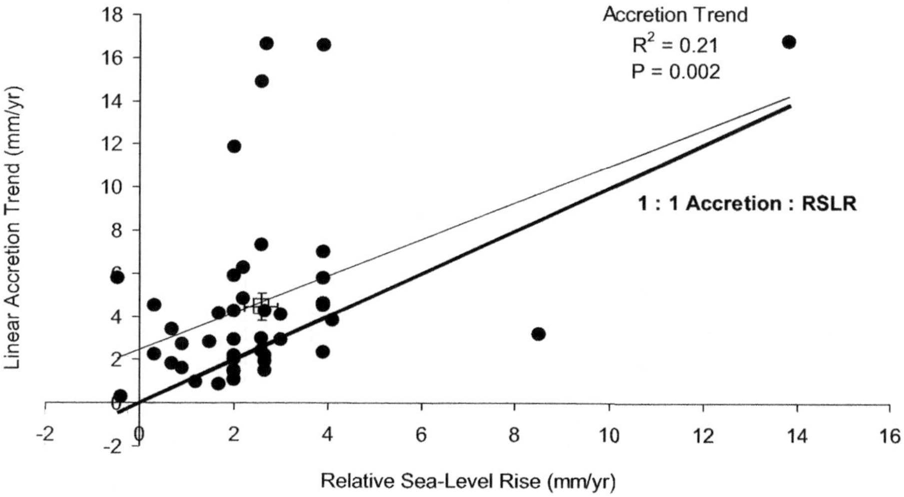

the variability is large, suggesting that the local process environment exerts a strong influence. Disaggregated by geomorphic setting, deltas are characterized by high accretion with only moderate elevation change (Fig. 12.6), resulting in high shallow subsidence (P=0.03). Embayments also showed high shallow subsidence, but at a lower average accretion rate (P=0.03). Back-barriers and estuaries generally have good concordance between accretion and elevation change. In terms of geographic region (Fig. 12.7), the United States Atlantic NE and Mediterranean Europe are characterized by high accretion and high elevation change. The Mediterranean estimates are high because the network sites are located in areas receiving riverine sediments (e.g., deltas). Previously published evidence from eastern England (Cahoon et al. 2000a) suggests that many NW European sites are likely to show similar patterns of behavior. The United States Atlantic SE and Gulf Coast are characterized by high accretion and low elevation change, the United States Pacific and Australian sites by low accretion and low elevation change. Lastly, there are no

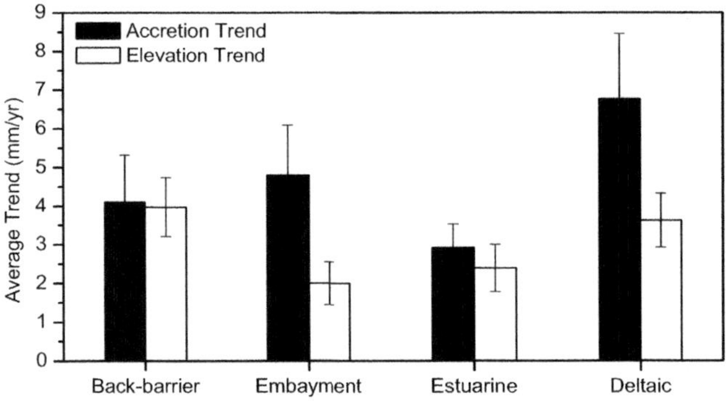

Fig. 12.6 Trends in accretion and elevation for 49 salt marsh sites in the SET network, disaggregated by geomorphic setting

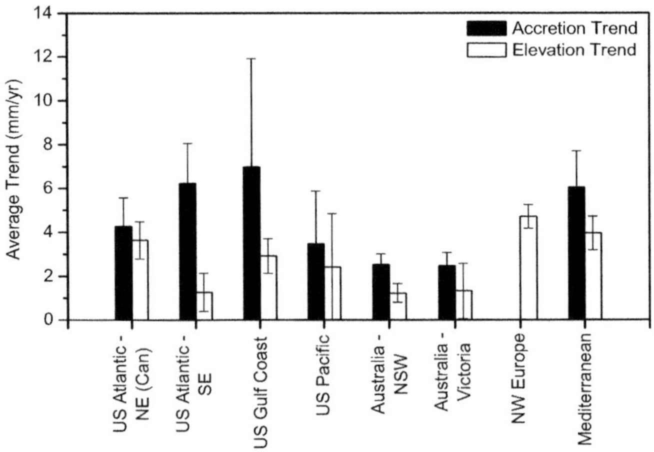

Fig. 12.7 Trends in accretion and elevation for 49 salt marsh sites in the SET network, disaggregated by geographic region

clear trends in the relations between either accretion or elevation change and tidal range in relation to geomorphic setting ($P>0.1$).

12.4.3 The Mangrove Forest SET Network

The SET mangrove network comprises 25 geographical locations and 72 coastal wetland settings in the Gulf of Mexico, the Caribbean, Central America, the Western Pacific, and Australia. Data analysis was conducted on 60 mangrove forest settings from 19 of the sites where sediment accretion and elevation change records were >1 year. Approximately one-half of the data records span 4 years or less.

Five geomorphic classes (adapted from Woodroffe 2002) are represented (back-barrier, embayment, estuarine, open coast, oceanic island), from five main geographic regions (Gulf of Mexico, Caribbean, Pacific Central America, SE Australia, Micronesia), and seven countries (United States, Mexico, Belize, Honduras, Costa Rica, the Federated States of Micronesia, Australia). The small oceanic islands in the Caribbean are microtidal, remote from terrigenous sediment input, experience only moderate rates of RSLR, and rely mostly on biogenic processes for elevation maintenance (Cahoon et al. 2003). The Micronesian islands are also remote from terrigenous influence, but have very high topography and rainfall, resulting in somewhat less dependence on biogenic processes. The mangrove settings from the Florida border of the Gulf of Mexico are classified as embayments, open coast, and estuarine. However, these settings are somewhat similar to the Caribbean islands in terms of tidal regime, sediment input, mangrove species composition, and local RSLR. The SE Australian coastline experiences a meso to macro-tidal regime; chosen settings are low-energy, such as bays, back-barriers, and riverine habitats. The geologically active Pacific coast of Central America features large, macrotidal estuaries with high sediment loads. The settings chosen in Costa Rica are experiencing substantial subsidence (Emery and Aubrey 1991). This wide diversity of habitats and environmental settings presents a first opportunity to study processes and vulnerabilities on a global scale.

Results

Accretion rates exceeded sea-level rise ($P=0.03$, $n=41$) and increased with increasing rates of sea-level rise ($P<0.0001$, $n=41$) across the network of settings. However, this accretion surplus did not necessarily translate into elevation gain, as most sites had elevation deficits with respect to RSLR ($P<0.001$, $n=55$). No linear trends were evident between elevation change rates and RSLR ($P=0.49$, $n=55$). The discrepancy between accretion and elevation change resulted in an overall significant positive shallow subsidence ($P=0.017$, $n=41$). Accretion also showed a significant positive relationship to tidal range (P=0.084, n=38). Estuarine settings showed a strong linear relationship between accretion and RSLR (Fig. 12.8; P<0.001, n=20), as did embayments, although the fit was less significant (P=0.08, n=8). Embayments were the only class showing increasing elevation change rates with increasing RSLR (P=0.07, n=8) and tidal range (P=0.07, n=8) and increasing accretion with increasing tidal range (P=0.08, n=8). These results suggest there exist belowground process controls on mangrove surface elevation dynamics.

No strong vegetation-dependent accretion/elevation trends were evident at a global scale from the network. *Avicennia* and *Rhizophora* provinces displayed similar trends with respect to sea-level rise and tidal range. The significantly positive relationship between accretion and soil bulk density (P=0.054, n=10), and the negative logarithmic relationship between accretion and percent organic matter (P=0.003, n=12) were both related to differences among geomorphic classes. Peat-based soils of the oceanic islands (Belize, Roatan) had the lowest bulk density, highest organic matter content, and the lowest accretion. Mineral, riverine Pacific Central American mangrove soils had the highest bulk density, lowest organic content, and the highest accretion. Here too, elevation failed to show any relationships to bulk density or organic content, suggesting belowground process controls on elevation dynamics.

Shallow subsidence was seen in many of the 41 settings for which data were available, yet no shared characteristics suggested a unifying mechanism. Subsidence was greatest in areas that received highest sedimentation: settings on

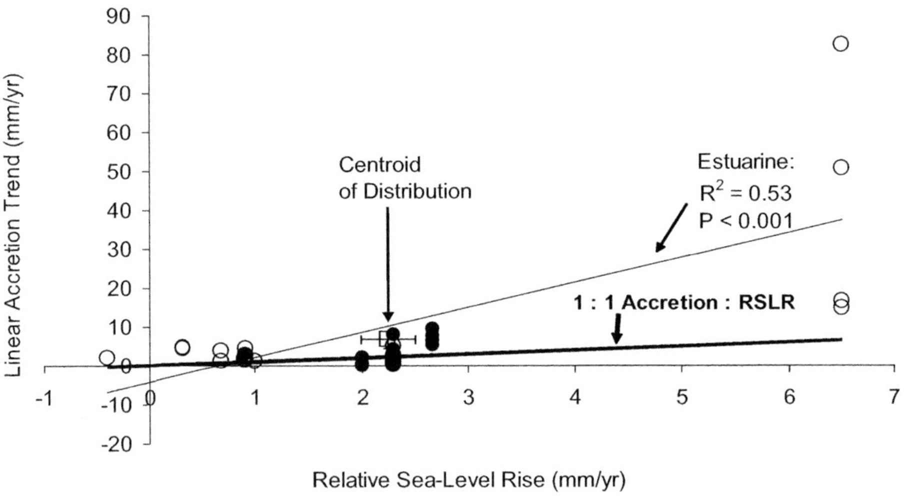

Fig. 12.8 Linear trend in accretion versus relative sea-level rise for mangrove sites in the SET network. The *heavy black line* represents the 1:1 relationship. The *thin black line* represents the slope for riverine mangroves. The *open box* with two-dimensional error bars represents the centroids of the distribution

the Pacific coast of Costa Rica accreted vertically at a rate of 41 mm $year^{-1}$, yet elevation increased at a mere 1 mm $year^{-1}$. This large difference is likely due to the compaction and decomposition of recent deposits.

The consistent elevation gain of Caribbean mangroves despite low sediment accretion underscores the importance of organic belowground productivity in these settings. Caribbean islands in particular had elevation 50% greater than accretion. This relationship is not specific to particular mangrove species, since the Caribbean islands share the same species as many of the other settings. Rather, the contribution of root growth may be related to nutrient status, hydrology, decomposition rates, disturbance regimes, and other site-specific variables.

Groundwater is known to directly influence mangrove surface elevation by altering soil volume (Smith and Cahoon 2003; Whelan et al. 2005). The sampling period throughout southeastern Australia (2000–2004) occurred during an intense El Niño-related drought. Associated negative SOI and rainfall reductions caused significant subsidence and lowered rates of surface elevation gain at settings in southeastern Australia. This may explain the low rate of surface elevation gain in these settings compared to the global network. While sea levels may be locally reduced under drought conditions, repeated drought cycles have the potential to significantly affect the structure and function of mangrove forests (Drexler and Ewel 2001).

12.5 Further Considerations

This review of the processes controlling marsh surface elevation change shows that biotic processes, both directly and indirectly, are likely to have a major influence on marsh vulnerability to future sea-level rise. The influence of biotic processes will vary depending upon: sediment supply, hydrology, dominant vegetation, and plant community structure. In addition, it seems clear that variations in tide, ET, and river discharge can influence surface elevation of some of the more organic wetland substrates over short time-scales. Empirical measures of the influence of tidal flooding and groundwater fluxes on wetland surface elevation are available from a limited number of wetlands encompassing a limited range of geomorphic settings and soil types. The responses of a wider range of wetlands to variations in soil water storage need to be determined to elucidate the importance of this driver of elevation change.

Interpreting the status of coastal wetlands across the range of geomorphic settings and localities encompassed by the SET network has identified several significant relationships among accretion, elevation change, and hydrologic drivers (e.g., tidal range, RSLR). Elevation change and accretion in wetlands where the delivery of sediment to the marsh was supported by local sediment supply and tidal/estuarine hydrology generally were seen to keep pace with RSLR. However, examining the data sets as a whole masks some important differences. For example, salt marshes are typically found at much higher intertidal levels in NW Europe than on the eastern seaboard of the United States. Furthermore, for minerogenic marshes, accretion rates and rates of surface elevation change decline over time as surfaces become progressively higher in the tidal frame. The nature of this decline over decadal to centennial timescales is strongly asymptotic towards Highest Astronomical Tide, although reaching an equilibrium level below this height (French 1993; Allen 2000). Such behavior complicates the assessment of marsh performance/sea-level relations across such varied sites. Whereas 'mature' marsh surfaces will exhibit an equilibrium level related to tidal parameters and thus respond to changes in sea level and frequency of tidal inundation, 'immature' marshes will show high, non-equilibrium rates of accretion and surface change as they build up rapidly from low positions in the tidal frame.

Importantly, examination of elevation change and accretion data available in the SET network has demonstrated the importance of time scale considerations both in data collection and interpretation. Due to the effects of local hydrology measurements of surface elevation for a given marsh should always be collected at the same stage of the tide, meteorological conditions, or season of the year to minimize bias in the data. Interpretations of elevation trajectories should be based on multi-year data records so as to incorporate as much temporal variability as possible but there is also the issue of the record length necessary to allow meaningful comparison with the sea-level record. To assess the long-term vulnerability of coastal wetlands, studies must incorporate the range of variability in biotic and abiotic surface and subsurface processes within the wetlands reviewed in this paper as well as both cyclic and episodic changes in local sea level.

The data available in the SET network can be used to predict the future vulnerability of coastal wetlands when used to drive numerical models that consider more than just future changes in sea level. How will predicted changes in surface runoff and groundwater, local sediment delivery to the coast, and coastal storm intensity interact with predicted accelerations in sea-level rise to affect wetland elevation change? Will a long-term reduction in groundwater flow result in a persistent reduction in wetland elevation that will exacerbate the effects of sea-level rise? How will changes in plant productivity and latitudinal shifts in vegetative communities with global change affect the role of biotic processes in wetland soil development? As well as future changes in climate drivers, some of the processes currently maintaining wetland elevation may not be sustainable in the long-term and over large areas. Thus, for example, marsh edge erosion may feed sediment onto adjacent surfaces. Whilst such processes may lead to local accretion and surface height gains locally, they may in fact be symptomatic of more long-term, regional patterns of marsh retreat and net loss of area.

This first analysis of accretion and elevation change data from the SET network has raised many questions. Future work should seek a better understanding of the relationships between abiotic and biotic factors in different coastal systems and how the combination of these factors results in characteristic trajectories of surface elevation change. Studies that focus on effects of specific factors such as nutrients, temperature, elevated CO_2, and flooding regimes, as well as their interactions, are needed to quantify and predict future wetland elevation change. Such studies should be coupled with a detailed interrogation of the existing database and a strategic extension of the SET-monitoring network to embrace regions and geomorphic settings as yet unsampled. The ultimate goal is to provide a wider understanding of the processes determining wetland vulnerability and sustainability worldwide.

Acknowledgements Jon French, John Day, Kerrylee Rogers, and Brian Perez contributed ideas and data to the symposium presentations at the INTECOL conference. In addition to the authors, the following participants in the SET network contributed data for our network analyses: Daniele Are, Andrea Rismondo, Francesco Scarton, John Day (Venice Lagoon, Italy), Linda Blum, Mark Brinson, Robert Christian (Va., USA), David Burdick (Maine, USA), Gail Chmura (Bay of Fundy, Canada), Carlos Coronado-Molina, Fred Sklar (Florida Everglades, USA), Chris Craft (Ga., USA), John W. Day, Jr. (Rhone Delta, France; Terminos Lagoon, Mexico), Patricia Delgado (Tempisque River, Costa Rica), Michael Erwin (Mass./N.J./Va., USA), Jon French (Essex/Suffolk/Norfolk, UK), Charles Hopkinson (Mass., USA), Carles Ibanez, John Day (Ebro Delta, Spain), James Morris (S.C., USA), Kerrylee Rogers (NSW/Victoria, Australia), Charles Roman (Mass., USA), Thomas J. Smith III (Florida Everglades, USA). Jim Lynch formatted and finalized the figures.

References

Abrami G, Day JW Jr (2003) Historical and morphological aspects of the evolution of the Venice Lagoon and the Po Delta: new criteria for the management of the area. Proc IAHR Symp River Coast Estuar Morphodyn 3:481–493

Allen JRL (2000) Morphodynamics of Holocene salt marshes: a review sketch from the Atlantic and Southern North Sea coasts of Europe. Quat Sci Rev 19:1155–1231

Boumans R, Day JW Jr (1993) High precision measurements of sediment elevation in shallow coastal areas using a sedimentation–erosion table. Estuaries 16:375–380

Cahoon DR (2003) Storms as agents of wetland elevation change: their impact on surface and subsurface sediment processes. (Proceedings of the international conference on coastal sediments 2003, CD-ROM) World Scientific/East Meets West, Corpus Christi

Cahoon DR, Turner RE (1989) Accretion and canal impacts in a rapidly subsiding wetland II. feldspar marker horizon technique. Estuaries 12:260–268

Cahoon DR, Reed DJ, Day JW Jr (1995) Estimating shallow subsidence in microtidal salt marshes of the southeastern United States: Kaye and Barghoorn revisited. Mar Geol 128:1–9

Cahoon DR, French J, Spencer T, Reed DJ, Moller I (2000a) Vertical accretion versus elevational adjustment in UK saltmarshes: an evaluation of alternative methodologies. In: Pye K, Allen JRL (eds) Coastal and estuarine environments: sedimentology, geomorphology and geoarchaeology. Geol Soc Lond Spec Publ 175:223–238

Cahoon DR, Marin PE, Black BK, Lynch JC (2000b) A method for measuring vertical accretion, elevation, and compaction of soft, shallow-water sediments. J Sediment Res 70: 1250-1253

Cahoon DR, Lynch JC, Hensel PF, Boumans R, Perez BC, Segura B, Day JW Jr, (2002a) High precision measurement of wetland sediment elevation: I. recent improvements to the Sedimentation-Erosion Table. J Sediment Res 72:730–733

Cahoon DR, Lynch JC, Perez BC, Segura B, Holland R, Stelly C, Stephenson G, Hensel P (2002b) High precision measurement of wetland sediment elevation: II. The rod surface elevation table. J Sediment Res 72:734–739

Cahoon DR, Hensel P, Rybczyk J, McKee KL, Proffitt CE, Perez BA (2003) Mass tree mortality leads to mangrove peat collapse at Bay Islands, Honduras after Hurricane Mitch. J Ecol 91:1093–1105

Cahoon DR, Ford MA, Hensel PF (2004) Ecogeomorphology of *Spartina patens*-dominated tidal marshes: soil organic matter accumulation, marsh elevation dynamics, and disturbance. In: Fagherazzi S, Marani M, Blum LK (eds) The ecogeomorphology of tidal marshes, coastal and estuarine studies, vol 59. American Geophysical Union, Washington, D.C., pp 247–266

Carpenter KE, Pye K (1996) Salt marsh change in England and Wales – its history and causes. (Technical Report W12) HR Wallingford, Marlow

Curco A, Ibanez I, Day J, Prat N (2002) Net primary production and decomposition of salt marshes of the Ebro delta, Catalonia, Spain. Estuaries 25:309–324

Day JW Jr, Rybczyk J, Scarton F, Rismondo A, Are D, Cecconi G (1999) Soil accretionary dynamics, sea-level rise and the survival of wetlands in Venice Lagoon: a field and modelling approach. Estuar Coast Shelf Sci 49:607–628

Drexler JZ, Ewel KC (2001) Effects of the 1997–1998 ENSO-related drought on hydrology and salinity in a Micronesian wetland complex. Estuaries 24:347–356

Emery KO, Aubrey DG (1991) Sea levels, land levels, and tide gauges. Springer, Berlin Heidelberg New York

French JR (1993) Numerical simulation of vertical marsh growth and adjustment to accelerated sea-level rise, North Norfolk, UK. Earth Surf Landform Proc 18:63–81

Furukawa K, Wolanski, E, Mueller H (1997) Currents and sediment transport in mangrove forests. Estuar Coast Shelf Sci 44:301–310

Gleason ML, Elmer DA, Pien NC, Fisher JS (1979) Effects of stem density upon sediment retention by salt marsh cord grass, *Spartina alterniflora* Loisel. Estuaries 2:271–273

Gornitz V (1991) Global coastal hazards from future sea level rise. Palaeogeogr Palaeoclim Palaeoecol 89:379–398

Hammar-Klose ES, Thieler ER (2001) Coastal vulnerability to sea-level rise: a preliminary database for the U.S. Atlantic, Pacific, and Gulf of Mexico coasts. (Digital Data Series DDS-68) US Geological Survey, Reston

Harrison EZ (1975) Sedimentation rates, shoreline modification, and vegetation changes on tidal marshes along the coast of Connecticut. MS thesis, Cornell University, Ithaca, 107 pp

Hoozemans FMJ, Marchand M, Pennekamp HA (1993) Sea level rise: a global vulnerability assessment – vulnerability assessments for population, coastal wetlands and rice production on a global scale, 2nd edn. Delft Hydraulics and Rijkswaterstaat, Delft/The Hague, 184 pp

Joye SB, Lee RY (2004) Benthic microbial mats: important sources of fixed nitrogen and carbon to the Twin Cays, Belize ecosystem. Atoll Res Bull 528:1–24

Kennedy VS, Twilley RR, Kleypas JA, Cowan JH Jr, Hare SR (2002) Coastal and marine ecosystems and global climate change: potential effects on US resources. Pew Center for Global Climate Change, Arlington, 52 pp

Krauss KW, Allen JA, Cahoon DR (2003) Differential rates of vertical accretion and elevation change among aerial root types in Micronesian mangrove forests. Estuar Coast Shelf Sci 56:251–259

Leonard LA, Reed DJ (2002) Hydrodynamics and sediment transport through tidal marsh canopies. J Coast Res Spec Iss SI36:459–469

Leonard LA, Hine AC, Luther ME (1995) Surficial sediment transport and deposition processes in a *Juncus roemerianus* marsh, west-central Florida. J Coastal Res 11:322–336

McFadden L, Vafeidis A, Nicholls R (2003) A coastal database for global impact and vulnerability analysis. (Proceedings of the international conference on coastal sediments 2003, CD-ROM) World Scientific/East Meets West, Corpus Christi

McKee KL, Faulkner PL (2000) Mangrove peat analysis and reconstruction of vegetation history at the Pelican Cays, Belize. Atoll Res Bull 468:46–58

Middleton BA, McKee KL (2001) Degradation of mangrove tissues and implications for peat formation in Belizean island forests. J Ecol 89:818–828

Mitsch WJ, Gosselink JG (2000) Wetlands. Wiley, New York

Morris JT, Sundareshwar PV, Nietch CT, Kjerfve B, Cahoon DR (2002) Responses of coastal wetlands to rising sea level. Ecology 83:2869–2877

Nicholls RJ, Hoozemans FMJ, Marchand, M, (1999) Increasing flood risk and wetland losses due to global sea-level rise: regional and global analyses. Global Environ Change 9:S69–S87

Nuttle WK (1988) The interpretation of transient pore pressures in salt marsh sediment. Soil Sci 146:391–402

Nuttle WK, Hemond HF (1988) Salt marsh hydrology: implications for biogeochemical fluxes to the atmosphere and estuaries. Global Biogeochem Cycles 2:91–114

Nuttle WK, Hemond HF, Stolzenbach KD (1990) Mechanisms of water storage in salt marsh sediments: the importance of dilation. Hydrol Proc 4:1–13

Nyman JA, DeLaune RD, Roberts HH, Patrick WH Jr (1993) Relationship between vegetation and soil formation in a rapidly submerging coastal marsh. Mar Ecol Prog Ser 96:269–279

Nyman JA, Delaune RD, Pezeshki S, Patrick WH Jr (1995) Organic matter fluxes and marsh stability in a rapidly submerging estuarine marsh. Estuaries 18:207–218

Paquette CH, Sundberg KL, Boumans RMJ, Chmura GL (2004) Changes in salt marsh surface elevation due to variability in evapotranspiration and tidal flooding. Estuaries 27:82–89

Perez BC, Cahoon DR (2004) Marsh vertical accretion and elevation change in hydrologically managed and unmanaged marshes within the McFaddin National Wildlife Refuge, Texas. (Final report to US Fish and Wildlife Service) McFaddin National Wildlife Refuge, Sabine Pass, Tex.

Perez B, Cahoon D, Segura B, Lynch J, Olney T (2003) Marsh elevation dynamics over various depth profiles in a Louisiana salt marsh. Bienn Conf Estuar Res Fed Conf Abstr 17:103

Poff LF, Brinson MM, Day JW Jr (2002) Aquatic ecosystems and global climate change: potential impacts on inland freshwater and coastal wetland ecosystems. Pew Center for Global Climate Change, Arlington, 45 pp

Pont D, Day J, Hensel P, Franquet E, Torre F, Rioual P, Ibanez C, Coulet E (2002) Response scenarios for the deltaic plain of the Rhone in the face of an acceleration in the rate of sea level rise, with special attention for Salicornia-type environments. Estuaries 25:337–358

Rogers K, Saintilan N, Heijnis H (2005) Mangrove and salt marsh monitoring at Westernport Bay, Victoria. Estuaries 28:551–559

Rooth JE, Stevenson JC, Cornwell J (2003) Increased sediment accretion rates following invasion by *Phragmites australis*: the role of litter. Estuaries 26: 475–483

Roy PS, Williams RJ, Jones AR, Yassini I, Gibbs PJ, Coates B, West RJ, Scanes PR, Hudson JP, Nichol S (2001) Structure and function of south-east Australian estuaries. Estuar Coast Shelf Sci 53:351–384

Rybczyk JM, Cahoon DR (2002) Estimating the potential for submergence for two subsiding wetlands in the Mississippi river delta. Estuaries 25:985–998

Smith TJ III, Cahoon DR (2003) Wetland sediment elevation in the Florida Everglades: response to surface water stage variation. (Proceedings of the international conference on coastal sediments 2003, CD-ROM) World Scientific/East Meets West, Corpus Christi

Stevenson JC, Ward LG, Kearney MS (1986) Vertical accretion in marshes with varying rates of sea level rise. In: Wolfe DA (ed) Estuarine variability. Academic, Orlando, pp 241–259

Swarzenski C, Perez BC, Faulkner S, Gambrell R, Michot T, Cahoon DR (2006) A field investigation into causes and potential consequences of widespread salt marsh dieback in coastal Louisiana in 2000: task 2.3. (Final report) Louisiana Department of Natural Resources, Baton Rouge (in press)

Whelan KRT, Smith TJ III, Cahoon DR, Lynch, JC, Anderson GH (2005) Groundwater control of mangrove surface elevation: shrink and swell varies with soil depth. Estuaries 28:833–843

Woodroffe C (2002) Coasts: form, process and evolution. Cambridge University Press, Cambridge

13 Connecting Arctic and Temperate Wetlands and Agricultural Landscapes: The Dynamics of Goose Populations in Response to Global Change

ROBERT L. JEFFERIES, RUDI H. DRENT, JAN P. BAKKER

13.1 Introduction

During the past 50 years, agricultural practices in Europe and North America have undergone immense change. The development of high-yielding varieties of crops and grasses and their requirement for large amounts of nitrogen and phosphorus fertilizers to sustain yields have revolutionized agriculture. In some countries, larger agricultural units of hundreds of hectares are replacing the small holdings formally held by individual farmers, the outcome of economies of scale, loss of subsidies and the need to remain competitive. In 2001, 43 % of the land area of the European Union (E.U.) was devoted to agriculture. However, in the United Kingdom, Denmark, the Netherlands, Belgium and Ireland, the area under agricultural production has fallen by 14–22 % from 1961 to 2001, although agricultural output on remaining areas has intensified, leading to an increase in total food production of more than 25 % compared with 30 years ago (Fox et al. 2005). This industrialization of agriculture, resulting in decisions by large organizations on land use and crop production at a regional scale, has transformed the agricultural landscape on the two continents. The present patterns of land use are changing rapidly, driven by economic dictates. For example, the recent enlargement of the E.U. will likely result in increased potato production in Poland, at the expense of production in western Europe, where arable land and grasslands are being converted for maize production (Van Eerden et al. 2005). In Louisiana, increased maize production also is replacing traditional rice cultivation (Abraham et al. 2005). Changes have taken place in the nutritional quality of a crop, in addition to the planting of alternative crops. The recent directive from the E.U. limiting the use of nitrogenous fertilizers means that the nutritional quality of permanent grasslands for herbivorous birds has started to decline after decades of

Ecological Studies, Vol. 190
J.T.A. Verhoeven, B. Beltman, R. Bobbink, and D.F. Whigham (Eds.)
Wetlands and Natural Resource Management

enhancement (Van Eerden et al. 2005). Similarly, in the United States, a reduction in waste maize, increasing areas of soybean production and a switch to genetically modified crops are leading to competition for resources between waterfowl species (Krapu et al. 2004). Collectively, these on-going changes indicate a very fluid and dynamic scenario, where the dictates of agricultural policy and economics drive land-use changes, many of which are unforeseen but which have important implications for wildlife management.

It is well recognised that individuals of different species of geese show a range of responses to these changes in land use that not only affect local populations of the species, but also trophic interactions and ecosystem functioning. In the case of migratory species, the outcomes are not limited just to agricultural landscapes, as alternative habitats where the populations reside and often breed are as much affected, because of migratory connectivity. Populations may be very opportunistic and individuals show considerable capacity for behavioural adjustment to environmental change, as they exploit the new conditions that can lead to an increase in their numbers. In other cases, populations of some species have declined in numbers where their preferred food supply has disappeared as a result of land-use changes, including drainage of wetlands and enrichment of inshore marine habitats. These changes in the numbers of migratory species, in turn, may be expected to affect trophic relations in the alternative habitats. The differences in the nature of the response to environmental change represent a continuum of responses from species which can be broadly classed as specialist to those which are generalist.

Arctic-breeding geese display many of the responses outlined above. They mostly nest in Arctic coastal freshwater wetlands, or in salt marshes and migrate south in the autumn along established flyways to their respective wintering grounds (Madsen et al. 1999). Traditionally, the salt marshes of western Europe and the United States have been the main wintering areas for these different species and are still widely used. Many temperate and sub-arctic salt marshes in North America harbour natural vegetation, whereas in Europe the marshes are grazed by livestock and hence the vegetation is regarded as semi-natural. However, salt-marsh vegetation is a vanishing resource. In the Netherlands, the time spent feeding on agricultural land by barnacle geese (*Branta leucopsis*), staging brent (brant) geese (*Branta bernicla bernicla*) and bewick's swan *(Cygnus bewickii)* is 80%, >33% and 50% respectively (Van Eerden et al. 2005); these and other species are increasingly exploiting agricultural crops as a resource. On both continents, this is particularly true of those species of geese that winter in terrestrial environments. In contrast, North American populations of the brant goose that winter in inshore marine habitats have failed to exploit agricultural land as a major food resource in winter. These birds are less catholic in their feeding preferences and the restricted diet of brant geese largely consists of *Zostera* spp and marine algae. As such, they represent one end of the continuum of responses shown by Arctic-breeding geese on their wintering grounds.

In this short overview of the impact of modern agriculture on geese that breed in the Arctic or sub-Arctic and winter in more southerly latitudes, we examine the responses of the different species of geese to agricultural change on the wintering grounds, flyways and breeding grounds in northern latitudes. We adopt a comparative approach in order to determine similarities and dissimilarities in the responses between the different species to direct and indirect effects of modern agriculture. The effects on northern ecosystems of the substantial increase in goose numbers, which has likely resulted from the availability of agricultural foods, are also evaluated. Finally, we assess whether the behavioural responses of geese to agricultural change will continue to lead to an increase in numbers, because of attempts to limit population sizes by wildlife management agencies. We have devised a flow chart to visualise the interactions we believe to be critical, linking events in the temperate zone so heavily influenced by man to their repercussions in the Arctic (Fig. 13.1). Note that the impact of climate change has hardly been mentioned so far, despite its obvious implications for goose populations.

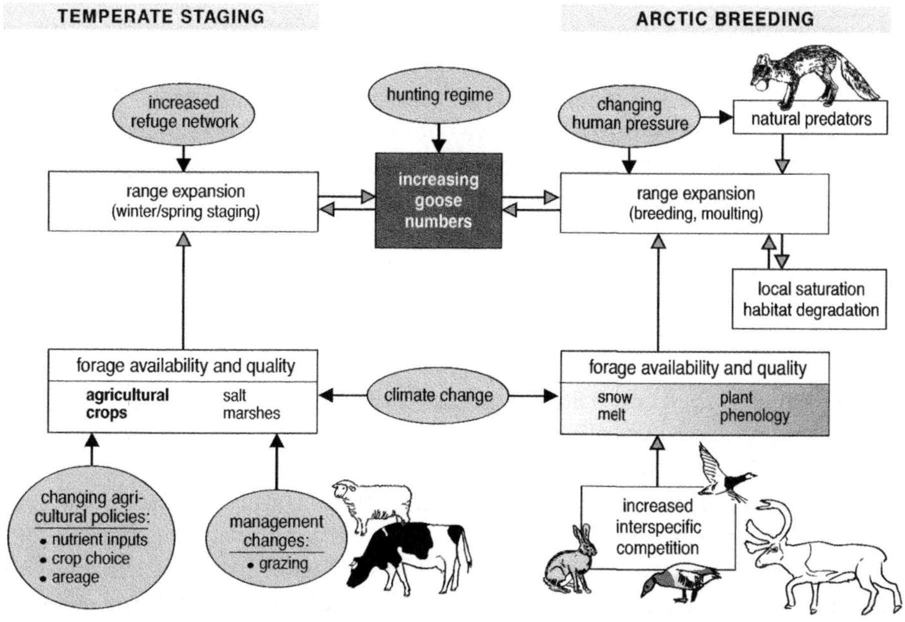

Fig. 13.1 Interaction matrix of factors that influences the current increases in goose numbers in Europe and North America. Temperate zone areas and breeding stations in the Arctic are shown on the left and right respectively. *Ovals* indicate factors directly or indirectly under human influence. Note that "the forage box" in the temperate zone includes semi-natural habitats, but these are no longer much used by the increasing numbers of geese of the different species

13.2 Links Between Modern Agriculture as a Food Source and the Increase in the Size of Arctic Goose Populations

The rapid expansion in the size of terrestrial goose populations in recent decades appears to be resource-based. Many bird species have a long-standing association with agriculture. In the Netherlands, for example, coastal wetlands have been modified since Roman times and waterfowl have adjusted to the ever-changing landscape. However, the substantial increases in goose numbers in recent decades is not just a consequence of land-use change and increased crop production but is linked to the ready availability of highly nutritious agricultural products. It has allowed the birds dependable access to protein and energy-rich foods on both continents. One of the most compelling sources of evidence of this link between geese and the quality of agricultural products are the studies by Van Eerden and his colleagues (see references in Van Eerden et al. 2005). The steady improvement in the food quality of grasslands since 1920 until recently has led to long-term shifts in the abundance of Anatidae (ducks, geese, swans) feeding in permanent grasslands in the Netherlands in winter. The use of improved varieties of grasses and heavier applications of nitrogen fertilizers have increased the crude protein content of the forage grasses from 15 % to 35 % on a dry weight basis over the period (Van Eerden et al. 1996). The timing of the initial use of agricultural crops by the different bird species was not the same and is related to the body mass of the birds. The larger species have a lower basal metabolic rate (W g^{-1} of body tissue) and hence are less demanding with repect to food quality than smaller species (Mattocks 1971). The larger species (e.g. mute swan, *Cygnus olor*) started feeding on these permanent grasslands decades before the steadily improving quality of the pastures enabled the smaller species (e.g. wigeon, *Anas penelope*) to take advantage of the new food source (Van Eerden et al. 1996).

Similarly in the United States, although the total area of maize production has not increased substantially since the 1950s, a sharp rise in yield took place between 1970 and 1975, associated with the planting of high-yielding crop varieties and the increased use of fertilizer. Winter counts of lesser snow geese (*Chen caerulescens caerulescens*) (Fig. 13.2) were relatively steady from 1965 to 1975 and averaged between 600 000 to 800 000 birds, but subsequently numbers increased to 2×10^6 by 1990. The coefficient of determination of numbers of lesser snow geese with corn yield from 1975 to 1990 gave an r^2 value of 0.89 (Jefferies et al. 2004). The corresponding values for wheat and rice are 0.90 and 0.91, respectively. In the region of Lac-St-Pierre on the north shore of the St. Lawrence River, west of Québec City, there was a dramatic increase in corn production from 3 % in 1966 to 49 % in 2001 of the area of cultivated land (Gauthier et al. 2005). Greater snow geese (*Chen caerulescens atlantica*) arrived earlier (approximate dates: 1971–1975, 4 April; 1990–1998,

Fig. 13.2 A group of blue and white lesser snow geese (*Chen caerulescens caerulescens* A.O.U.) indicating the different color morphs found in this species. Photo Bruce Batt, Ducks Unlimited

25 March) on their northward spring migration co-incident with this change. The birds altered their migration route so that they moved up the Hudson/Richelieu valley in New York State instead of flying along the Atlantic coastline to reach Lac-St-Pierre, where at least 51 % of the population feed on spilled grain in the fields of maize stubble (Gauthier et al. 2005). They move to the traditional staging sites in the salt marshes of the estuary of the St. Lawrence River at a later date. The net outcome of this change is that, in recent years, the birds leave for the Arctic with 9–12 % more body fat compared with that in 1979–1980, much of it derived from maize, as indicated by the isotopic signature of the fat (maize has the C_4 type of photosynthesis and the δ C^{13}/C^{12} ratio is different from that of a C_3 plant; Gauthier et al. 2003).

In all of these examples, crop selection by agricultural interests and improvements in the nutritional quality of forage grasses released a bottleneck that led to birds changing their choice of preferred food as new sources of acceptable resources became available on a dependable basis. Often, the maintenance of suitable grassland pasture for waterbirds is dependent on livestock grazing, both on the wintering grounds and along the migration routes (Fox et al. 2005; Van Eerden et al. 2005). In addition, as new food sources have become available, changes in both migration route and the timing of migration have followed as the birds exploit these foods.

13.3 Hunting Practices, Availability of Refuges, Agricultural Food Supplies and the Size of Goose Populations

The rapid increase in the size of Arctic goose populations in recent decades that winter in agricultural habitats is for the most part not characterized by a steady geometric increase in numbers, but rather by phases of rapid growth and periods of stasis, or even a temporary decline in numbers. The changes in growth rates result from the impact on population growth of biotic and abiotic variables, other than the availability and quality of agricultural foods. The variables include episodic weather events, climatic change, changes in hunting practices, interspecific competition and the establishment of refuges. Each affected Arctic goose populations in different ways, depending on the individual circumstances, but the overall effect has led to geographical changes in distributions on both the wintering and summer breeding grounds and changes in the spatial and temporal patterns of migration, as indicated below.

13.3.1 Hunting Practices in Agricultural Landscapes and the Size of Goose Populations

The impact of hunting practices on the size of goose populations in the different countries is not easy to assess, particularly in Europe where some countries, such as Belgium, have a total ban on the hunting of geese while others, such as Denmark and the United Kingdom, permit the hunting of designated goose species during defined periods. This means that species can be hunted at some locations along the autumnal flyway, but not others and likewise parts of their wintering range are not free from hunting. Given this geographical mosaic of hunting regulations across Europe (Madsen et al.1999) and the propensity of goose populations to shift migration routes and the location of wintering grounds, it is difficult to tease apart the effects of hunting on population growth, compared with the influence of other factors. Until recently, these difficulties were compounded by the lack of long-term data bases of population trends of the different species and the number of birds shot in a given year in the different countries. However, in individual cases, the effect of a change in hunting practices on population growth is apparent. The population size of the pink-footed goose (*Anser brachyrhynchus*), which breeds in Svalbard and mostly winters in the Low Countries, has benefited from a total ban on shooting (The Netherlands, 1976; Belgium, 1981). Numbers increased from 12 000–20 000 in the mid-1960s to 40 000–50 000 in 2003 (Madsen et al.1999; Fox et al. 2005). The Greenland white-fronted goose (*Anser albifrons flavirostris*) is another species that has benefited from protection from winter hunting. In 1982, this population fell to an estimated 16 600 individuals, which led to its protection from winter hunting at its most important wintering site in south-eastern. Ireland. The rate of increase following protection strongly suggested that hunting had limited the size of the local population (Fox et al. 2003). The increase continued until 1999, when the estimated size of the population reached 35 600 birds, but since then numbers have fallen, probably because of a long-term decline in reproductive success unrelated to events in winter (Fox et al. 2003, 2005).

On the North American continent, the effect of fall hunting on the population growth of different species of geese also has been examined. The growth rate of the greater snow goose population was higher in the last three decades of the twentieth century than in previous decades. Although native subsistence hunting and recreational hunting took place in Canada during the 1960s and 1970s, the hunting of this species was halted in the United States. The re-opening of hunting in the United States in 1975 (Menu et al. 2002) temporarily halted population growth, but from 1980 to 1999 rapid population growth occurred, even though hunting regulations were relaxed further (i.e. increased bag limits). Hence, changes in hunting regulations cannot explain the increase in numbers (Gauthier et al. 2005). Similarly, the mid-continent population of the lesser snow goose was relatively stable in numbers from about 1950 to about 1970, but increased four-fold from 1970 to peak numbers in 1998. In spite of considerable relaxation of the hunting regulations, the proportion of the population harvested failed to rise at the same rate as the overall growth of the mid-continent population, indicating that hunting was ineffective at limiting population growth and that the increase in number of birds was not the outcome of a decline in hunting pressure (Abraham et al. 2005).

13.3.2 The Synergistic Link Between Refuges and Agriculture: Effects on Wintering and Migrating Goose Populations

The presence of refugia from hunting along migration routes and on the wintering grounds may be expected to have contributed to population increases of geese in recent decades, but the evidence is somewhat equivocal. Most of the extensive network of United States National and State Wildlife Refuges used by the mid-continent population of lesser snow geese, for example, were established between 1930 and 1970 and are thought to have contributed to the large increase in numbers of this population since 1970. The refuges were created in order to protect and restore wetlands for breeding and migrating waterfowl (Bellrose 1980). Once established, they were used by geese migrating from northern latitudes to Texas and Louisiana and this quickly led to a cessation of the long-distance fall migration of geese that was characteristic of much of the twentieth century (Cooch 1955; Johnsgard 1974). The refuges function as a sanctuary from hunting for the population, which feeds on resources both in the reserves and in adjacent agricultural fields. There is good evidence that, for the Canada goose, the system led to spectacular increases in population in the United States (for a review, see Owen 1980a). At some refuges (e.g. Bosque del Apache, New Mexico) goose movements between agricultural land and adjacent refuges are jointly controlled by agricultural and wildlife officers, in order to manage goose populations as efficiently as possible in relation to various interests (farming, wildlife conservation, tourism, hunting; Abraham et al. 2005). When establishing refuges, it is important to assess whether the vegetation is natural or semi-natural. In the latter case, the cessation of livestock grazing can result in a tall canopy and leaves of low forage quality (Van Eerden et al. 2005). For example, the number of spring-staging brent geese will likely be reduced by a factor of four to eight when all salt marshes in the Wadden Sea are no longer grazed (Bos 2002). The geese will probably start feeding on inland agricultural land, leading to conflicts between agricultural and conservation interests. A possible way out is financially compensating the farmers when they do not disturb foraging geese, as is practiced on some Wadden Islands (Bos 2002, Bos and Stahl 2003).

This stepping-stone model of birds frequently stopping while on migration also describes the movement northwards in spring, as the birds follow the retreating snowline in a series of stop-overs. The flexible tactics enable birds to acquire resources for reproduction in addition to migration, in contrast to birds that migrate directly to the breeding sites, where unpredictable weather conditions may lead to inadequate resources in spring, delaying breeding. The extensive network of reserves throughout the continent, many of which are in intensive agricultural areas, in effect provides multi-channel migration routes enabling birds to switch routes, depending on local circumstances. Where birds stage at only one or two sites during migration, this creates a geographic bottleneck and there is less opportunity for a shift in migration routes in the event of adverse conditions, as in the case of greater snow geese and their staging sites in the estuary of the St. Lawrence River. Total protected areas for this sub-species are about ten times smaller on the staging than on the wintering areas (Gauthier et al. 2005).

13.3.3 The Direct and Indirect Effects of Weather Patterns and Climate Change on Wintering and Migrating Goose Populations

As mentioned above, there is compelling evidence that the increase in the size of Arctic goose populations in recent decades has been driven by the dependable availability of agricultural foods on the wintering grounds and along flyways. Where populations of a species, such as the brant goose in North America and the Arctic red-breasted goose (*Branta ruficollis*) in Europe, do not feed in agricultural fields but use natural vegetation as their food source, the populations are potentially much more vulnerable to climatic events and the vagaries of plant growth. In the case of the brant goose, the heavy reliance on the sea grass *Zostera marina* and, to a lesser extent, the marine alga *Ulva* sp. means that, if the crop of these two species is diminished, a fall in brant numbers is predicted to occur. The populations of the various sub-species of brant tend to concentrate in autumn and spring at a single major staging area which is rich in *Zostera*, but sea temperatures can limit *Zostera* growth (Ward et al.

2005). Although massive die-offs of *Zostera* do not occur on the Pacific coast, unlike the north Atlantic coasts, the changes in levels of irradiance and sea-water temperature, probably linked to ENSO, can lead to a decline in the spatial extent and abundance of *Z. marina*, particularly along the coast of Baja California, Mexico. Wintering brant geese reduce their numbers at southern stations in Mexico associated with a decline in *Zostera*, but the population increases at more northern stations, such as along the Alaska coastline where higher temperatures reduce the period of ice cover (Ward et al. 2005). Nevertheless, the total wintering population of black brant on the Pacific coast of North America has not yet recovered to the levels of the mid-1960s, in sharp contrast to other goose populations (see Fig. 13.3). Note that the closely related dark-bellied brent wintering in northwestern Europe has responded strongly to the combined influence of protection and the exploitation of agricultural foods since the early 1970s.

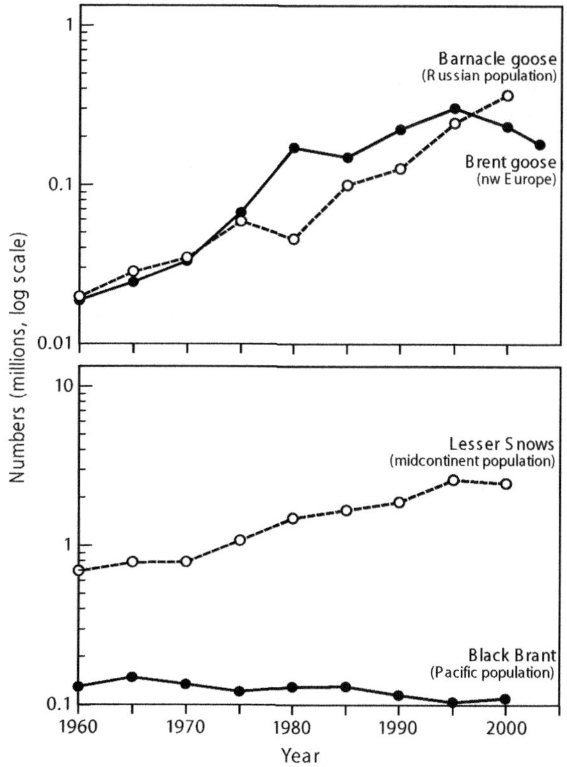

Fig. 13.3 Contrasting trajectories of strongly increasing goose populations. *Top* Barnacle geese of the Russian population counted in northwestern Europe in winter are compared with dark-bellied Brent wintering in northwestern Europe during recent decades. *Bottom* The midcontinent population of lesser snow geese [with low numbers (< 5 %) of Ross's geese; from Abraham et al. 2005] are compared with a goose population showing little change over the same period (total black brant population for the west coast of North America from Ward et al. 2005).
Note that the dark-bellied brent (closely related to black brant) wintering in northwestern Europe initially strongly increased over this period, following the habitat shift onto agricultural crops, a move the black brant has yet to make. Note the logarithmic scale of goose numbers which renders rates of change directly comparable

Similar links between the North Atlantic Oscillation Index and Atlantic brant geese wintering north of core wintering areas in New Jersey since 1960 are absent. During the severe winters of 1976–1977 and 1977–1978, when sea ice developed in coastal areas off the states of New Jersey and New York, the Atlantic brant were unable to feed on *Zostera* or *Ulva*. For the first time on the North American continent, the birds were recorded feeding on cultivated grasses and clover on golf courses. However, unlike the populations of European brent geese, the North American brant geese did not switch to agricultural foods. In addition to climatic events affecting the growth of *Zostera* species, other causes can lead to marked declines in the availability of this species. In 1931–1932, the pathogenic slime mould, *Labyrinthula zosterae*, caused extensive die-off of *Z. marina* in the north Atlantic. This led to long-term shifts in brant/brent distributions and migration patterns on both sides of the Atlantic; and the overall population declined 80–90 %. One consequence was that the birds started feeding on *Ulva* and in Europe there was increasing consumption of salt-marsh vegetation, followed by use of agricultural crops. On the east coast of North America, cultural eutrophication also led to a decline in *Zostera* and an increase in *Ulva*. These changes in the abundance of a key resource (*Zostera*) brought about by direct and indirect biotic and abiotic effects indicate the vulnerability of brant populations to sudden shifts in the availability of natural vegetation.

Changes in crop choice by geese in response to weather patterns can influence their northern range in winter. Before the mid-1980s, pink-footed geese in Europe stayed south of areas with sub-zero air temperatures. They fed in permanent grassland and moved north or south depending on the prevailing temperature. More recently, this pattern changed, as the birds now feed on winter cereals which are more widely grown. The leaves of winter wheat maintain a high nutrient content at sub-zero temperatures, unlike grass leaves, and the birds are able to balance their energy budget by feeding on this crop at temperatures below the freezing point (Fox et al. 2005). Similarily, winter feeding on *Eriophorum angustifolium* and *Rhynchospora alba* by Greenland white-fronted geese in oceanic blanket mires and raised bogs in north-western Europe depends upon a mean January temperature higher than 3 °C and the absence of ground frost, enabling the geese to consume the lower stem base of these plant species. Further north, the peatlands are frozen and, further south, high temperatures and low rainfall mitigate against peatland formation (Fox et al. 2005). Hence, there is a narrow geographic window that allows the birds to feed and roost in this type of habitat at this time of year (Fox et al. 2005).

All Arctic-breeding species, at least in Europe, stage during the spring migration at sites where the air temperature is between 3 °C and 6 °C, at which temperatures the onset of the spring growth of grasses occurs. The time spent at staging sites by barnacle and brent geese corresponds to the period when the birds are able to achieve a positive protein accumulation feeding on the spring growth of grasses, consistent with the green-wave hypothesis (c.f. Owen 1980b). The onset of spring growth is marked by a daily mean cumulative temperature above 0 °C of 180 °C (Van Eerden et al. 2005). In the event of increased global warming, it might be expected that the northerly spring migration will occur earlier. However, the presence of the boreal forest with its near absence of grass and croplands limits the ready accessibility of forage plants that respond to the warmer temperatures early in the season. It is significant that, although greater snow geese arrive earlier at their staging sites in Québec, the date of departure to the Arctic has not changed (Gauthier et al. 2005). All of these examples indicate the subtle interplay between the direct and indirect effects of climate on plant growth and the foraging responses of geese that result in winter-range extensions and changes in the migration routes of birds (Van der Graaf 2006).

13.4 Habitat Changes in Response to Population Growth of Geese

13.4.1 Effects of the Geese on Temperate Salt-Marsh Vegetation

As early as 1947, Lynch et al. described the grubbing of below-ground organs of salt-marsh plants by lesser snow geese (and muskrat) in the coastal marshes of Texas and Louisiana, where the birds winter. The grubbing by certain goose species is very much a function of their bill size and shape. Severe grubbing in the Gulf States leads to the development of open-water mudflats, termed "eatouts" where the birds roost. The damage to vegetation by the geese can be greatly accelerated by livestock trampling in these Gulf marshes (Bateman et al. 1988). Similar eatouts occur in the Atlantic marshes where greater snow geese feed on rhizomes of *Spartina alterniflora* (Smith and Odum 1981). However, these eatouts are a small percentage of the total area of marshes on the Gulf coast and the Atlantic seaboard and they have not increased in area in recent decades, largely due to the availability of alternative sources of food on farmland (Gauthier et al. 2005). In the estuary of the St. Lawrence River, Québec, greater snow geese also grub for the rhizomes of *Scirpus pungens* in brackish marshes at staging sites (Giroux and Bédard 1987; Giroux et al. 1998). In the heavily used areas on the south shore of the estuary, the above-ground *Scirpus* biomass was lower than that in the lightly used areas, but full recovery occurred within three years if areas were exclosed. In contrast on the north shore at Cap Tourmente, there has been a 47 % decrease in stem density of *Scirpus* since 1971, although the total use of the reserve has declined over this period (Gauthier et al. 2005). In the Dutch Wadden Sea, greylag geese (*Anser anser*) grub stands of *Spartina anglica* and *Scirpus maritimus* (Bakker et al. 1999). This species has been a resident in the Netherlands since 1965. The birds are not colonial, but up to 60 000 birds collect at a moulting site in the Oostvaarderplassen, where they can severely damage the reed beds of *Phragmites australis* by their foraging activities (Van Eerden et al. 2005). Spring-staging barnacle geese in the Dollard estuary in the north-east Netherlands damage swards of *Puccinellia maritima* that are replaced by secondary pioneer vegetation. However, this occurs only in the presence of livestock grazing (Esselink 2000).

On the Dutch island of Schiermonnikoog, barnacle geese and brent geese stage during winter and spring. During spring, the biomass of the vegetation on the salt marsh is increasing despite the foraging of geese (Van der Graaf 2006,Van der Graaf et al. 2005). The barnacle geese migrate to the island of Gotland in the Baltic for a fuelling stop on the way to the Arctic. However, since the early 1980s, increasing numbers breed on Gotland (Larsson et al. 1988). The biomass of the vegetation on the salt marsh on this island does not increase, in contrast to that on Schiermonnikoog. This may be due to the tenfold higher grazing pressure on Gotland (Van der Graaf et al. personal communication). On Schiermonnikoog, the salt-marsh grass, *Festuca rubra,* represents a common resource for barnacle and brent geese and brown hares (*Lepus europaeus*) in spring, when it forms more than 50 % of the diet of all three herbivores (Van der Wal et al. 1998). Although hares and geese forage for this grass competitively, the mutual relationships between barnacle and brent geese are more complex, as brent geese in spring prefer to forage on vegetation previously grazed by barnacle geese (Stahl 2001). Ydenberg and Prins (1981) showed that the nitrogen content of leaf tips in grazed plots was consistently higher than that in the tips of leaves from ungrazed plots. Further experiments at this colony using captive geese (Bazely, McCleery and Prins, unpublished data) demonstrated that grazing by barnacle geese can lead to a higher cumulative net above-ground production of *Festuca rubra* than that in ungrazed swards. Collectively, these studies strongly suggest that brent geese profit from increased production of new tissue with a high nutrient load induced by barnacle geese grazing the swards earlier in the spring (Van der Graaf 2006). This type of facilitation maintains the protein content of the forage at a higher level than might be expected in late spring after the barnacle geese have started their migration northwards, but before the brent geese have left the island. However, there is little compelling evidence from studies in temperate regions that the various foraging activities of different species of Arctic breeding geese result in permanent long-term changes in plant communities, either in agricultural ecosystems or in (semi-)natural wetlands. Short-term changes in the abundances of different species may be expected, but the vegetation subsequently recovers. This is a reflection that, although local damage may occur, brought about by high densities of geese, the birds are transient and move elsewhere, allowing the vegetation to recover. It is interesting to speculate whether the traditional salt-marsh habitats and freshwater wetlands used by large aggregations of geese in winter would historically have remained intact, if goose numbers in the past were as high as they now are for the different species. The present-day agricultural landscapes alleviate the foraging pressure of these birds on (semi-)natural landscapes. Within the Arctic, of course, agricultural landscapes are effectively absent and this absence of an alternative source of forage results in considerable loss of natural vegetation where high densities of geese occur during the breeding season, as discussed below. This further suggests that goose numbers in the past, before the advent of modern agriculture and before records were kept, were lower than they are at present, simply because of the limited carrying capacity of natural wetlands and grasslands in winter and in summer.

Fig. 13.4 Salt marsh on the Arctic coast of Russia, nesting ground of the barnacle goose (*Branta leucopsis*) here foraging during the pre-egg stage along the melting snow edge on 8 June. There is no indication that this small-billed goose species causes environmental degradation in this habitat despite intensive utilization of this marsh during one hundred days every year over the past decade (the mean laying date that season was 12 June). Photo Götz Eichhorn, University of Groningen

Fig. 13.5 Barnacle goose (*Branta leucopsis*) female at the nest of the colony area depicted in Fig. 13.4. Photo Götz Eichhorn, University of Groningen

13.4.2 Effects of Geese on Arctic Coastal Vegetation

The adverse effects of geese on coastal vegetation in the Arctic depend on the densities of birds at the various sites, as well as their foraging behaviour (grazing, grubbing, shoot-pulling of sedges) which is related to bill size and shape. Coastal sites in the southern Arctic are potentially more vulnerable to disturbance, because they serve as both breeding and staging sites for the different sub-populations of geese, some of which breed locally while others breed at higher latitudes. If wintery conditions persist further north in spring, it may delay the departure of birds from these southerly staging sites, adding to the demand for resources. Under extreme circumstances, the birds that normally nest at high latitudes may nest at these more southerly locations, swelling the numbers of the local populations. The birds need to replenish protein and fat for body maintenance and reproduction, following expenditure of energy during migration. Birds that fly long distances over open water on the spring migration particularly need to regain lost resources. Greater snow geese fly approximately 2900 km over the boreal forest from the St. Lawrence River staging sites in Québec to Bylot Island, off the coast of Baffin Island, Nunavut, hence their need for "topping-up" on arrival at the breeding grounds. The daily food intake of the female greater snow geese on the breeding grounds is high: the birds do not have sufficient fat reserves at the onset of incubation to sustain a complete fast during this period (Boismenu et al. 1992). Similar considerations apply to Arctic geese in northern Europe that migrate over long distances to breeding sites, such as the sub-population of barnacle geese which breeds on Spitzbergen (Drent et al. 2003). In contrast, the Hudson Bay populations of the lesser snow geese migrate over 1200 km to the southern shores of the Bay, but there is still a considerable amount of feeding by birds, either in coastal wetlands in the final stages of migration, or on the breeding grounds. The intensity of the feeding partly depends on local weather conditions in early spring (Jefferies et al. 2003). The combination of high densities of birds and their established use of particular wetlands with thawed ground, where the birds feed intensively at a time when there is little or no above-ground growth of vegetation, is a recipe for habitat degradation (Fig. 13.6).

Only the mid-continent population of the lesser snow goose that breeds in the eastern Canadian Arctic, particularly in the Hudson Bay Lowland, southern Baffin Island and Queen Maud Gulf in the central Arctic, has a noticeable effect on coastal wetland plant assemblages and soils at a large spatial scale (Jefferies et al. 2003), which can be detected with the use of LANDSAT imagery (Jano et al. 1998, Jefferies et al. 2006). The current population size is about 7?10^6 birds and they nest in colonies where the density can be in excess of 2000 nests km^{-2}, so considerable impact on the vegetation may be anticipated at these densities. The sequence of change resulting in vegetation loss in intertidal marshes and the establishment of an alternate stable state is described elsewhere in detail (for a review and a list of key references, see Jefferies et al. 2003). The loss is triggered by snow geese that grub for roots and rhizomes of summer forage graminoids in thawed ground during early spring (Fig. 13.6). However, it is the subsequent abiotic changes, including the development of hypersalinity, loss of organic matter and compaction of sediments, that limit the potential for recolonization of exposed soils. The limitation is compounded by biotic factors such as the loss of the soil seed bank, the asexual nature of the dominant salt-marsh grass *Puccinellia phryganodes* (sterile triploid) and the infrequent flowering of a sedge, *Carex subspathacea*. This spread of destruction leading to the alternate stable state is characterized by an ascending spatial scale. Intertidal swards initially undergo increasing fragmentation at spatial scales of 1 m^2 or less. Because of the cumulative impact of grubbing, the patches of exposed sediment coalesce into larger units, marked by adverse soil conditions and the inability of vegetative units of these graminoid species to re-establish. Over time, the vegetation mosaic is lost, as exposed sediments develop with further coalescing. Re-establishment of vegetation in hypersaline compacted sediments of poor nutrient status is long-term, especially given the reproductive biology of the salt-marsh graminoids and the ever-presence of geese that re-set the successional clock. Colonization occurs more rapidly in unconsolidated soft sediments that

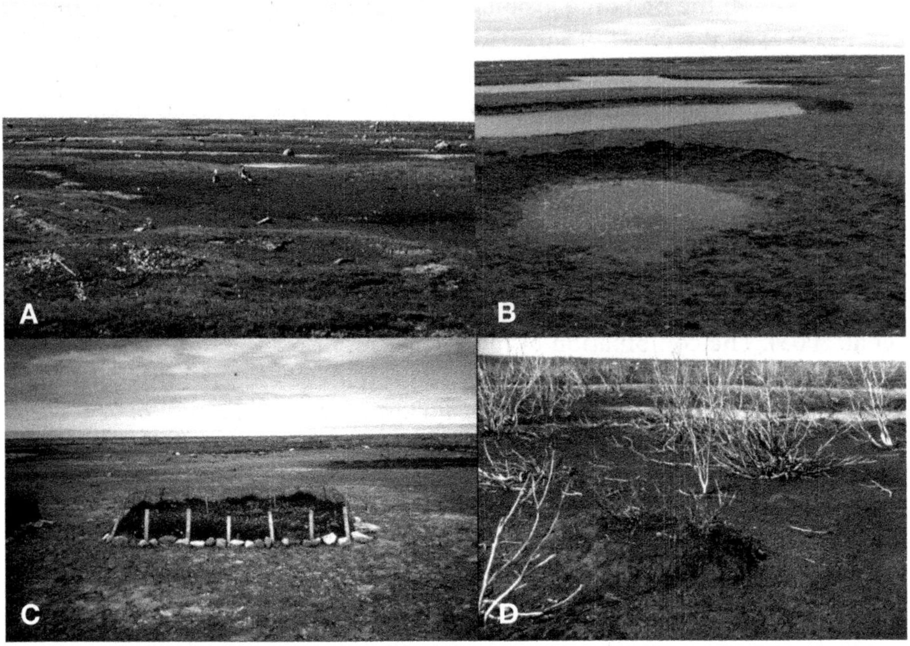

Fig. 13.6 A The salt marsh at La Pérouse Bay, Manitoba in 1977 before the period of heavy grubbing. Note a pair of snow geese (*Chen caerulescens caerulescens* A.O.U.) in the foreground. Photo Ken Abraham, Ontario Ministry of Natural Resources. **B** The effects of grubbing by lesser snow geese (*Chen caerulescens caerulescens* A.O.U.) in early spring on the intertidal salt marsh at La Pérouse Bay, Manitoba. Grubbing takes place before above-ground growth of vegetation and after the surface layer of soil has thawed. Photo Hudson Bay Project Team. **C** The effects of grubbing of the vegetation in the intertidal marsh at La Pérouse Bay in 1991. Loss of vegetation has occurred exposing sediment over most of the marsh. The exclosure was established in 1991 by Dawn Bazely and Robert Jefferies. It indicates that in the absence of grubbing the vegetation remains intact. Photo Hudson Bay Project Team. **D** The effects of grubbing in the supratidal marsh at La Pérouse Bay leads to hypersaline conditions and the death of willow bushes (*Salix brachycarpa*) and exposure of the surface organic layer. Photo Hudson Bay Project Team

result from the deposition of eroded consolidated sediment. However, while high densities of geese continue to forage in these coastal marshes, long-term establishment in soft sediments is highly precarious. Effectively, there is loss of an ecological sere. A similar loss of vegetation occurs in the fresh-water marshes, although the biotic and abiotic processes are different (Jefferies et al. 2003). Erosion of peat following loss of vegetation can lead to exposure of glacial gravels and marine clays and can alter the trajectory of succession (Handa et al. 2002).

At other sites within the Arctic beyond the coastal lowlands of Hudson Bay where lesser snow geese and Ross's geese (*Chen rossii*) nest, such as at Karrak

Lake, Nunavut, south of Queen Maud Gulf in the central Canadian Arctic, similar adverse effects of high densities of nesting geese are evident in freshwater marshes (Didiuk et al. 2001). However, elsewhere in the Arctic, there is little evidence of sustained loss of vegetation from the foraging activities of other goose species (Fig. 13.4 and Fig. 13.5). Incipient damage to, and even loss of, vegetation can occur at small spatial scales (<50 cm), or there may be shifts in the relative abundance of species, but the swards recover quickly. In the future, if the different populations continue to increase or if birds are less dispersed within the breeding range than at present, then there is the possibility that loss of vegetation could be more long-term than at present, particularly in the case of those species which grub vegetation.

One of the consequences of increases in the size of goose populations and range expansions is the extent to which inter-specific interactions are occurring, such that a subordinate species may lose access to primary feeding sites because of competition from an invader (Fox et al. 2005). The long-term decline in Greenland white-fronted geese caused by the replacement rate falling below 1.0 may be a consequence of the spread of Canada geese (*Branta canadensis interior*) from Québec into west Greenland that has been taking place since the 1980s. The Canada geese, which winter in the eastern United States on agricultural and recreational land, have displaced the whitefront, the endemic species, from territories they once occupied. Ironically, both species profited from modern agriculture but, as a consequence of range extensions, this led to inter-specific competition and a decline in the reproductive success of the Greenland white-fronted goose (Fox et al. 2005). In a similar way, the smaller brent goose may be outcompeted by the bigger barnacle goose along the Dutch (or entire Wadden Sea) coast, with negative repercussions on subsequent breeding.

13.5 Anthropogenic Constraints on Population Growth

Although agricultural subsidies act to sustain population growth by providing dependable resources for migrating and wintering geese, there are other anthropogenic effects that act to restrain population growth. Many of these effects are recent and the long-term consequences of the effects on the populations are uncertain. A very important requirement is the need for careful monitoring of the survival and reproductive success of populations at all stages of the annual cycle, in order to assess the effects. Regrettably, data of this type in the Arctic are often unavailable, or else are based on intensive studies at one site.

Within the North American continent, a large-scale biomanipulation experiment is underway in an attempt to control the numbers of greater and lesser snow geese that have profited from the use of agricultural resources. A spring hunt of lesser snow geese was introduced by the United States Fish and Wildlife Service and the Canadian Wildlife Service in 1999 and 2001, respectively, in an attempt to reduce numbers and allow vegetation in northern habitats to recover from the effects of intense foraging. A similar hunt of greater snow geese was introduced in Québec in 1999 to limit vegetation damage in wintering and staging habitats. The objective is to maintain the population growth of the geese below the replacement rate of 1.0. Hence, at the continental scale, two experiments are in progress with different management objectives; and the outcomes at this stage on population numbers are unknown. In the case of the lesser snow goose, official wildlife reports show the number of birds harvested increased from 0.6×10^6 in 1998 to 1.4×10^6 in 2001; and the decline in the winter count since 1999 may be the outcome of this increase in harvest rate, but there are many uncertainties. In western Europe, although goose numbers have risen in some populations associated with their foraging activities in agricultural landscapes, hunting is either banned or is severely restricted in the different countries.

Another effect of the spring hunt is the increased harassment of birds at staging sites in North America. In the case of the greater snow goose, the reproductive output of radio-collared birds declined sharply after the introduction of the spring hunt (Mainguy et al. 2002). Harassment may limit the ability of the birds to acquire fat and protein at staging sites on agricultural land (Bélanger and Bédard 1989; Mosbech and Glahder 1991; Madsen and Fox 1997; Madsen 1998). Adult female geese that were shot on arrival on the breeding grounds contained less abdominal fat than birds before the introduction of the spring hunt (Mainguy et al. 2002). In the case of the greater snow goose, the spring hunt led to a decline in the population to a number below the target level, hence the remedial action was successful.

Similar to the greater snow goose, pink-footed geese in Europe stage at well established sites of limited area that represent a geographical bottle-neck where the high densities of birds are vulnerable to disturbance. Madsen (2001) and Drent et al. (2003) reported that the harassment of pink-footed geese by farmers on the island of Vesterålen off the Norwegian coast limits the accumulation of abdominal fat by birds and leads to a low reproductive output. Ringed birds with a low abdominal index of fat accumulation in spring return in autumn without goslings. There is also evidence that migrating barnacle geese may be unable to acquire sufficient amounts of high-quality protein when the birds feed on agricultural grasses, rather than salt-marsh grasses, at a staging site in Helgeland, Norway (Prop and Black 1998). This adversely affects their reproductive success. In non-agricultural habitats, the birds' rate of protein acquisition and their reproductive output were much higher. Although agricultural crops contain high amounts of protein, their amino acid composition may be deficient in key amino acids for goose nutrition, resulting in poor protein storage and reproductive failure (Prop and Black 1998). Significantly, in the Netherlands, barnacle geese start their spring migration when the salt-marsh vegetation on which they feed is no longer capable of providing adequate nitrogen to maintain a positive daily gain in storage protein. The heavy feeding in these marshes is coincident with the first flush of spring growth when the leaves are protein-rich, ensuring a high intake of nitrogen per bite of tissue (R. Drent, unpublished data). The timing of the feeding events and the subsequent migration is broadly coincident with the predictions of the green-wave hypothesis (Owen 1980b, Van der Graaf et al. 2006).

13.6 Conclusion

The use of intensive farming methods in recent decades, including the addition of large amounts of nitrogenous fertilizer to agricultural land, has resulted in the production of crops of high nutritional quality. In turn, this has led to changes in the foraging behaviour and dispersion of Arctic breeding geese that feed in agricultural landscapes, both on the wintering grounds and along migration routes in Europe and in North America. It is evident that highly dynamic and unpredictable patterns of movement of geese have resulted. In North America, the declining food supply for geese associated with recent developments in agriculture, is likely to create a resource bottleneck and to result in competition between different species of waterfowl for the diminishing quantities of waste agricultural foods and weeds (Krapu et al. 2004). This, in turn, may well affect adult survival and reproductive success on the Arctic breeding grounds (Drent et al. 2006). Particularly worrying are the indirect effects of agriculture, such as the recent inter-specific competition between Canada geese and Greenland white-fronted geese on the breeding grounds – a consequence of population expansion and range extension. Episodic events, such as the freeze in western Europe during the winter of 1963, resulted in Greenland white-fronted geese establishing new feeding grounds on agricultural lands that they continue to use today (E. Kuijken, personal communication). Changes in land use and the amounts of fertilizer applied to land also have resulted in abrupt shifts in the foraging behaviour of waterfowl. Two species in particular, the barnacle goose and the lesser snow goose, appear to be extremely opportunistic in their ability to exploit agricultural landscapes. Populations of barnacle geese have started breeding on Gotland and in Estonia, where they feed in coastal hay meadows and on cereals (Larsson et al. 1988; Leito 1996). These were and still are used as migratory stop-over sites in spring. In Zeeland in the Netherlands, which is a wintering area, barnacle geese started nesting in 1984 (Meininger and Van Swelm 1994) and, in 2003, there were almost 3000 pairs breeding (Van Eerden et al. 2005). In the case of the lesser snow goose, several pairs bred in the marshes of Lake Winnipeg, Manitoba, but a population never established. Extensions of the wintering range of this species are likely to produce increased mixing of the sub-populations (i.e. Mid-Continent, Central and Pacific populations). Hence, small groups of both these species are breeding in southern locations and failing to migrate north each spring; and, in the case of the lesser snow goose, the range extensions indicate that birds winter over much of the southern and central United States and northern Mexico. Effectively, both species have become invasive.

These rapidly changing scenarios stress the need for careful recording of foraging behaviour and range extensions of the different species and how these relate to population dynamics and interspecific interactions. This is no mean task, especially in northern latitudes; but the rewards are considerable, both in our understanding of ecological processes and in providing sound management advice to safeguard the integrity of the Arctic environment.

Acknowlegements Our colleagues who presented papers at the symposium in Utrecht in July, 2004, kindly allowed us access to MSS that are undergoing review at this time and which cover the presented material. We hope that we have given a fair and balanced overview of the material. We are particularly grateful to Kenneth Abraham, Mennobart van Eerden, Tony Fox, Gilles Gauthier and Austin Reed for insightful discussions during the meeting. R.L.J. gratefully acknowledges support from the Natural Sciences and Engineering Council of Canada, and R.D. from the Netherlands Arctic Programme of the Netherlands Organisation for Scientific Research NWO.

References

Abraham KF, Jefferies RL, Alisauskas RT (2005) The dynamics of landscape change and snow geese in mid-continent North America. Global Change Biol 11:841–856

Bakker L, Van der Wal R, Esselink P, Siepel A (1999) Exploitation of a new staging area in the Dutch Wadden Sea by greylag geese *Anser anser*, the importance of food-plant dynamics. Ardea 87:1–13

Bateman HA, Joanen T, Stutzenbaker CD (1988) History and status of midcontinental snow geese on their Gulf Coast winter range. In: Weller MW (ed) Waterfowl in winter. University of Minnesota, Minneapolis, pp 495–515

Bélanger L, Bédard J (1989) Responses of greater snow geese to human disturbance. J Wildl Manage 53:713–719

Bellrose FC (1980) Ducks, geese and swans of North America. Stackpole, Harrisberg

Boismenu C, Gauthier G, Larrochelle J (1992) Physiology of prolonged fasting in greater snow geese (*Chen caerulescens atlantica*). Auk 109: 511–521

Bos D (2002) Grazing in coastal grassland – Brent geese and facilitation by herbivory. PhD thesis, University of Groningen, Groningen

Bos D, Stahl J (2003) Creating new foraging opportunities for dark-bellied brent geese *Branta bernicla* and barnacle geese *Branta leucopsis* in spring, insights from a large-scale experiment. Ardea 91: 153–165

Cooch FG (1955) Observations on the autumn migration of Blue Geese. Wilson Bull 67:171–174

Didiuk AB, Alisauskas RT, Rockwell RF (2001) Interactions with arctic and subarctic habitats. In: Moser TJ (ed) The status of Ross's Geese Arctic. (Goose joint venture special publication. US Fish and Wildlife Service, Washington, and Canadian Wildlife Service, Ottawa) US Fish and Wildlife Service, Washington, D.C., pp 19–32

Drent RH, Both C, Green M, Madsen J, Piersma T (2003) Pay-offs and penalties of competing migratory schedules. Oikos 103:274–292

Drent RH, Fox AD, Stahl J (2006) Travelling to breed. J. Ornithol 147: 122–134

Esselink P (2000) Nature management of coastal salt marshes – interactions between anthropogenic influences and natural dynamics. PhD thesis, University of Groningen, Groningen

Fox AD, Glahder CM, Walsh AJ (2003) Spring migration routes and timing of Greenland white-fronted geese – results from satellite telemetry. Oikos 103: 415–425

Fox AD, Madsen J, Boyd H, Kuijken, E, Norriss DW, Tombre IM, Stroud DA (2005) Effects of agricultural change on abundance, fitness components and distribution of two Arctic-nesting goose populations. Global Change Biol 11:881–893

Gauthier G, Bety J, Hobson KA (2003) Are greater snow geese capital breeders? New evidence from a stable-isotope model. Ecology 84:3250–3265

Gauthier G, Giroux J-F, Reed A. Béchet A. Bélanger L (2005) Interactions between land use, habitat use and population increase in greater snow geese: what are the consequences for natural wetlands? Global Change Biol 11:856-869

Giroux J-F, Bédard J (1987) The effects of grazing by greater snow geese on the vegetation of tidal marshes in the St. Lawrence estuary. J Appl Ecol 24:773–788

Giroux J-F, Gauthier G, Costanzo G, Reed A (1998) The impact of geese on natural habitats. In: Batt BDJ (ed) The Greater Snow Goose: report of the Arctic goose habitat working group. (Goose joint venture special publication. US Fish and Wildlife Service, Washington, and Canadian Wildlife Service, Ottawa) US Fish and Wildlife Service, Washington, D.C., pp 32–57

Handa IT, Harmsen R, Jefferies RL (2002) Patterns of vegetation change and the recovery potential of degraded areas in a coastal marsh system of the Hudson Bay Lowland. J Ecol 90:86–99

Jano AP, Jefferies RL, Rockwell RF (1998) Foraging geese, vegetation loss and soil degradation in an Arctic salt marsh. J Ecol 78:93–99

Jefferies RL, Rockwell RF, Abraham KF (2003) The embarrasssment of riches: agricultural food subsidies, high goose numbers, and loss of Arctic wetlands – a continuing saga. Environ Rev 11:193–232

Jefferies RL, Rockwell RF, Abraham, KF (2004) Agricultural food subsidies, migratory connectivity and large-scale disturbance in Arctic coastal systems: a case study. Integr Comp Biol 44:50–59

Jefferies RL, Jano AP, Abraham KF (2006) A biotic agent promotes large-scale catastrophic change in the coastal marshes of Hudson Bay. J Ecol 94: 234-242.

Johnsgard PA (1974) Song of the north wind: a story of the snow goose. Anchor/Doublesday, New York

Krapu GL, Brandt DA, Cox RR (2004) Less waste corn, more land in soybeans, and the switch to genetically modified crops:trends with important implications for wildlife management. Wildl Soc Bull 32:127–136

Larsson K, Forslund P, Gustafsson L, Ebbinge BS (1988) from the high Arctic to the Baltic: the successful establishment of a barnacle goose population on Gotland, Sweden. Ornis Scand 19:182–189

Leito A (1996) The Barnacle Goose in Estonia. Eston Mar 1:1–103

Lynch JJ, O'Neil T, Lay DW (1947) Management significance of damage by geese and muskrats to Gulf Coast marshes. J Wildl Manage 11:50–76

Madsen J (1998) Changing trade-offs between predation risk and food intake:gaining access to feeding patches during spring-fattening in pink-footed geese *Anser brachyrhynchus*. In: Mehlum F, Black JM, Madsen J (eds) Research on arctic geese. (Proc Svalbard Goose Symp, Nor Polar Inst, Oslo) Nor Polarinst Skr 200:305–311

Madsen J (2001) Spring migration strategies in pink-footed geese *Anser brachyrhynchus* and consequences for spring fattening and fecundity. Ardea 89:43–55

Madsen J, Fox AD (1997) Impacts of hunting disturbance on waterbird populations: the concept of flyway networks of disturbance-free areas. Gibier Faune Sauvage 14:201–209

Madsen J, Cracknell G, Fox T (1999) Goose populations of the western Palearctic. (Wetlands international publication 48) National Environmental Research Institute, Copenhagen

Mainguy J, Bêty J, Gauthier G, Giroux J-F (2002) Are body condition and reproductive effort of laying snow geese affected by the spring hunt? Condor 104:156–162

Mattocks JG (1971) Goose feeding and cellulose digestion. Wildfowl 22:107–113

McLaren JR, Jefferies RL (2004) Initiation and maintenace of vegetation mosaics in an Arctic salt marsh. J Ecol 92:648–660

Meininger PL, Van Swelm ND (1994) Brandganzen *Branta leucopsis* als broedvogel in het Deltagebied. Limosa 67:1–5

Menu S, Gauthier G, Reed A (2002) Changes in survival rates and population dynamics of greater snow geese over a 30-year period: implications for hunting regulations. J Appl Ecol 39:91–102

Mosbech A, Glahder C (1991) Assessment of the impact of helicopter disturbanceon moulting pink-footed geese *Anser brachyrhynchus* and barnacle geese *Branta leucopsis*. Ardea 79:233–238

Owen M (1980a) The role of refuges in wildfowl management. In: Wright EN (ed) Bird problems in agriculture. BCPC, London, pp 144–156

Owen M (1980b) Wild geese of the world. Batsford, London

Prop J, Black JM (1998) Food intake, body reserves and reproductive success of barnacle geese *Branta leucopsis* staging in different habitats. In: Mehlum F, Black JM, Madsen J (eds) Research on arctic geese. (Proc Svalbard Goose Symp) Norwegian Polar Institute, Oslo, pp 175–193

Smith TJ III, Odum WE (1981) The effects of grazing by snow geese on coastal marshes. Ecology 62:98–106

Stahl J (2001) Limits to the co-occurrence of avian herbivores. PhD thesis, University of Groningen, Groningen

Van Eerden MR, Zijlstra M, Van Roomen M, Timmermann A (1996) The response of Anatidae to changes in agricultural practice: long-term shifts in the carrying capacity of wintering waterfowl. Gibier Faune Sauvage 13:681–706

Van Eerden MR, Drent RH, Stahl J, Bakker JP (2005) Connecting seas:western Palaearctic continental flyway for water birds in the perspective of changing land use and climate. Global Change Biol 11:894–909

Van der Graaf AJ (2006) Geese on a green wave: flexible migrants in a changing world. PhD thesis, University of Groningen, Groningen

Van der Graaf AJ, Stahl J, Bakker JP (2005) Compensatory growth of *Festuca rubra* after grazing–Can migratory herbivores increase their own harvest during staging? Funct Ecol 19: 961–969

Van der Graaf AJ, Stahl J, Klimkowska A, Bakker JP, Drent RH (2006) Surfing on a green wave–how plant growth drives spring migration in the barnacle goose. Ardea 94 (in press)

Van der Wal R, Kunst P, Drent RH (1998) Interactions between hare and brent goose in a salt marsh system: evidence for food competition? Oecologia 117:227–234

Ward DH, Reed A, Sedinger JS, Black JM, Derksen DV, Castelli PM (2005) Effects of anthropogenic and climate events on habitats and population dynamics of nearctic brant. Global Change Biol 11:869–881

Ydenburg RC, Prins HHT (1981) Spring grazing and the manipulation of food quality by barnacle geese. J Appl Ecol 18:443–453

14 Eurasian Mires of the Southern Taiga Belt: Modern Features and Response to Holocene Palaeoclimate

T. MINAYEVA, W. BLEUTEN, A. SIRIN, E.D. LAPSHINA

14.1 Introduction

Terrestrial ecosystems play an important role in the global carbon cycle; and peatlands, wetlands with a high capacity for carbon storage, are an important component of the carbon pool of the biosphere. In Northern Hemisphere subarctic zones, Northern American boreal and Euroasian taiga zones, an estimated 3.5×10^6 km^2 is covered by various types of peatlands that store 220–460 Pg carbon (1 Pg C = 10^{15} g C). This is about 60 % of the C pool in the atmosphere and 30 % of the total C store in soils of the world (Joosten and Clark 2002). Carbon storage in boreal and other northern peatlands is estimated at 114–882 Pg C (Post et al. 1982; Armentano and Menges 1986; Gorham 1991; Zoltai and Martikainen 1996). According to the State land inventory, peatlands cover over 8 % of Russia (Sirin and Minaeva 2001) and, together with shallow peat paludified lands, they could make up over 20 % of the country territory (Vompersky et al. 1996). Russian peatlands alone are estimated to contain between 113.5 Pg C (Vompersky et al. 1996) to more than 200 Pg C (Botch et al. 1995), which could make from 20 % to 50 % of the carbon stored in world peats. While peatlands cover over 3 % of the earth's land surface, they have not been fully taken into account in the global climate change models and scenarios. Most efforts to predict global responses to climate change focus on oceans and forests (Fan et al. 1998) and the role of peatlands is underestimated (Ramsar Resolution 2006). Unlike many other types of ecosystems, peatlands have the ability to store carbon for long (i.e., geologic) periods of time but their ability to store carbon is being influenced by direct and indirect human activities and by climate change.

Large-scale draining of peatlands results in their destruction and replacement by farmland, forests and other human land-use types. Unlike natural

Ecological Studies, Vol. 190
J.T.A. Verhoeven, B. Beltman, R. Bobbink, and D.F. Whigham (Eds.)
Wetlands and Natural Resource Management

peatlands that are sinks for atmospheric carbon, managed peatlands become strong carbon sources caused by oxidation of the peat and peat (forest) fires. As a result of human activities, it is estimated that carbon sequestration in peatlands no longer occurs on 500×10^3 km^2 (20 %) of the non-tropical peatlands in the world. Global mire and peat resources are currently decreasing at approximately 0.1 % per year. Peatland losses have been caused by their conversion to agriculture (50 %), forestry (30 %), peat extraction (10 %) and urbanization (5 %). Peatland losses have been largest in Europe, where some 60 % of the area that formerly was a carbon sink no longer accumulates peat. Joosten and Clarke (2002) estimate that 10–20 % of the original mire area does not even exist anymore as peatlands.

In addition to human activities, the impacts of climate change also influence the ability of peatlands to sequester carbon. The objective of this chapter is to describe peat accumulation rates of mires during Holocene climate fluctuations, to support the prediction of mire response to future climate changes. The data that we report are based on a study of Russian mires within the Southern Taiga Belt in the Northern Hemisphere. The study focuses on this region because it contains a high percentage of intact peatlands that are known to have high peat accumulation rates, which makes them substantial sinks for atmospheric carbon. Mires in this region thus have the potential to respond to climate change but they may also be vulnerable to climate warming.

We begin with a description of the main features of mires in the Southern Taiga Belt of Eurasia, followed by a description of mires, peat stratigraphy and carbon accumulation at study sites in European Russia and in Western Siberia. In the discussion, we focus on changes in carbon sequestration that might result from climate change, based on estimates of variations in carbon accumulation data during the Holocene period as determined through analyses of peat cores collected at the study sites.

14.2 Peatlands of the Southern Taiga Belt of Northern Eurasia

14.2.1 The Features of the Southern Taiga Bioclimate

The Southern Taiga Belt in Eurasia is mainly located between latitudes 56° N to 60° N. In the western part of Eurasia, this zone exceeds further north than 60° N due to oceanic influences. In European Russia, located closer to the Ural mountains, the zone narrows to between 57° N and 59° N. The Southern Taiga Belt occupies a vast area in Western Siberia, while in Central and Eastern Siberia the extent of the zone narrows due to continental climatic conditions. The zone shifts in a south-eastern direction in the area of Baikal Lake and

Table 14.1 Climate data from key study areas

Key area	West Dvina research station	Central forest reserve	Vasygan
Weather station	Zapadnaya Dvina	Lesnoy Zapovednik	Pudino, Sherstobitovo
Latitude, N	56°10′	56°30′	57°34′, 57°14′
Longitude, E	32°10′	32°60′	79°26′, 78°52′
Elevation (m a.s.l.)	190	240	120
Average annual air *t*°C	+4.1	+4.2	−1.5
T°C, January	−8.8	−9.0	−18.8
T°C, July	+17.2	+16.5	+17.0
Annual precipitation (mm)	758	730	518
Average duration of snow cover (days)	130	134	176
Period of snow cover	26/11–30/03	21/11–03/04	31/10–18/04

from there continues to the east between 52° N and 53° N along the Amur valley up to Sakhalin (Lavrenko 1947, 2000). The main feature of the vegetation in the zone is the dominance of coniferous forests and open peatlands that include a variety of mire-types, including domed raised bogs and forested peatlands (Tolmachev 1954). According to Neushtadt (1957), the Taiga Belt could be also considered also as a 'peatland' Belt – here is the maximum of peatland distribution and peat accumulation (Nikonov 1955).

Characteristic features of the Southern Taiga are a temperate-continental climate in the western part of European Russia and a continental climate in West Siberia (Table 14.1). A general climatic feature is that precipitation is higher then evapotranspiration on an annual basis, resulting in conditions that are ideal for peat accumulation. Warm summer temperatures support high annual primary production; and long and cold winters support low decomposition rates and subsequent peat accumulations.

14.2.2 Peatland Distribution and Main Types

Peatlands in Northern Eurasia have been impacted little by anthropogenic activities and thus are mostly still in a natural condition. Most of the peatlands of the Russian Federation (Fig. 14.1) are found in plains and lowlands where there are temperate-continental and oceanic climates. The numerous peatlands in the arctic zone developed during the Holocene are not expanding under current climatic conditions. In contrast, peatlands in the Southern Taiga are currently undergoing a high rate of paludification.

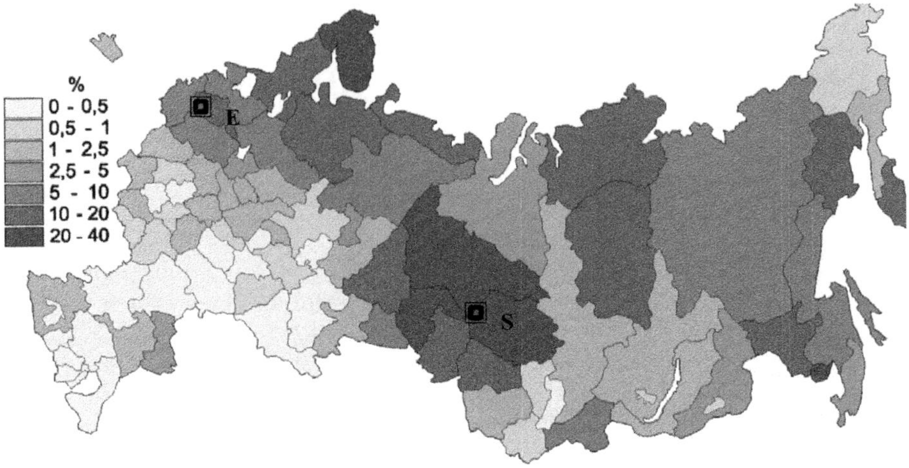

Fig. 14.1 Location of the study areas and percentage of peat area of the provinces of the Russian Federation (State Land Inventory data: Sirin and Minaeva 2001). Study sites: *E* European Russia, *S* Western Siberia

Peatlands in the Southern Taiga Belt include a variety of mire types including: domed raised bogs, sedge spring and slope fens, sedge-reed floating mats in shallow waters, valley sedge fens, forested bogs and fens – including vast areas of larch fens ('mar') in Eastern Siberia. In the Siberian part of the Southern Taiga Belt, one can even find relict arctic types of peatlands known as aapa- and palsa-mires. Raised sphagnum bogs are the most widespread type of peatland in the Southern Taiga Belt; and they account for 100 000 ha in the European part of the Belt and millions of hectares in the Siberia part. Because of their large size, Siberian mire massifs are not homogenous but are complexes of raised bogs with hummock hollow complexes within large transitional fens, patches of spring fens and both mesotrophic and eutrophic floating mats. The largest mires in the Siberian and European parts of the Belt are forested. Because many of the mires are forested, they are often mapped as terrestrial (i.e., non-wetland) vegetation types; thus the area of forested mire is an underestimate of their actual extent. The forested mires occupy various topographic positions within watersheds, including gentle slopes or valleys. They also vary in their nutrient status, ranging from ombrotrophic mires that receive only precipitation to more nutrient-rich mires and valley fens that are fed by groundwater. The area of forested mires in the European Southern Taiga (especially in the Russian plain) could account for as much as 25 % of all forests on the slopes in a watershed and 5 % of the forested area in valleys (Sirin and Minayeva 2001). In West Siberia, forested mires are presented by oligotrophic types dominatd by *Pinus sylvestris* ('ryam') and eutrophic valley fens dominated by *P. sibirica* and *Picea obovata* ('sogra').

14.2.3 Main Features of Peatland Development

The relationship between patterns of mire development and the geographical position and bioclimatic zone in West Siberia and the North-West European part of Russia have been described by Neushtadt (1957, 1985). Paludification in both regions started in the early Holocene or Boreal period (10 000–12 000 BP) according to the Blytt–Sernander scheme (see Section 14.2.4). The oldest Siberian peat contains more wood than similar-aged deposits from the European part of Russia, where peat layers are dominated by hypnum–sedge or sedge–reed. This proves that paludification in European Russia frequently started by the filling (e.g., terrestrialization) of lakes with undecomposed plant remains. In Siberia, this peat accumulation started by forest flooding which was caused by raised groundwater levels.

In the European region, trees eventually colonized the developing peatlands, while in West Siberia both open and treed communities occurred. In European cores, a peat layer with a high percentage of undecomposed wood fossils (referred to as the boundary horizon) is frequently present but it is absent from cores from Siberia. During the period of development of this boundary horizon in Europe, Siberian mires accumulated partially decomposed layers of sphagnum peat. During the last stage of Siberian mires development, a thick layer of *Sphagnum fuscum*–peat accumulated, reaching 4–5 m. The European bogs are dominated by sphagnum mosses communities in the most recent layers, which do not usually exceed 2.0–2.4 m in depth and were formed by *Sphagnum fuscum* or *S. magellanicum* and *S. angustifolium*.

14.2.4 Main Features of Climate During the Holocene

The Holocene palaeoclimate in Eurasia has been described in many publications (e.g., Khotinsky 1977; Klimanov 1984; Velichko et al. 1991). Here, we use the Blytt–Sernander scheme to describe the changes in climate that occurred during that period Holocene. The scheme was developed by Blytt (1882) and Sernander (1910) and further adopted by Khotinsky (1969), who divided the Holocene into five time-periods before present: 12 000–9800 = Pre-Boreal (PB), 9800–7700 = Boreal (B), 7700–4600 = Atlantic (AT), 4600–2700 = Sub-Boreal (SB), 2600–present = Sub-Atlantic (SA).

Khotinsky (1977) described the relationship between variations in Holocene climates and the Earth warm supply factor. The most striking climatic changes were increases in humidity and a decrease in temperature in the second half of the Holocene in the Western parts of Eurasia and a decrease of humidity with corresponding temperature rise during the Atlantic period of the Holocene in South Siberia.

Palaeoclimate reconstruction curves given for different regions within longitudinal gradient over Eurasia show deep changes in the Holocene climate (Fig. 14.2). Deviation of the mean annual palaeotemperatures demonstrates many similarities but also deep contrasts in the Holocene palaeoclimate for different sectors of the Northern Eurasia Taiga zone within the oceanic–continental gradient (Klimanov and Sirin 1997). The Holocene climate optimum, a relative warm period during the AT proved from pollen analyses of peat cores by the Northward extension of tree species growing in a warmer climate, is well observed for the European part as well as for the northern parts of Siberia. The Holocene climate optimum is not so evident for the Southern Taiga part of West Siberia and East Siberia, where a brief period of cold in the beginning of the AT followed by a warming period – also associated with a significant increase in precipitation and humidity – did not occur. The result was a period of partial drying (e.g., aridization) in the southern part of the Taiga zone during a period of global warming.

The current period of the Holocene is characterized by a shift in the globe radiation balance that has resulted in climatic warming since the beginning of the industrial age (after 1750). This recent global warming is caused by industrially induced increases in the concentrations of natural greenhouse gases (CO_2, CH_4, N_2O, O_3) and artificial products like halocarbonates and aerosols (Houghton et al. 1996). Since the late nineteenth century, the mean global temperature has increased 0.3–0.6 °C, with maximum warming in northern altitudes between 40° N and 70° N (Nicholls et al. 1996). With respect to increasing greenhouse gases and aerosols, IPCC models for the period 1990–2100 assess a rise in global mean temperature between 0.9–3.5 °C, with highest

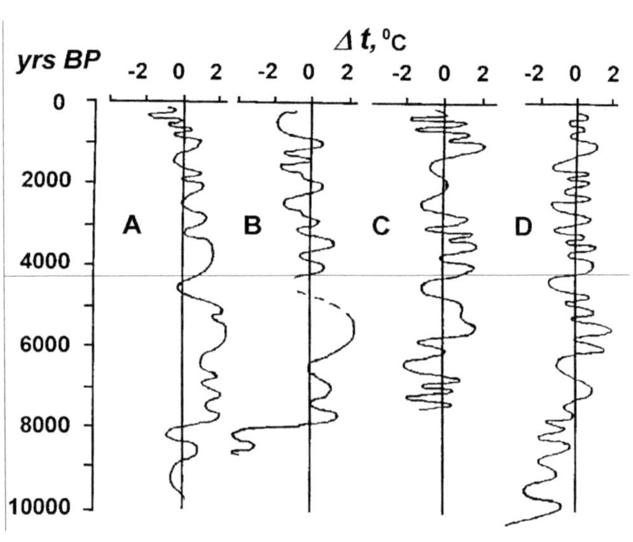

Fig. 14.2 Deviation of the mean annual palaeotemperatures during the Holocene compared with today at different regions in the Northern Eurasia Taiga zone, from west to east: *A* central European Russia, *B* Urals, *C* Western Siberia, *D* Eastern Siberia. The time-scale is based on uncalibrated ^{14}C dates. (revised after Klimanov and Sirin 1997)

warming in the mid- and high northern latitudes and an increase in winter precipitation, with a decrease in summer precipitation.

14.2.5 Peat Accumulation Dynamics

Inequality in peat stratigraphy and fluctuations in peat growth were registered and their possible connection to climate changes in the past were suggested already at the beginning of mire and peatland research (Barbier 1981). Advances in radiocarbon and other methods of peat-dating as well as the development of knowledge on palaeoclimate reconstruction provide a basis for further ideas in this field. Higher resolution of ^{14}C dating along a peat core could help to build more precise curves of peat age versus depth (Tolonen 1987; Gorham and Janssens 1992; Warner et al. 1993), but even frequent sampling with regular steps (Tolonen 1987; Belyea and Warner 1996) or a consideration of peat stratigraphy does not always fit intervals with different palaeoclimates. Not many studies are questioning the peat growth changes in different periods of the Holocene (Elina 1991; Franzén 1992; Elina et al.1995; Glebov et al. 1996, 1997; Vasiliev et al. 2001).

Variations in peat accumulation rates for peatlands in Russia have been described by several authors. using the example of 28 mires located between 50° N and 71° N and 25° E and 136° E, significant spatial and temporal differences in peat accumulation rates were demonstrated for the entire Holocene and within the Sub-Atlantic period in particular. It was shown that many Eurasian mires grew faster in cold periods and less in warmer periods during the past three millennia (Klimanov and Sirin 1997). Vompersky et al. (2000) demonstrated differences in the peat accumulation of the Sub-Atlantic period from the average Holocene peat accumulation for more than 150 cores from different parts of Eurasia, from Ukraine to Kamchatka. Vasiliev (2000) presented data on the decrease in peat accumulation during the Holocene climatic optimum (6000–4500 years BP) in Southern Taiga zone in Siberia compared to its increase during the same period in Europe or in northern territories of Siberia. He explains that fact by a more dry climate in the southern part of Siberia during the Holocene climate optimum. Even if the mire vegetation did not change in response to short-term (century level) climatic changes, there were differences in rates of peat accumulation (Rauber 2002).

14.3 Mire Development and Peat Accumulation Dynamics in the Key Areas During the Holocene

14.3.1 Study Sites in European Russia

The two study areas in the European part of Russia (West Dvina Institute of Forest Science Russian Academy of Science Research Station and Central Forest Biosphere Nature Reserve) have similar climates but differ in geomorphic settings. The West Dvina Station (56°10' N, 32°10' E; 25 km^2) is situated in the interfluve between the Zapadnaya Dvina (Daugava) and Velesa rivers on the northern edge of the West Dvina lowland. Elevations of the study area vary between 170 m and 190 m a.s.l., with moraine ridges up to 250 m a.s.l. The Central Forest Biosphere Nature Reserve (56°30' N, 32°60' E; 210 km^2) is situated 80 km north of the West Dvina Station on the southern edge of the Valday Hills. Elevations range from 220 m to 300 m a.s.l. The upper 80–100 m bedrock consists of limestone, dolomite, marl, sandstone and layers of blue clay of Devonian age. The Quaternary deposits covering the Devon layers at the West Dvina Station include moraine tills and fluvioglacial sands of the Valday (Würm) glaciation, resulting in a mosaic of landscape units, including valleys of different size between alternating sandy dunes and almost flat loamy morainic upland. The Central Forest Reserve is located in an area called the 'limestone terrace' that was not covered by Valday glaciation. Quaternary deposits in the study area were formed by uniform loamy material of Moscow (Riss) moraine deposits, in some cases covered by fluvioglacial deposits of Valday moraine.

The moderate-continental climate here is influenced by the Atlantic ocean: the long-term average annual precipitation is 730–750 mm and the average temperature is up to 4.1–4.3 °C (Table 14.1).

The vegetation of both study sites is typical of the Southern Taiga zone throughout Europe and Asia. Coniferous forests, dominated by *Picea abies* (L.) Karst., are widespread, but on poor sandy soils in the dune landscapes and in places with high groundwater levels they are replaced by forests dominated by *Pinus sylvestris* L. At both study sites, primary deciduous trees [*Betula pendula* var. *pubescens* and *Alnus glutinosa* (L.) Gaertn.] occur in forests in river valleys. Deciduous forests dominated by *B. pendula* var. *pubescens*, *Populus tremula* and *A. incana* replace spruce stands after windthrows, fires or cuttings. All depressions, upper elevations of watersheds, valleys of all sizes and natural stream beds are covered with peat deposits. The main mire types are raised bogs, which are formed either in loamy depressions (former lakes) or on the sandy deposits in positions with a high groundwater level. Concentric raised bog vegetations with hollow–hummock complexes and dwarf pine forests are developed on weak slopes and water-divides. Slopes of raised bogs are usually occupied by pine stands dominated by *Pinus sylvestris* f. *uliginosa* with ground-layer vegetation varying, depending on the slope and hydrology. In natural conditions, the edges of raised bogs (called 'lag') have transitional vegetation types between bog and fen. The vegetation of lags depends on the hydrology and minerotrophic conditions and varies from *Carex* and *Phragmites* domination to forests of black alder or pine and spruce forests that form by paludification.

Forested peatlands on higher elevations and on slopes are dominated by mixed stands of spruce, birch and pine, with *Carex* spp, *Eriophorum* spp and dwarf shrubs (*Chamaedaphne calyculata, Andromeda polifolia, Ledum palustre, Vaccinium uliginosum, V. myrtillus, V. vitis-idaea*), or sometimes more mesotrophic vegetation (*Carex rostrata, Calamagrostis canescens, Menyanthes trifoliata*). In the valley peatlands, the vegetation is typically dominated by *Alnus glutinosa*. In similar areas, spruce carrs occur but are rare. In former pastures and artificially maintained meadows, herbaceous species dominate. In the valleys, the springs and mesotrophic lakes have *Phragmites–Sphagnum* or *Carex–Sphagnum* transitional fens.

The Usvyatsky Mokh study site at the West Dvina Research station is situated in a glacial depression on fluvio-glacial sands. Vegetation is a typical raised bog, with vegetation zonation as described in the previous section. Today, the peat covers 1.7 km^2 with a maximum thickness of 5 m in the center. The peat structure shows that the site was originally dominated by a fen, followed by a transitional mire and finally an oligotrophic bog. The bog has reached the stage of a domed mire (Fig. 14.3A). The entire bog has been continually drained since 1972 with a distance of 100–120 m between ditches and a ditch depth of 0.7–1.4 m, lowering the water level in the middle of ditch spacing by approximately 0.2 m (Vompersky et al. 1988; Sirin et al. 1991). As a result of the drainage, the typical hollow–hummock structure in the center of the bog has been degraded. The bog currently is totally covered by a *Pinus sylvestris* stand that ranges in height from 20 m to 2–8 m on the margins and the central part, respectively. The vegetation in the bog center is dominated by *Eriophorum–Chamaedaphne*–dwarf pine, surrounded by a belt of *Sphagnum–Chamaedaphne*–dwarf pine. In the central parts, dwarf shrubs are presented mainly by *Andromeda polifolia* and *Chamaedaphne calyculata* and mosses by *Sphagnum fuscum* and *Sphagnum magellanicum*. In the marginal communities of these central parts, the dominant dwarf shrubs are *Ledum palustre* and *Vaccinium* spp, while *Sphagnum fuscum* is replaced by *Sphagnum angustifoluim* and *Sphagnum capillifolium*.

The Sopki study area at the Central Forest Nature Reserve is a 2.3 km^2 forested bog that occupies a saucer-like depression on the gentle slope (<1‰) of a watershed divide that is at an elevation of 261 m a.s.l. Peat depth varies from 0.3 m to 2.0 m, mostly 0.6 m. The surface is very flat, sometimes even concave (Fig. 14.3B). The mineral subsoil consists of loamy moraine deposits with occasional inclusion of sandy deposits, indicating that the site was a for-

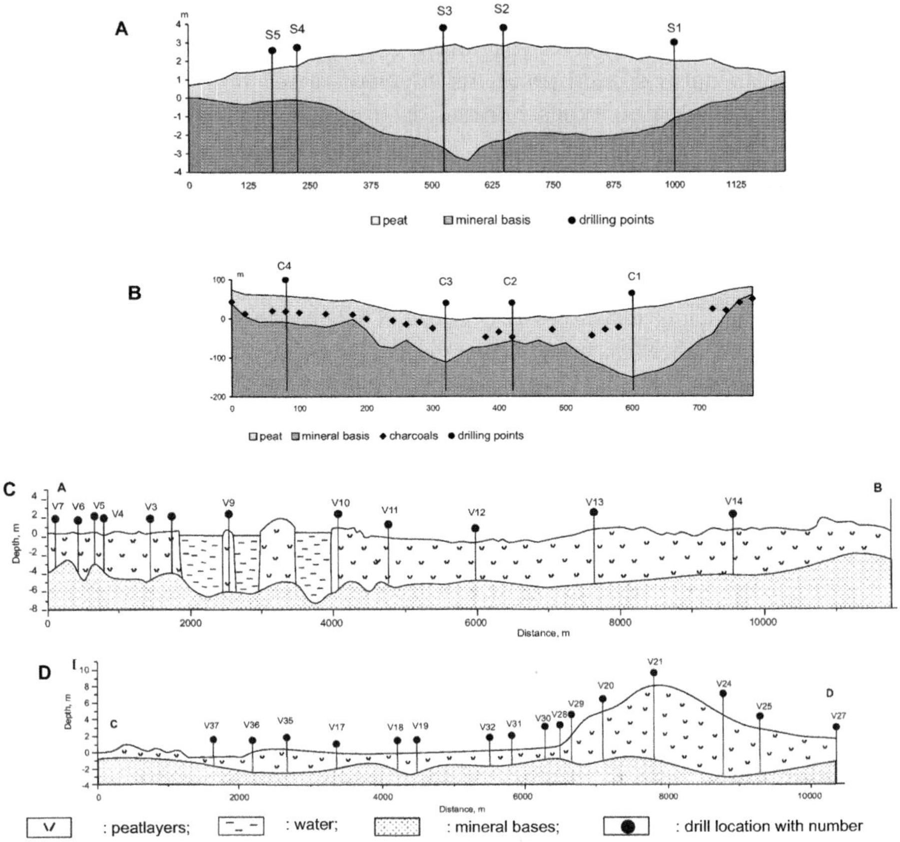

Fig. 14.3 Cross-section of peat deposits in the study areas. Central European Russia: **A** study site 'Usviatsky Mokh' (Zapadnaya Dvina Field Experimental Station), **B** study site 'Sopki' (Central Forest State Nature Reserve). West Siberia (the Great Vasyugan Bog): **C** West Vasyugan key area ('Malaya Icha'), **D** East Vasyugan key area ('Uzas')

mer streambed. The vegetation is dominated by *Picea*, *Pinus* or *Betula* stands that occur on sites that are in different stages of paludification. Most vegetation is oligotrophic, with *Sphagnum* mosses, *Chamaedaphne* dwarf shrubs, *Eriophorum* and sedges (*Carex globularis*, *Carex canescens* and patches of *Carex rostrata*). The moss layer is represented by *S. angustifolium* and *S. magellanicum* with large patches of mesotrophic species – *S. flexuosum* and even *S. fallax* in depressions. *Betula* stands have *Polytrichum commune* and *Eriophorum vaginatum* and they occur on a former large 40-year-old fire scar in the eastern part of the bog. The water level across the study area is uneven and depends on the distribution of precipitation (Minayeva et al. 2004).

14.3.2 Study Sites in Western Siberia

Two key study areas, typical for the Southern Taiga Subzone, are located within the western part of the Great Vasyugan Bog (Bleuten and Lapshina 2001). The Vasyugan bog is the largest, almost completely pristine peat bog complex in the world, with a total area of more than 5×10^4 km^2. The Uzas study area (57°15' N, 76°95' E) is located 150 km east of the other study area, known as the Malaya Icha site (56°55' N, 78°30' E).

The relief at both study areas is flat, with the surface having a slight inclination from north to the south. Elevations varies from 129.5 m to 136.0 m a.s.l. at the Uzas site and 125–127 m to 132 m at the Malaya Icha site (Fig. 14.3C, D).

The hydrography at both sites is typical for large mire massifs. Water flows slowly over and through the active peat-forming vegetation layer, gradually concentrating in brooklets at elevations of 120–130 m a.s.l. By the time the brooklets emerge from the peatland and flow into larger streams in deciduous woodlands, the streams have incised to a depth of 1–2 m. Both sites have many lakes. The surface area of individual lakes varies from several square meters to 2 km^2. The large lakes with mineral substrates are called primary ones because these lakes have never been terrestrialized. The smaller, secondary lakes developed in the hollows of the ridge–hollow complexes do have peat substrates. The climatic conditions of the two sites are presented in Table 14.1, based on data from the Sherstobitovo and Pudino weather stations.

The mire landscapes are vegetated by ombrotrophic dwarf pine–dwarf shrub–*Sphagnum* bogs (in Siberia called 'ryam'), ridge–hollow and ridge–hollow–pool complexes, which are raised as small or huge (0.01–25.0 km^2) islands (0.5–8.0 m) above the vast homogeneous or patterned minerotrophic through-flow fens. Various types of forested shrub fens and herbaceous–sedge tussock swampy forests are the third general category of mire vegetation (Lapshina 2003).

The vegetation associated with ombrotrophic mire landscapes is relatively uniform throughout all the forested zone of West Siberia. Forests are characterized by a dwarf pine layer, which is 0.5–1.5 m (ave. 2.5 m) high. There are well developed low shrubs (*Chamaedaphne calyculata, Ledum palustre, Oxycoccus microcarpus*) and some herbs (*Rubus chamaemorus, Drosera rotundifolia*). *Sphagnum fuscum* is the dominant moss. The open wet hollows and floating mats are characterized by the occurrence of *Carex limosa, Scheuchzeria palustris, Rhynchospora alba, Eriophorum vaginatum* and *E. russeolum*. Important mosses in the open hollows and floating mats are *Sphagnum balticum, S. jensenii, S. papillosum*.

The patterned fens are characterized by the development of low narrow peat ridges (strings), oriented across the direction of water movement. On the fens with very slight gradients, the spatial structures tend to lose their orientation and become irregular shapes. The vast wet areas (flarks) and floating mats are dominated by sedges (*Carex lasiocarpa, C. diandra, C. limosa, C.*

chordorrhiza) and brown moss species (*Scorpidium scorpioides*, *Homatocaulis vernicosus*, *Campylium polygamum*, *C. stellatum*, *Meesia triquetra*). *Utricularia intermedia*, *U. minor*, *Drosera anglica* and *Menyanthes trifoliata* occurs in these communities with high frequency. The vegetation on the ridges depends on their height above the wet areas. On the wettest and lowest ridges, *Betula nana*, *Carex elata* ssp. *omskiana* usually dominate. The vegetation on the somewhat higher and drier ridges is characterized by *Chamaedaphne calyculata*, *Ledum palustre* and *Potentilla palustris* in the field layer, with scattered birch, a few pine trees 1.5–5.0 m high. *Sphagnum warnstorfii* and *Tomentypnum nitens* dominate in moss layer.

The sedge (*Carex lasiocarpa*)–*Sphagnum* (*S. obtusum*) and sedge (*C. rostrata*)–cotton grass (*Eriophorum vaginatum*)–*Sphagnum* (*S. angustifolium*) communities are very widespread in transitional zones between ombrotrophic bogs and minerotrophic sedge brown moss fens.

Dwarf shrub (*Betula nana*, *Chamaedaphne calyculata*) fens that are interspersed with birch and pine (3–5 m high) are very common over wide areas bordering headwater streams on the southern slopes of the Great Vasyugan Bog. A number of sedges and herbs species may be present in such communities, but *Carex lasiocarpa*, *Potentilla palustris*, *Equisetum fluviatile* are the most characteristic. The moss layer consists of *Sphagnum warnstorfii*, *S. centrale*, *S. angustifolium*, *S. contortum*, *Calliergon* spp, *Warnstorfia* spp and *Lophocolea heterophylla*.

Forested swamps occur at the margins of the mires where there are mineral soils. The vegetation composition of the forested swamps (in Siberia called 'sogra') is usually quite varied and species-rich because of the great variety of microhabitats and their relatively high degree of contact with mineral substrates of varying chemical composition. Forested areas are dominated mainly by *Pinus sylvestris* and *Betula pubescens*. *Pinus sibirica* and *Picea obovata* are much more common towards the edge of the peatlands. In the ground layer, tussock sedges (*Carex cespitosa*, *C. appropinquata*, *Carex elongata*, *C. disperma*, *C. loliacea*) and herbs (*Thelypteris palustris*, *Menyanthes trifoliata*, *Calla palustris*, *Cicuta virosa*) occur with high frequency. *Calliergon* spp, *Drepanocladus* spp and *Mnium* spp are the dominant mosses in waterlogged hollows. Dwarf shrubs, herbs and moss species are constantly present on the hummocks.

14.3.3 Study Methods

The stratigraphy and other peat characteristics were obtained by coring along transects, crossing the middle of mire massifs or a part of it (see Fig. 14.3 for relative locations of cores at each study site). The peat cores were sampled for analyses of macrofossils, organic matter content, ash content, bulk density, decomposition degree and radiocarbon (^{14}C) dating. European cores were sampled at intervals of 2–5 cm and Siberian cores at 10 cm intervals.

The peat macrofossil composition was determined by Dr. O. Uspenskaya (The All-Russian Institute of Horticulture, Russian Academy of Agricultural Sciences) for the European sites and Dr. E. Muldiyarov (Laboratory of Biogeocoenology, Tomsk State University) for the Siberian sites. The bulk density was determined from the weight of peat samples of known volume, dried at 105 °C. The ash content was determined by oxidative burning at temperatures between 950 °C and 1000 °C. Carbon content was analyzed with a CHN analyzer (Elementar Vario EL III). Absolute dating of peat samples for European sites was carried out at the Institute of Geology RAS (laboratory head L.Sulerzhitsky) using the standard ^{14}C dating method. For Siberian sites, the ^{14}C age of 88 peat samples, including 22 basal samples of peat, was determined at the Institute of Geography of The Russian Academy of Science (IGRAS) in Moscow. Calibration was performed with CALIB 4.1 (Stuiver and Reimer 1993). The dynamics of Holocene accumulation rates were calculated from peat depth, peat age, bulk density and dry matter carbon content.

14.3.4 Holocene Peat Dynamics

Stratigraphy and Vegetation Cover Changes in the European Study Sites

Palaeoecological information from the central core of the 'Usvyatsky Mokh' study area (S3, Fig. 14.4) indicated that paludification was initiated 9500±1400 years BP (uncalibrated ^{14}C dates). There were no fossils of aquatic species in the core, but the vegetation must have been typical of very wet habitats because of the dominance of *Menyanthes trifoliata* and green mosses of the *Drepanocladus* group. The next stage of mire development – the sedge–hypnum transitional fen with dominance of *Carex diandra* – was relatively short and changed to a longer phase dominated by transition sedge–sphagnum fens with *Carex inflata, Carex lasiocarpa, Scheuchzeria palustris, Sphagnum obtusum, Sphagnum squarrosum* and *Sphagnum teres.* The next stage was characterized by the presence of pine and *Sphagnum fuscum* and mesotrophic sedge–sphagnum vegetation. That suggests that it was formed as hummock–hollow complex. The next distinct period was a homogenous, long period of pine forest with cotton grass and sphagnum mosses. The last 2000 years, based on macrofossils, was mainly dominated by *Sphagnum angustifolium* along with *Sphagnum magellanicum*. The last 2000 years were characterized by high rates of peat accumulation and slight deviations from time to time that were related to climate fluctuations to drier or wetter conditions (Rauber 2002).

Paludification at the 'Sopki' study site started 6500 BP (Fig. 14.5), beginning in a shallow lake as indicated the presence of macrofossils of *Potamogeton* sp.

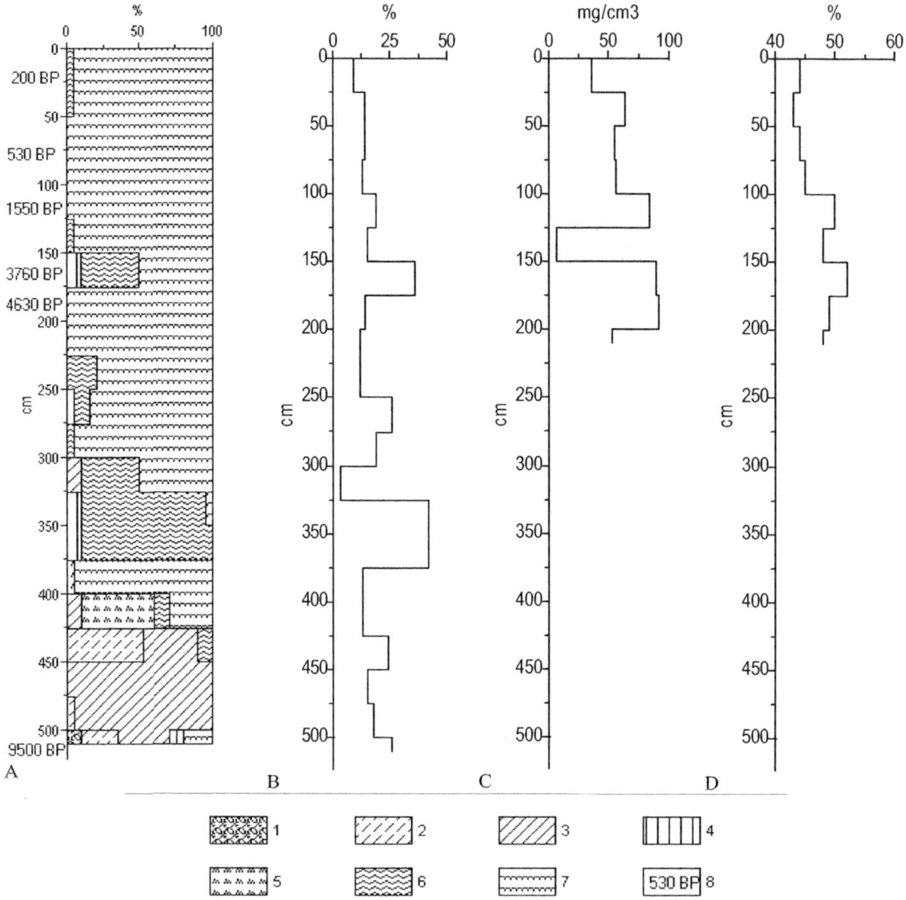

Fig. 14.4 Peat stratigraphy in the central peat core (S3) in study site 'Usviatsky Mokh' (Central European Russia). **A** Composition of macrofossil plant remains (%): *1* semi-aquatic vegetation, *2* spring fen species, *3* transition sedge–sphagnum bog species, *4* pine-wood fossils, *5* dwarf shrubs with *Sphagnum fuscum* community, *6* cotton-grass community, *7* dwarf shrubs with *Sphagnum magellanicum* community, *8* uncalibrated ^{14}C dates. **B** Degree of peat decomposition (%). **C** Peat bulk density (mg cm^{-3}). **D** Carbon content (%)

and remnants of floating mats with *Drepanocladus*, *Menyanthes trifoliata* and *Carex diandra*. After 500 years, a reed swamp developed with *Sphagnum teres* and *Drepanocladus* as the dominant mosses. The reed swamp was replaced by a birch stand. Peat accumulation during the initial period of paludification was low (0.15 mm $year^{-1}$ and 19.1 g C m^{-2} $year^{-1}$), partially due to the loss of peat during the frequent fires that occurred during that period of time. The high frequency of fires was marked by numerous charcoal layers in the peat.

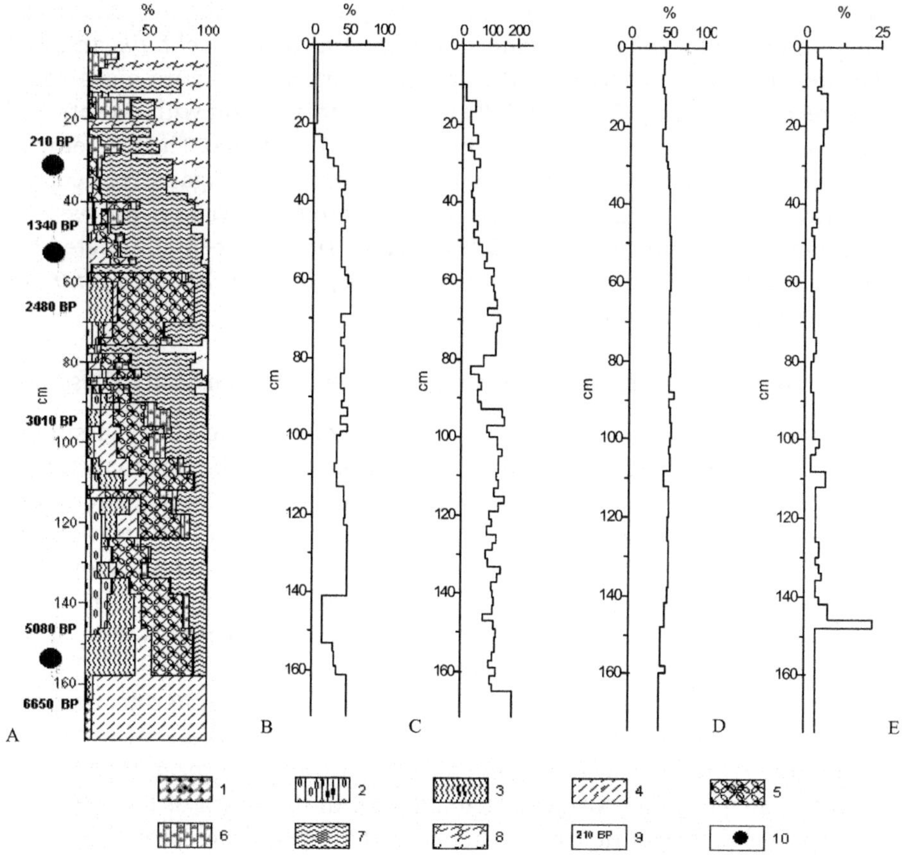

Fig. 14.5 Peat stratigraphy in peat core (C1) at study site 'Sopki' (Central European Russia). **A** Composition of macrofossil plant remains (%): *1* aquatic species, *2* species of alder forest, *3* species of reed communities, *4* species of herb–willow mesotrophic bogs and spring fens, *5* species of wooded mesotrophic bog, *6* species of spruce forest, *7* species of pine sphagnum bog with dwarf shrubs, *8*slightly decomposed tow of *Sphagnum flexuosum* and *Sphagnum fallax*, *9* uncalibrated ^{14}C dates, *10* charcoals. **B** Degree of peat decomposition (%), **C** peat bulk density (mg cm^{-3}), carbon (**D**) and organic (**E**) matter content (%)

By 5200 BP, the birch stand was replaced by a pine forests, which has dominated to the present. Until 2000 BP, the forest was a mesotrophic pine stand mixed with black alder and spruce with reeds, *Menyanthes trifoliata* and *Sphagnum centrale* in the herb layer. From time to time, the site changed to more oligotrophic conditions, indicated by the presence of *Eriophorum* sp. and *Carex rostrata*. The abundance of *Betula* and *Alnus* decreased slightly, beginning about 4000 years ago, being replaced by *Picea* accompanied by *Sphagnum girgensohnii*. Fire which occurred 3000 years ago was followed by a 300-year period of fast growth of the bog (0.5 cm annual linear peat increment and 52.8 g C m^{-2} year^{-1}), indicating the presence of oligotrophic vegetation dominated by *Sphagnum magellanicum*, *Andromeda polifolia*, and *Eriophorum vaginatum*. Thereafter, the system returned to a more mesotrophic pine–reed community. Between 3500 and 2500, this period with mesotrophic communities was again interrupted by a fire 2200 years ago. The fire returned the site development to a more oligotrophic trend once again. Immediately after fire, *Carex rostrata,* some herbs and *Sphagnum magellanicum* were the dominant plants. During the last 200 years, the site has been dominated by a pine forest mixed with stands of dwarf shrubs, *Eriophorum vaginatum* and *Sphagnum angustifolium*.

Stratigraphy and Vegetation Cover Changes in the Siberian Study Sites

The Great Vasyugan Bog consists of a number of separate mire massifs that grew in size and tended to fuse during the Holocene epoch. Peat analyses suggest that there were three general sequences of mire development across the landscape, based on the presence of three distinct types of peat stratigraphy: (1) three-layered peat deposits with centers of origin in well developed flat bog massifs, (2) homogeneous *Sphagnum* deposits near the borders of primary lakes and (3) mostly fen–peat deposits in areas where fens were in contact with minerotrophic groundwater.

The initial stages of peat accumulation occurred in numerous deep and shallow lake depressions or poorly drained peaty basins, where the water table was always at the surface. According to radiocarbon dates, peat deposition in the most ancient center of peat accumulation in the south of the Ob–Irtysh watershed plain began in Pre-Boreal and early Boreal times about 9000–9500 years BP (Fig. 14.3C, D; cores V9, V21). The peat is underlain by gyttja and aquatic (*Typha*) peat, both produced during the time that the system was aquatic. The thickness of lake sediments is commonly in excess of 0.5 m, but seldom exceeds 2.0–2.5 m in locations that were lakes or where pools now occur or were present in the past. The thickness of peat deposits in the initial center of origin is commonly 6.5–7.0 m above the mineral soil.

The peat cores indicate that the vast areas of the Great Vasyugan Bog that were adjacent to the initial lakes began as emergent wetlands that formed in depressions. Peat accumulation in those habitats was initiated at the beginning of the Atlantic Period, about 8000–8500 years BP (Fig. 14.3C; V10–V14).

There are generally three distinct strata in the peat deposits of the oldest mire massifs of the Great Vasyugan Bog. The basal peat layer is usually well decomposed fern, sedge–fern or herbaceous (*Thelypteris palustris*, *Equisetum* sp., *Menyanthes trifoliata*) peat about 1 m thick (Fig. 14.6). About 5500–4500 years BP, the open herbaceous fen was invaded by dwarf birch (*Betula nana*) and Ericaceae shrubs (not further identified), depositing a dwarf shrub peat layer that often also had a high percentage of cotton grass (*Eriophorum* sp.) remains. These fens in turn were overgrown by *Sphagnum* communities (*S.*

Fig. 14.6 Peat stratigraphy in peat core (V21, *Sphagnum fuscum* peat deposit on the top of domed bog) at study site 'Malaya Icha' (Great Vasyugan Bog, East Vasyugan key area, West Siberia). **A** Composition of macrofossil plant remains (%): *1 Sphagnum fuscum*, *2 S. angustifolium*, *3 S. teres*, *4* low sedges (*Carex limosa*), *5 Eriophorum vaginatum*, *6* brown mosses, *7* wood, *8* calibrated ^{14}C dates. **B** Carbon (g C dm^{-3}). **C** Organic matter (g dm^{-3}). **D** Degree of peat decomposition (%)

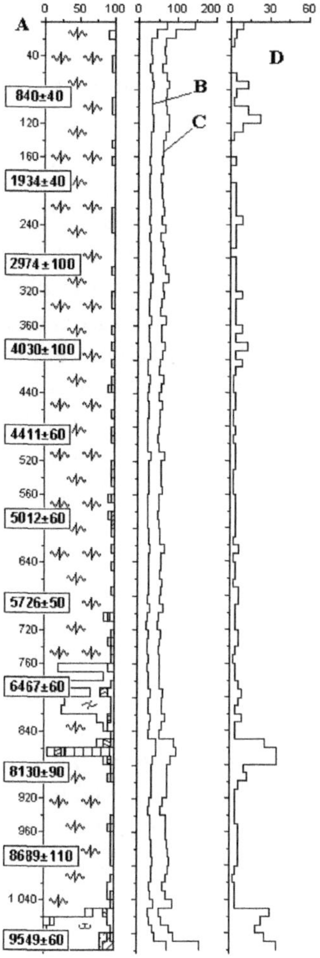

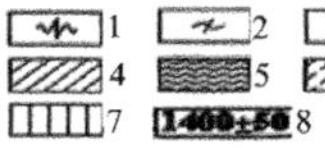

magellanicum, *S. angustifolium*) with scattered pine trees, indicating a transition to a somewhat elevated bog. This layer in the peat profile ranges in thickness from 0.5 m to 1.0 m and involves predominantly dwarf shrub–cotton grass, *Sphagnum*–dwarf shrubs or *Sphagnum*–woody peat, whose deposition coincided with growth of climate coldness at the end of Atlantic and the beginning of Sub-Boreal time.

The reduction in groundwater influence and the increasing importance of precipitation caused a sharp change in environmental conditions. As a result, a typical raised bog, with dwarf pine–ericaceous–*Sphagnum* (*S. fuscum*) vegetation, was established at the beginning of the Sub-Boreal period. During the Sub-Atlantic time, the development of wet hollows and numerous pools formed ombrotrophic mire complexes in the place of homogeneous raised bogs ('ryam'). The peat sequences show that the surface layer, composed of poorly to moderately decomposed *Sphagnum* moss materials, may reach 3–5 m in thickness.

The simplest sequence of peat development was recorded in peat cores taken on a domed bog (Fig. 14.3C; V21). The wetland began here about 9500 years BP (calibrated $^{14}C = 9549 \pm 60$) with a brief period of open *Sphagnum teres* fen, perhaps in the form of a floating mat, depositing 40 cm of peat. The macrofossils indicate that typical bog vegetation, dominated by *Sphagnum fuscum* ('ryam'), occupied the site at the beginning of the Boreal about 9000 years BP and has been maintained until the present without any variation (Fig. 14.6). The whole thickness of peat is 11 m above the mineral soil.

Peat macrofossils suggest that this type of peatland was initiated as wet sedge–grass meadow. These were followed during the wettest epochs of Atlantic time by open nutrient-rich herbaceous fens covered by fern, sedge–fern or sedge–bogbean–fern vegetation composed by *Thelypteris palustris*, *Carex* spp, and *Menyanthes trifoliata*, depositing about 1.0–1.5 m of peat.

As the peat deposits increased, these fens were followed by open homogeneous or patterned brown moss and sedge–brown moss fens, sometimes with shrubby patches and scattered birch trees, which form the mire complex landscapes that have been maintained over vast areas until the present.The development of mire complexes includes lateral expansion over the surface of adjacent fens. Peat macrofossils from these areas indicate that herbaceous fens initially occupied the wet depressions, followed by sedge–brown moss fens, which in turn have been replaced by mire complexes with oligotrophic *Sphagnum fuscum* hummock sites among the minerotrophic sedge–(*Carex rostrata*, *C. elata* ssp. *omskiana*)–*Sphagnum* (*Sphagnum teres*)–brown moss (*Warnstorfia* spp, *Scorpidium scorpioides*) fen, that exists at present.

14.3.5 Peat and Carbon Accumulation Rates

The study results show significant changes of the average value of annual peat and carbon accumulation during the Holocene period (Fig. 14.7). For European key sites, the annual peat increase in raised bog varies from 0.4 mm $year^{-1}$ to 0.5 mm $year^{-1}$, with a minimum value during Sub-Boreal period (Fig. 14.7A). Peat increase in the forested bog varied from 0.3 mm $year^{-1}$ to 0.45 mm $year^{-1}$ and with a reverse trend compared to the raised bog. Maximum values were observed for the Sub-Boreal period, which exceeded peat growth in the Sub-Atlanticum and Atlanticum by near to one-half (Fig. 14.7B).

Because of the bulk density changes, organic matter content and carbon accumulation values in the raised bog ('Usviatsky Mokh'), the peat accumulation rates varied during different Holocene periods, with a consistent decrease from the Atlanticum to the Sub-Boreal and Sub-Atlanticum periods.

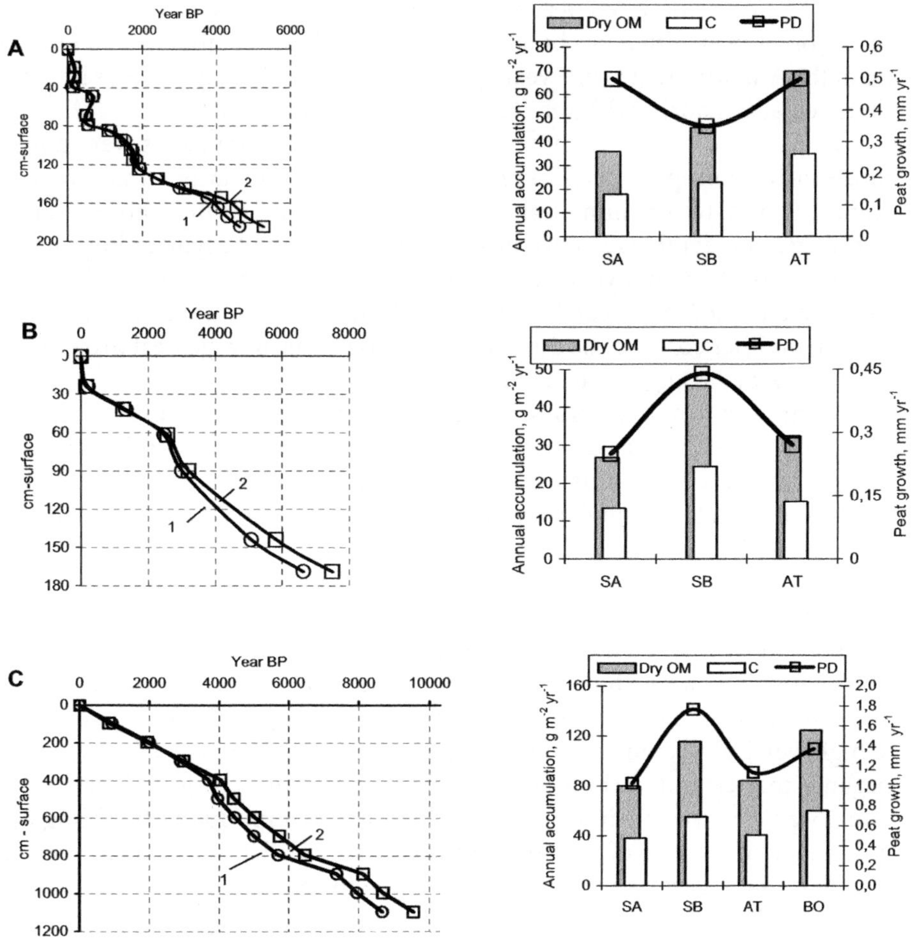

Fig. 14.7 Relationship between age (*1* uncalibrated, *2* calibrated ^{14}C dates), peat depth and accumulation of dry organic matter (OM), carbon (C) and peat accumulation (PD) in different periods of the Holocene (*BO* Boreal, *AT* Atlanticum, *SB* Sub-Boreal, *SA* Sub-Atlanticum). Study sites: **A** 'Usviatsky Mokh' and **B** 'Sopki' (Central European Russia), **C** 'Malaya Icha' (West Siberia)

During those transitions, the rate of carbon accumulation varied from 35 g C to 23 g C and 18 g C m^{-2} $year^{-1}$. Low dry organic matter and carbon accumulation values during the Sub-Atlanticum were due to low bulk density values (mean 0.07 g cm^{-3}) of mainly poorly decomposed (mean R=12 %) *Sphagnum* peat. In the Sub-Boreal and Atlanticum periods, the mainly cotton-grass peats had decomposition rates of 21 % and 25 % and bulk densities of 0.13 g cm^{-3} and 0.14 g cm^{-3}.

Carbon accumulation in the forested bog ('Sopki') was influenced by a sequence of peat fires, which took please periodically (as marked by charcoals). The times between peat fires were characterized by an intensive accumulation of peat with a low bulk density. The carbon accumulation increased from the Atlantic (15 g C m^{-2} $year^{-1}$) to the Sub-Boreal (24 g C m^{-2} $year^{-1}$) and then decreased during the Sub-Atlantic to 13 g C m^{-2} $year^{-1}$.

The peat and carbon accumulation rates in the Siberian sites were double those in the European sites, with annual peat accumulations rates varying from 1.0 mm to 1.8 mm $year^{-1}$, with a maximum in the Sub-Boreal. In Fig. 14.7C, the depth:age curve of the Icha core in the Vasyugan bog show a lowered rate of 0.6 mm $year^{-1}$ at the beginning of the Atlanticum and a maximum rate of 2.62 mm $year^{-1}$ at the transition between Atlanticum and Sub-Boreal. In other cores of the Southern Taiga Belt, the average peat growth was 0.57 mm $year^{-1}$ (range 0.34–1.37 mm $year^{-1}$).

The highest peat dry organic matter content (OM) and carbon accumulation rates were measured for the Boreal period (124 g OM m^{-2} $year^{-1}$ and 60 g C m^{-2} $year^{-1}$). In this period, relatively dense (0.08 g cm^{-3}) peat was formed at a high accumulation rate (1.37 mm $year^{-1}$). During the Atlantic, the annual accumulation of dry organic matter carbon decreased (84 g OM m^{-2} $year^{-1}$ and 40 g C m^{-2} $year^{-1}$. The presence of cotton grass (*Eriophorum vaginatum*) and wood remains in the peat layers formed in this period indicate a warmer and dryer climate, conditions which promote decomposition more then productivity. The wetter and cooler Sub-Boreal showed thick layers of loose peat with a low density (0.06–0.07 g cm^{-3}) but a high vertical increment (0.95–2.62 mm $year^{-1}$). The rate of annual dry organic matter and carbon accumulation increased (116 g OM m^{-2} $year^{-1}$ and 55 g C m^{-2} $year^{-1}$) in this period due to a high productivity of bog and a decrease in the rate of peat decomposition. A smaller annual peat production and lower peat density cause a decrease in the annual dry organic matter and carbon accumulation (80 g OM m^{-2} $year^{-1}$ and 38 g C m^{-2} $year^{-1}$) during the Sub-Atlantic period.

14.4 Discussion and Conclusions

The data for carbon sequestration measured in this study, in general, correspond with the values obtained by other authors (Table 14.2). The annual input into long-term carbon accumulation (LORCA) is estimated at $90–110 \times 10^{12}$ g C $year^{-1}$ for the world's mires (Sjörs 1980; Silvola 1986; Gorham 1991) and at $50–75 \times 10^{12}$ g C $year^{-1}$for northern mires (Armentano and Menges 1986), which corresponds with a carbon accumulation rate of 14–29 g C m^{-2} $year^{-1}$ (Gorham 1995). Carbon accumulation by Russian mires of the former Soviet Union is estimated to be 50×10^{12} g C $year^{-1}$ (Vompersky 1994; Botch et al. 1995).

Table 14.2 Apparent carbon accumulation by Northern mires according todifferent authors

Region	g C m^{-2} $year^{-1}$	Reference
Canada Boreal	29.9	Turunen (1999)
Canada Sub-Arctic	23.5	
Canada	10.0–35.0	Ovenden (1990)
North America Interior	29.0	Gorham (1991)
Finland, Estonia and USA East Coast	4.6–85.8 (mean 19.9±10.7)	Korhola (1995)
Russia	22.4	Vompersky (1994)
Former USSR	31.0	Botch et al. (1995)

Based on the data presented for the world's mires (Table 14.2), there is clearly considerable geographic uniformity in peat accumulation rates over regions of Eurasia. Comparison of the peat accumulation rates for the Sub-Atlanticum clearly indicates that there are some areas in which there are now increased rates of peat accumulation (Klimanov and Sirin 1997). In the European part of Russia, areas with increased rates of peat accumulation are connected with the Southern and Middle Taiga, starting from Estonia and north-western European Russia to its central part. In this region, average values of peat accretion range over 1.0–1.5 mm $year^{-1}$. The rate of peat accumulation decreases in the south and the north to a range of 0.3–0.03 mm $year^{-1}$ in the Belorussian and Ukrainian poles'ya and 0.55–0.35 mm $year^{-1}$ in the Kola Peninsula and Arkhangelsk district.

A clear gradient in the peat accretion rate of the Western Siberian Taiga zones was established (Table 14.3). Beyond these zones the rates decrease; toward the north (at Salekhard) down to 0.05 mm $year^{-1}$ and toward the south (Northern Kazakhstan) to 0.35 mm $year^{-1}$ (Klimanov and Sirin 1997).

Data presented in Table 14.3 suggest that peat growth could depend even more on mire origin that on its geographical location. Highest growth rates have been established in ice marginal valleys, where the wetlands developed at sites that were in contact with mineral-rich groundwater, which enhances productivity. In floodplains, high rates of peat accumulation also occurred where the wetlands were in contact with groundwater or were flooded with mineral-rich sediments. The average peat accumulation rate at the eight sites, including bogs, fens and forested swamps ('sogra'), in Southern Taiga and Sub-Taiga zones of West Siberia varied from 0.35±0.03 mm $year^{-1}$ to 1.13±0.02 mm $year^{-1}$ and the LORCA values from 19.0±1.1 g C m^{-2} $year^{-1}$ to 69.0±4.4 g C m^{-2} $year^{-1}$ (Borren et al. 2004).

Similarly large differences in peat growth were found for the central European Taiga part, as shown in Fig. 14.8. The average long-term carbon accumulation rate was 34.8 g C m^{-2} $year^{-1}$ (range 13–65 g C m^{-2} $year^{-1}$). Peat accretion

Table 14.3 Peat growth in Western Siberia by vegetation–climate zone (Borren et al. 2004). *N* numbers of cores analyzed

Climatic zone	*N*	Peat accumulation rate (mm $year^{-1}$)		
		Average	Minimum	Maximum
Northern Taiga	16	0.39	0.10	0.78
Middle Taiga	27	0.56	0.17	1.34
Southern Taiga	29	0.74	0.36	1.27
Sub-Taiga	12	0.80	0.45	1.32
Ice marginal valleys	9	1.09	0.35	1.67
Flood pains	12	0.98	0.56	1.64

and carbon accumulation in raised bogs formed on moraine planes, as taken for the Little Climatic Optimum (which can be regarded as a climatic analogue of today's conditions) was twice higher than in forested swamps and fens and nearly five times higher than in raised bogs formed on outwash sands. The better drainage of the latter makes them more vulnerable to climatic changes and thus causes lower peat accumulation rates. During other palaeoclimatic periods, however, the contrast between different mires was less notable.

Our studies clearly demonstrate substantial changes in peat growth and carbon accumulation in response to variations in palaeoclimatic conditions. However, variation between different mires was limited. According to the data of the above-mentioned research into the peat accumulation dynamics for different mires of the West Dvina Field Station, the LORCA was approximately 30–40 g C m^{-2} $year^{-1}$. Looking at distinct palaeoclimatic periods makes the differences between mire ecosystems more apparent. Sites that had little or no contact with groundwater were more sensitive to palaeoclimate fluctuations. The Malaya Icha site in West Siberia is located on the divide between two watersheds and was expected to be a suitable site for evaluating the effects of climate change on peat growth and carbon accumulation because of the absence of non-precipitation sources of water (e.g. river floods, groundwater).

Peat accumulation rates at the site, especially during the early Holocene, varied due to changes in vegetation composition (i.e., succession). Later during the Holocene, the bog ecosystem (dominated by *Sphagnum fuscum)* did not change. Because compaction was negligible within the peat profile and the decomposition rate was not important in undrained peatlands (Borren et al. 2004), changes in sequestration rate (Fig. 14.7C) can be attributed to climate changes.

High productivity in the Southern part of Forest Zone can be explained by the relative long and warm growing season of this strict continental climate, where the spring and autumn seasons are very short (1–2 weeks) In contrast, the relatively long period with temperatures far below zero (4–5 months) results in a comparatively low substrate temperature, low microbiological activity and therefore low decomposition rate. Together, this may result in the measured high rates of carbon sequestration. The decrease in carbon sequestration during the Atlanticum (Fig. 14.7C) most probably results from the higher decomposition in the warmer climate.

Within the Holocene, there was a steady growth of peatland area at all of our study sites, resulting in an expansion of peat deposits and increased carbon storage. The temporal factor in carbon sequestration manifests itself not only in palaeoclimatic periods or at the millennium scale. External factors such as changes of solar activity, dry and wet periods may affect the peatland contribution to carbon cycle. Many Eurasian mires tended to grow faster in cold periods and to slow down in warmer periods during the last three millennia (Klimanov and Sirin 1997), as shown in Fig. 14.8. Even if the vegetation of specific mires did not react to minor climatic changes (according to macro-

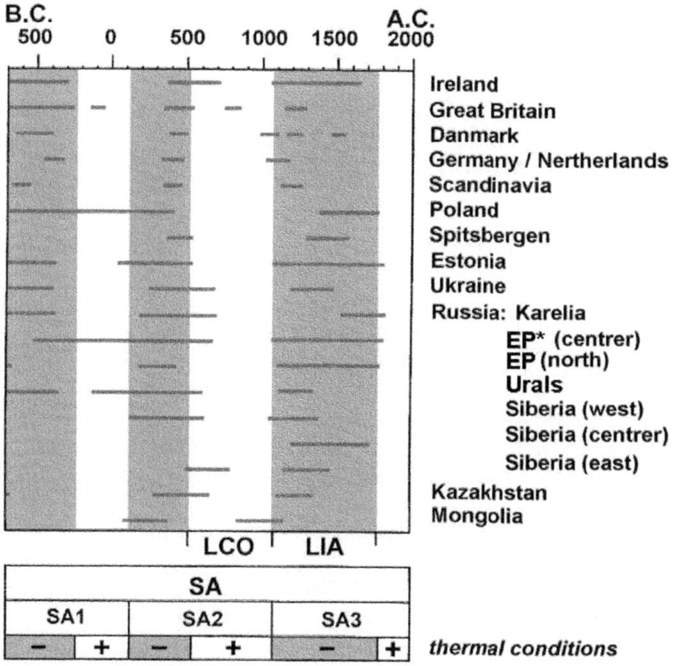

Fig. 14.8 Periods of intensive peat accumulation in different regions of Eurasia during the Sub-Atlanticum, 2600 BP (uncalibrated ^{14}C dates) to present, and its sub-periods (*SA1, SA2, SA3*). *LCO* Little Climatic Optimum, 800–1100 AD, *LIA* Little Ace Age, 1550–1850 AD. *EP** European part. Note: + and – indicate thermal conditions deviation from actual climate (after Klimanov and Sirin 1997)

fossil analysis data), such climate changes were reflected in the values of carbon accumulation (Rauber 2002).

A future climate change resulting in a temperature rise (IPCC 2001) can enhance primary production and decomposition, with a net negative effect on carbon sequestration, in particular when the climate becomes dryer. However, if precipitation increases more than evapotranspiration and temperature rises simultaneously, a rapid peat growth and higher carbon sequestration may occur.

The analyses of the selected mires in the Eurasian Taiga zone demonstrated a clear response of peat accumulation/carbon sequestration by climate change. The other way around, taiga mires may negatively feed-back on climate warming, depending on the change in wetness associated with future climate conditions. As no climate wetness predictions for Siberia are available, quantification of this feedback remains impossible at this time.

Acknowledgements We thank L.A.Sulerzhitsky for ^{14}C dating of peat samples, O.N. Uspenskaya and E.Y. Mouldiyarov for analyzing the macrofossil content of the peat samples and Christine Rauber and Igor Glushkov for sharing field data. We acknowledge partial financial from the Wetlands International Russia Programme project 1094-002 UNEP-GEF "Integrated Management of Peatlands for Biodiversity and Climate Change" and the CASUS project funded by EU-INTAS (project 03-51-6294).

References

Armentano TV, Menges ES (1986) Patterns of change in the carbon balance of organic soil wetlands of the temperate zone. J Ecol 74:755–774

Barbier KE (1981) Peat stratigraphy and climatic change. A palaeoecological test of the theory of cyclic peat bog regeneration. Balkema, Rotterdam, 219 pp

Belyea LR, Warner BG (1996) Temporal scale and the accumulation of peat in a *Sphagnum* bog. Can J Bot 74:366–377

Bleuten W, Lapshina ED (eds) (2001) Carbon storage and atmospheric exchange by West Siberian peatlands. (FGUU scientific reports 2001-1) FGUU, Tomsk, 165 pp

Blytt A (1882) Die Theorie der wechselnder kontinentalen und isularen Klimate. Engler Bot. Jahrb 2

Borman FH, Likens GE (1979) Pattern and process in a forested ecosystem. Springer, Berlin Heidelberg New York, 253 pp

Borren W, Bleuten W, Lapshina ED (2004) Holocene peat and carbon accumulation rates in the southern taiga of western Siberia. Quat Res 61:42–51

Botch MS, Kobak KI, Vinson TS (1995) Carbon pools and accumulation in peatlands of the former Soviet Union. Global Biogeochem Cycles 9:37–46

Elina G (1991) Methods for reconstruction of humidity regime in terms of regularities of paludification and mire dynamics in the Holocene. In: USSR Academy of Sciences (ed) Studies of mire ecosystems of Fennoscandia. (Materials of the Soviet–Finnish Symposium, 28–31 May 1990) Karelian Research Centre USSR Academy of Sciences, Petrozavodsk, pp 51–60

Elina GA, Arslanov HA, Klimanov VA (1995) The vegetation and climatochronology of Holocene in Lavozero lowland of Kola peninsula (after spore-pollen diagrams of palsa-mire) (in Russian). Bot Zh 80:1–16

Fan S, Gloor M, Mahlman J, Pacala S, Sarmiento J, Takahashi T, Tans P (1998) A large terrestrial carbon sink in North America implied by atmospheric and oceanic carbon dioxide data and models. Science 282:442–447

Franzén LG (1992) Can Earth afford to lose the wetlands in the battle against the increasing greenhouse effect? Proc Int Peat Congr 9:1–18

Glebov FZ, Karpenko LV, Klimanov VA (1996) Palaeoecological analyses of peat core on the Ob and Vasyugan watershed (in Russian). Sib Ecol J 6:497–504

Glebov FZ, Karpenko LV, Klimanov VA, Mindeeva TN (1997) Palaeoecological characteristics of Holocene between Ob and Vasyugan on the data of peat section "Vodorazdel" (in Russian). Ecology 6:412–418

Gorham E (1991) Northern peatlands: role in the carbon cycle and probable responses to climate warming. Ecol Appl 1:182–195

Gorham E (1995) The biogeochemistry of northern peatlands and its possible responses to global warming. In: Woodwell GM, Mackenzie FT (eds) Biotic feedbacks in the global climatic system. Will the warming feed the warming? Oxford University Press, Oxford, pp 169–187

Gorham E, Janssens JA (1992) The palaeorecord of geochemistry and hydrology in northern peatlands and its relation to global change. Suo 43:117–126

Houghton JT, Meira Filho LG, Callander BA, Harris N, Kattenberg A, Maskell K (eds) (1996) Climate change 1995: the science of climate change. (Contribution of Working Group I to the Second Assessment Report of the IPCC) Cambridge University Press, Cambridge, 572 pp

IPCC (2001) Climate change 2001: the scientific basis. The carbon cycle and atmospheric carbon dioxide. Cambridge University Press, Cambridge, pp 183–237

Joosten H, Clarke D (2002) The wise use of mires and peatlands – background and principles including a framework for decision-making. International Mire Conservation Group/International Peat Society, Saarijärvi, 304 pp

Khotinsky NA (1969) The correlation of the Holocene deposits and absolute chronology of Blytt–Sernander (in Russian). In: INQA (ed) Holocene. (VIII Congress of INQA) Nauka, Moscow

Khotinsky NA (1977) Holocene of the North Eurasia (in Russian). Nauka, Moscow, 198 pp

Klimanov VA (1984) Palaeoclimatic reconstructions based on the information statistical method, in Late Quaternary environments of the Soviet Union (translated from Russian). University of Minnesota, Minneapolis, 297 pp

Klimanov VA, Sirin AA (1997) The dynamics of peat accumulation by mires of Northern Eurasia during the last three thousand years. In: Trettin CC (ed) Northern forested wetlands, ecology and management. Lewis /CRC, Boca Raton, pp 319–330

Korhola A (1995) Holocene climatic variations in southern Finland reconstructed from peat-initiation data. Holocene 5:43–58

Lapshina ED (2003) Flora of mires of south-east of West Siberia (in Russian). Tomsk University, Tomsk, 296 pp

Lavrenko EM (1947) Principles and units of geobotanical regionalization (in Russian). In: Geobotanical regionalization of the USSR. Nauka, Moscow, pp 9–13

Lavrenko EM (2000) Selecta (in Russian). St Petersburg University, St Petersburg, 672 pp

Minayeva T, Glushkov I, Sulerzhicky L, Uspenskaya O, Sirin A (2004) On temporal aspects of shallow peat accumulation in boreal paludified forests: data from case studies in Central European Russia. Proc Int Peat Congr 12:150–155

Neushtadt MI (1957) The forest history and palaeogeorgaphy of USSR in Holocene (in Russian). Academy of Sciences of USSR, Moscow, 404 pp
Neushtadt MI (1985) Mire formation processes in the Holocene (in Russian). Izv Acad Sci USSR Ser Geogr 1:39
Nicholls N, Gruza GV, Jouzel J, Karl TR, Ogallo LA, Parker DE (1996) Climate change 1995: the science of climate change. (Contribution of Working Group I to the Second Assessment Report of the IPCC) Cambridge University Press, Cambridge, pp 133–192
Nikonov MN (1955) Regional peatlands zonation in relation to their economics (in Russian). Proc Inst For Wood 31:49–63
Ovenden L (1990) Peat accumulation in northern wetlands. Quat Res 33:377–386
Post WM, Emanuel WR, Zinke PJ (1982) Soil carbon pools and world life zones. Nature 298:156–159
Ramsar Resolution (2006) Resolutions VIII.3, VIII.17. Available at http://www.ramsar.org/key_res_viii_index_e.htm
Rauber C (2002) Stability of raised bogs to climatic changes – a case study. Shaker, Aachen, 168 pp
Sernander R (1910) Die Schwedischen Torfmoore als Zeugen postglazialer Klimaschwankungen. Veraenderungen des Klimas seit dem Maximum der letzten Eiszeit. Exekutivkomm. Int. Geologenkongress 11, Stockholm
Silvola U (1986) Carbon dioxide dynamics in mires reclaimed for forestry in eastern Finland. Ann Bot Fenn 23:59–67
Sirin AA, Minaeva TY (eds) (2001) Peatlands in Russia: towards an analysis of sectorial information (in Russian). GEOS, Moscow, 190 pp
Sirin AA, Vompersky SE, Nazarov NA (1991) Influence of forest drainage on river runoff regime: main concepts and examples from central part of the USSR European territory. Ambio 20: 334–339
Sjörs H (1980) Peat on earth: multiple use or conservation? Ambio 9:303–308
Smith LC, MacDonald GM, Velichko AA (2004) Siberian peatlands a net carbon sink and global methane source since the early Holocene. Science 303: 353–356
Stuiver M, Reimer PJ (1993) Extended ^{14}C database and revised CALIB radiocarbon calibration program. Radiocarbon 35:215–230
Tolmachev AI (1954) On the history of dark Taiga origination and development (in Russian). Academy of Sciences of USSR, Moscow, 156 pp
Tolonen K (1987) Natural history of raised bogs and forest vegetation in the Lammi area, Southern Finland studied by stratigraphical methods. Ann Acad Sci Fenn Ser A 144:46
Turunen J (1999) Carbon accumulation of natural mire ecosystems in Finland – applications to boreal and subarctic mires. PhD thesis, University of Joensuu,Joensuu, 30 pp
Vasiliev SV (2000) Peat accumulation rate in West Siberia (in Russian). In: Karelian Scientific Centre (ed) Dynamics of mire ecosystems of Northern Eurasia in Holocene (Proceedings of International Symposium). Karelian Scientific Centre RAS, Petrozavodsk, pp 56–59
Vasiliev SV, Titlyanova AA, Velichko AA (eds) (2001) West Siberian peatlands and carbon cycle: past and present. (Proceedings of the International Field Symposium, Noyabrsk, 18–22 August 2001). Agenstvo Sibprint, Novosibirsk, 250 pp
Velichko A, Frenzel B, Pecsi M (eds) (1991) Atlas of palaeoclimates and palaeoenvironments of the northern hemisphere, late Pleistocene–Holocene. Budapest, Budapest
Vompersky SE (1994) Role of mires in the cycle of carbon (in Russian). In: Nauka N (ed) Biogeocoenotical peculiarities of mires and their rational use. Nauka, Moscow, p 5
Vompersky SE, Sirin AA, Glukhov AI (1988) Formation and regime of flow during forest drainage (in Russian). Nauka, Moscow, 168 pp

Vompersky SE, Tsyganova OP, Valyaeva NA (1996) Peat-covered wetlands of Russia and carbon pool of their peat. In: Peatlands use – present, past and future. Proc Int Peat Congr 10:381–390

Vompersky SE, Tsyganova OP, Glukhova TV, Valyaeva NA (2000) The vertical peat increment of the mires in Russia on radiocarbon data (in Russian). In: Karelian Scientific Centre (ed) Dynamics of mire ecosystems of Northern Eurasia in the Holocene. (Proceedings of International Symposium) Karelian Scientific Centre RAS, Petrozavodsk, pp 53–55

Warner BG, Clymo RS, Tolonen K (1993) Implications of peat accumulation at Point Escuminac, New Brunswick. Quat Res 39:245–248

Zoltai SC, Martikainen PJ (1996) Estimated extent of forested peatlands and their role in the global carbon cycle. In: Apps MJ (ed) Forest ecosystems, forest management and the global carbon cycle. (NATO ASI Ser. I. Global environmental change, vol 40) Springer, Berlin Heidelberg New York, pp 47–58

Subject Index

Ecological Studies

Volumes published since 2002

Volume 153
UV Radiation and Arctic Ecosystems (2002)
D.O. Hessen (Ed.)

Volume 154
Geoecology of Antarctic Ice-Free Coastal Landscapes (2002)
L. Beyer and M. Bölter (Eds.)

Volume 155
Conserving Biological Diversity in East African Forests: A Study of the Eastern Arc Mountains (2002)
W.D. Newmark

Volume 156
Urban Air Pollution and Forests: Resources at Risk in the Mexico City Air Basin (2002)
M.E. Fenn, L. I. de Bauer, and T. Hernández-Tejeda (Eds.)

Volume 157
Mycorrhizal Ecology (2002, 2003)
M.G.A. van der Heijden and I.R. Sanders (Eds.)

Volume 158
Diversity and Interaction in a Temperate Forest Community: Ogawa Forest Reserve of Japan (2002)
T. Nakashizuka and Y. Matsumoto (Eds.)

Volume 159
Big-Leaf Mahogany: Genetic Resources, Ecology and Management (2003)
A. E. Lugo, J. C. Figueroa Colón, and M. Alayón (Eds.)

Volume 160
Fire and Climatic Change in Temperate Ecosystems of the Western Americas (2003)
T. T. Veblen et al. (Eds.)

Volume 161
Competition and Coexistence (2002)
U. Sommer and B. Worm (Eds.)

Volume 162
How Landscapes Change: Human Disturbance and Ecosystem Fragmentation in the Americas (2003)
G.A. Bradshaw and P.A. Marquet (Eds.)

Volume 163
Fluxes of Carbon, Water and Energy of European Forests (2003)
R. Valentini (Ed.)

Volume 164
Herbivory of Leaf-Cutting Ants: A Case Study on *Atta colombica* in the Tropical Rainforest of Panama (2003)
R. Wirth, H. Herz, R.J. Ryel, W. Beyschlag, B. Hölldobler

Volume 165
Population Viability in Plants: Conservation, Management, and Modeling of Rare Plants (2003)
C.A Brigham, M.W. Schwartz (Eds.)

Volume 166
North American Temperate Deciduous Forest Responses to Changing Precipitation Regimes (2003)
P. Hanson and S.D. Wullschleger (Eds.)

Volume 167
Alpine Biodiversity in Europe (2003)
L. Nagy, G. Grabherr, Ch. Körner, D. Thompson (Eds.)

Volume 168
Root Ecology (2003)
H. de Kroon and E.J.W. Visser (Eds.)

Volume 169
Fire in Tropical Savannas: The Kapalga Experiment (2003)
A.N. Andersen, G.D. Cook, and R.J. Williams (Eds.)

Volume 170
Molecular Ecotoxicology of Plants (2004)
H. Sandermann (Ed.)

Volume 171
Coastal Dunes: Ecology and Conservation (2004)
M.L. Martínez and N. Psuty (Eds.)

Volume 172
Biogeochemistry of Forested Catchments in a Changing Environment: A German Case Study (2004)
E. Matzner (Ed.)

Volume 173
Insects and Ecosystem Function (2004)
W. W. Weisser and E. Siemann (Eds.)

Volume 174
Pollination Ecology and the Rain Forest: Sarawak Studies (2005)
D. Roubik, S. Sakai, and A.A. Hamid (Eds.)

Volume 175
Antarctic Ecosystems: Environmental Contamination, Climate Change, and Human Impact (2005)
R. Bargagli

Volume 176
Forest Diversity and Function: Temperate and Boreal Systems (2005)
M. Scherer-Lorenzen, Ch. Körner, and E.-D. Schulze (Eds.)

Volume 177
A History of Atmospheric CO_2 and its Effects on Plants, Animals, and Ecosystems (2005)
J.R. Ehleringer, T.E. Cerling, and M.D. Dearing (Eds.)

Volume 178
Photosynthetic Adaptation: Chloroplast to Landscape (2005)
W.K. Smith, T.C. Vogelmann, and C. Chritchley (Eds.)

Volume 179
Lamto: Structure, Functioning, and Dynamics of a Savanna Ecosystem (2006)
L. Abbadie et al. (Eds.)

Volume 180
Plant Ecology, Herbivory, and Human Impact in Nordic Mountain Birch Forests (2005)
F.E. Wielgolaski (Ed.) and P.S. Karlsson, S. Neuvonen, D. Thannheiser (Ed. Board)

Volume 181
Nutrient Acquisition by Plants: An Ecological Perspective (2005)
H. BassiriRad (Ed.)

Volume 182
Human Ecology: Biocultural Adaptations in Human Cummunities (2006)
H. Schutkowski

Volume 183
Growth Dynamics of Conifer Tree Rings: Images of Past and Future Environments (2006)
E.A. Vaganov, M.K. Hughes, and A.V. Shashkin

Volume 184
Reindeer Management in Northernmost Europe: Linking Practical and Scientific Knowledge in Social-Ecological Systems (2006)
B.C. Forbes, M. Bölter, L. Müller-Wille, J. Hukkinen, F. Müller, N. Gunslay, and Y. Konstantinov (Eds.)

Volume 185
Ecology and Conservation of Neotropical Montane Oak Forests (2006)
M. Kappelle (Ed.)

Volume 186
Biological Invasions in New Zealand (2006)
R.B. Allen and W.G. Lee (Eds.)

Volume 187
Managed Ecosystems and CO_2: Case Studies, Processes, and Perspectives (2006)
J. Nösberger, S.P. Long, R. J. Norby, M. Stitt, G.R. Hendrey, and H. Blum (Eds.)

Volume 188
Boreal Peatland Ecosystem (2006)
R.K. Wieder and D. H. Vitt (Eds.)

Volume 189
Ecology of Harmful Algae (2006)
E. Granéli and J.T. Turner (Eds.)

Volume 190
Wetlands and Natural Resource Management (2006)
J.T.A. Verhoeven, B. Beltman, R. Bobbink, and D.F. Whigham (Eds.)

Volume 191
Wetlands: Functioning, Biodiversity Conservation, and Restoration (2006)
R. Bobbink, B. Beltman, J.T.A. Verhoeven, and D.F. Whigham (Eds.)

Plant Ecology

GPSR Compliance

The European Union's (EU) General Product Safety Regulation (GPSR) is a set of rules that requires consumer products to be safe and our obligations to ensure this.

If you have any concerns about our products, you can contact us on ProductSafety@springernature.com

In case Publisher is established outside the EU, the EU authorized representative is:

Springer Nature Customer Service Center GmbH
Europaplatz 3
69115 Heidelberg, Germany

Batch number: 10755284

Printed by Printforce, the Netherlands